DYNAMICS AND CONTROL
OF STRUCTURES

DYNAMICS AND CONTROL OF STRUCTURES

LEONARD MEIROVITCH
Department of Engineering Science and Mechanics
Virginia Polytechnic Institute and State University
Blacksburg, Virginia

A Wiley-Interscience Publication
JOHN WILEY & SONS
New York / Chichester / Brisbane / Toronto / Singapore

Copyright © 1990 by John Wiley & Sons, Inc.

All rights reserved. Published simultaneously in Canada.

Reproduction or translation of any part of this work
beyond that permitted by Section 107 or 108 of the
1976 United States Copyright Act without the permission
of the copyright owner is unlawful. Requests for
permission or further information should be addressed to
the Permissions Department, John Wiley & Sons, Inc.

Library of Congress Cataloging in Publication Data:

Meirovitch, Leonard.
   Dynamics and control of structures / Leonard Meirovitch.
     p.   cm.
   "A Wiley-Interscience publication."
   Includes bibliographical references.
   1. Structural control (Engineering). 2. Structural Dynamics
I. Title.
TA.9.M45   1989   89-22710
624.1'7--dc20   CIP
ISBN 0-471-62858-1

Printed in the United States of America

10 9 8 7 6 5 4 3 2

To Jo Anne

CONTENTS

PREFACE   xi

INTRODUCTION   1

1. NEWTONIAN MECHANICS   3

    1.1 Newton's Second Law, 3
    1.2 Impulse and Momentum, 6
    1.3 Moment of a Force and Angular Momentum, 7
    1.4 Work and Energy, 8
    1.5 Systems of Particles, 11
    1.6 Rigid Bodies, 16
    1.7 Euler's Moment Equations, 20

2. PRINCIPLES OF ANALYTICAL MECHANICS   21

    2.1 Degrees of Freedom and Generalized Coordinates, 21
    2.2 The Principle of Virtual Work, 22
    2.3 D'Alembert's Principle, 25
    2.4 Hamilton's Principle, 28
    2.5 Lagrange's Equations of Motion, 33
    2.6 Hamilton's Canonical Equations. Motion in the Phase Space, 38
    2.7 Lagrange's Equations of Motion in Terms of Quasi-Coordinates, 41

3. CONCEPTS FROM LINEAR SYSTEM THEORY   45

- 3.1 Concepts from System Analysis, 45
- 3.2 Frequency Response, 50
- 3.3 Response by Transform Methods. The Transfer Function, 53
- 3.4 Singularity Functions, 55
- 3.5 Response to Singularity Functions, 59
- 3.6 Response to Arbitrary Excitation. The Convolution Integral, 61
- 3.7 State Equations. Linearization About Equilibrium, 63
- 3.8 Stability of Equilibrium Points, 68
- 3.9 Response by the Transition Matrix, 74
- 3.10 Computation of the Transition Matrix, 76
- 3.11 The Eigenvalue Problem. Response by Modal Analysis, 79
- 3.12 State Controllability, 81
- 3.13 Output Equations. Observability, 83
- 3.14 Sensitivity of the Eigensolution to Changes in the System Parameters, 84
- 3.15 Discrete-Time Systems, 86

4. LUMPED-PARAMETER STRUCTURES   93

- 4.1 Equations of Motion for Lumped-Parameter Structures, 93
- 4.2 Energy Considerations, 98
- 4.3 The Algebraic Eigenvalue Problem. Free Response, 99
- 4.4 Qualitative Behavior of the Eigensolution, 108
- 4.5 Computational Methods for the Eigensolution, 113
- 4.6 Modal Analysis for the Response of Open-Loop Systems, 123

5. CONTROL OF LUMPED-PARAMETER SYSTEMS. CLASSICAL APPROACH   129

- 5.1 Feedback Control Systems, 129
- 5.2 Performance of Control Systems, 132
- 5.3 The Root-Locus Method, 137
- 5.4 The Nyquist Method, 141
- 5.5 Frequency Response Plots, 146
- 5.6 Bode Diagrams, 150
- 5.7 Relative Stability. Gain Margin and Phase Margin, 158
- 5.8 Log Magnitude-Phase Diagrams, 161
- 5.9 The Closed-Loop Frequency Response. Nichols Charts, 163
- 5.10 Sensitivity of Control Systems to Variations in Parameters, 169

5.11 Compensators, 171
5.12 Solution of the State Equations by the Laplace Transformation, 175

6. CONTROL OF LUMPED-PARAMETER SYSTEMS. MODERN APPROACH   179

6.1   Feedback Control Systems, 180
6.2   Pole Allocation Method, 183
6.3   Optimal Control, 191
6.4   The Linear Regulator Problem, 193
6.5   Algorithms for Solving the Riccati Equation, 195
6.6   The Linear Tracking Problem, 201
6.7   Pontryagin's Minimum Principle, 203
6.8   Minimum-Time Problems, 205
6.9   Minimum-Time Control of Linear Time-Invariant Systems, 208
6.10 Minimum-Fuel Problems, 222
6.11 A Simplified On-Off Control, 229
6.12 Control Using Observers, 232
6.13 Optimal Observers. The Kalman–Bucy Filter, 240
6.14 Direct Output Feedback Control, 252
6.15 Modal Control, 255

7. DISTRIBUTED-PARAMETER STRUCTURES. EXACT AND APPROXIMATE METHODS   269

7.1   Boundary-Value Problems, 270
7.2   The Differential Eigenvalue Problem, 274
7.3   Rayleigh's Quotient, 280
7.4   The Rayleigh–Ritz Method, 282
7.5   The Finite Element Method, 289
7.6   The Method of Weighted Residuals, 298
7.7   Substructure Synthesis, 300
7.8   Response of Undamped Structures, 304
7.9   Damped Structures, 309

8. CONTROL OF DISTRIBUTED STRUCTURES   313

8.1   Closed-Loop Partial Differential Equation of Motion, 314
8.2   Modal Equations for Undamped Structures, 315

8.3   Mode Controllability and Observability, 317
8.4   Closed-Loop Modal Equations, 318
8.5   Independent Modal-Space Control, 320
8.6   Coupled Control, 323
8.7   Direct Output Feedback Control, 333
8.8   Systems with Proportional Damping, 336
8.9   Control of Discretized Structures, 338
8.10 Structures with General Viscous Damping, 348

9. A REVIEW OF LITERATURE ON STRUCTURAL CONTROL 353

9.1   Issues in Modeling and Control Design, 354
9.2   Methods, Procedures and Approaches, 364
9.3   Aircraft and Rotorcraft Control, 373
9.4   Control of Civil Structures, 375
9.5   Sound Radiation Suppression, 376
9.6   Maneuvering of Space Structures. Robotics, 377
9.7   Control of Space Structures and Launch Vehicles, 379
9.8   Miscellaneous Applications, 380
9.9   Experimental Work and Flight Test, 381

REFERENCES 383
AUTHOR INDEX 411
SUBJECT INDEX 417

PREFACE

This book represents a blend of the various disciplines involved in the control of structures, namely, analytical mechanics, structural dynamics and control theory. Although tailored to the particular needs of the subject, the coverage of the material from each of the disciplines is sufficiently detailed so as to permit a broad and complete picture of the field of control of structures. Moreover, the integrated treatment of the various areas should give the reader an appreciation of how these areas fit together. The book is conceived as a professional reference book, although it should prove quite suitable as a textbook for graduate courses or short courses on the subject. To help the reader form an idea about the book, a chapter-by-chapter review follows.

The first chapter represents a brief discussion of Newtonian mechanics. Its purpose is to introduce certain important concepts from dynamics. Both systems of particles and rigid bodies are discussed. Chapter 2 includes some pertinent subjects from analytical mechanics, including Lagrange's equations, Hamilton's canonical equations and Lagrange's equations in term of quasi-coordinates. Chapter 3 presents various concepts from linear system theory of particular interest in control theory. Chapter 4 is in essence a short course in the vibration of lumped-parameter systems, with emphasis on computational techniques. Chapter 5 contains an introduction to classical control theory. Chapter 6 represents a short course in modern control theory. Chapters 5 and 6 are concerned entirely with the control of lumped-parameter structures. Chapter 7 is concerned with dynamics of distributed structures. The chapter includes important discretization procedures, such as the finite element method and substructure synthesis. Chapter 8 considers the control of distributed-parameter structures, as well as the control of discretized (-in-space) structures. Chapter 9 represents an extensive up-to-date review of literature on the control of civil, mechanical, aircraft and spacecraft structures.

The author wishes to express his appreciation to M. Y. Chang, M. E. B. France, M. K. Kwak and Y. Sharony for their valuable suggestions. Special thanks are due to Mark A. Norris for working out some of the numerical examples. Last but not least, the author wishes to thank Vanessa M. McCoy and Susanne M. Davis for typing the manuscript.

LEONARD MEIROVITCH

Blacksburg, Virginia
December 1989

DYNAMICS AND CONTROL
OF STRUCTURES

INTRODUCTION

In the last several decades, a new and exciting idea has been attracting increasing attention in many areas of engineering. The idea is that the performance of structures can be improved by the use of active control, which in essence permits the design of lighter structures than those without control. The implication of active control is that desirable performance characteristics can be achieved through feedback control, whereby actuators apply forces to a structure based on the structure response as measured by sensors. In contrast, passive control improves the performance of a structure through the use of materials or devices enhancing the damping and stiffness characteristics of the structure.

Control of structures involves a number of areas, in particular, analytical dynamics for efficient derivation of the equations of motion, structural dynamics for modeling and analysis and control theory for control system design. Good structural control demands a good understanding of all the areas involved, as conflicting requirements imposed by the different areas make some compromise a virtual necessity. These conflicts arise from the fact that structures are essentially distributed-parameter systems and the bulk of the control theory was developed for discrete systems, in particular for systems of relatively low order. It is common practice to approximate the behavior of distributed structures by discrete models derived through some discretization (-in-space) process. Faithful models, however, tend to require a large number of degrees of freedom, well beyond the capability of control algorithms. Hence, the compromise mentioned above amounts to developing a suitable reduced-order model so that the control system designed on the basis of the reduced-order model will perform well when applied to the actual distributed structure.

This book attempts to present the various disciplines involved in the control of structures in an integrated fashion. Both lumped-parameter and distributed-parameter structures are discussed. Moreover, both classical control and modern control are presented. The book should permit the reader to become acquainted with all aspects of control of structures, including the derivation of the equations of motion, structural modeling and control design.

CHAPTER 1

NEWTONIAN MECHANICS

Newtonian mechanics is based on three laws enunciated by Newton in 1687. The laws were formulated for a single particle and they can be extended to systems of particles and rigid bodies. They postulate the existence of inertial systems in which the laws of motion are valid, where an inertial space is either fixed or moving with uniform velocity relative to "fixed stars."

1.1 NEWTON'S SECOND LAW

Of the three laws of Newton, the second law is the most important. *Newton's second law* can be stated as follows: *A particle acted upon by a force moves so that the force vector is equal to the time rate of change of the linear momentum vector.* The mathematical expression of Newton's second law is

$$\mathbf{F} = \dot{\mathbf{p}} \tag{1.1}$$

where $\mathbf{F}$ is the *force vector* and

$$\mathbf{p} = m\mathbf{v} \tag{1.2}$$

is the *linear momentum vector* (Fig. 1.1), in which m is the *mass* of the particle, assumed to be constant, and $\mathbf{v}$ is its *absolute velocity*, i.e., the velocity relative to the inertial space.

NEWTONIAN MECHANICS

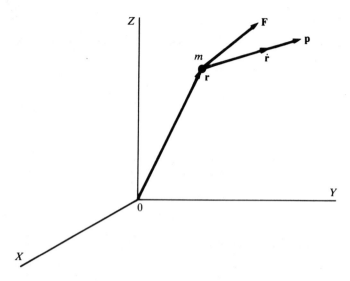

Figure 1.1

Example 1.1 The system shown in Fig. 1.2a consists of a mass m subjected to a force $F(t)$ while connected to a wall through a spring and dashpot. The spring resists elongations with a force equal to the product of the spring stiffness and the elongation. The implication is that the spring exhibits linear behavior. The spring stiffness is denoted by k, where k is commonly known as the spring constant. The dashpot represents a viscous damper that resists the velocity of separation between the two ends with a force equal to the product of a constant, known as the coefficient of viscous damping, and the velocity. The coefficient of viscous damping is denoted by c. Derive the equation of motion for the system by means of Newton's second law.

The motion takes place in the horizontal direction, so that there is only one component of motion, namely, the displacement $x(t)$. In view of this, we can dispense with the vector notation. The basic tool in applying

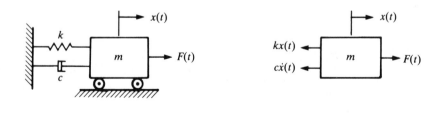

a.   b.

Figure 1.2

Newton's second law is the free-body diagram. This is a drawing of the isolated mass with all the forces acting upon it. In isolating the mass, one must cut through components, such as the spring and the dashpot. In the process, forces internal to the spring and dashpot become external. The free-body diagram for the system is shown in Fig. 1.2b. Hence, using Eq. (1.1) and recognizing that the left side represents the resultant force acting on m, we can write

$$F(t) - c\dot{x}(t) - kx(t) = \dot{p}(t) = m\dot{v}(t) = m\ddot{x}(t) \tag{a}$$

Rearranging, we obtain the equation of motion

$$m\ddot{x}(t) + c\dot{x}(t) + kx(t) = F(t) \tag{b}$$

Example 1.2 Derive the equation of motion for the simple pendulum shown in Fig. 1.3a.

Figure 1.3b shows the corresponding free-body diagram, in which $\mathbf{u}_t$ is a unit vector in the tangential direction t, $\mathbf{u}_n$ is a unit vector in the normal direction n and $\mathbf{k}$ is a unit vector perpendicular to the plane of motion. We observe that the unit vectors $\mathbf{u}_t$ and $\mathbf{u}_n$ are not constant, as their direction depends on the angle θ. In using Eq. (1.1), we need the time rate of change of the momentum, which involves the acceleration of the point mass m. To derive the acceleration, we write the radius vector from 0 to m in the form

$$\mathbf{r} = -L\mathbf{u}_n \tag{a}$$

The acceleration is then obtained by taking the second derivative of $\mathbf{r}$ with

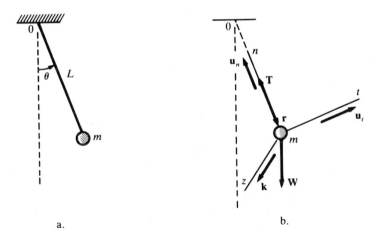

Figure 1.3

respect to time. In taking derivatives with respect to time, we recognize that the unit vectors $\mathbf{u}_t$ and $\mathbf{u}_n$ rotate with the angular velocity

$$\boldsymbol{\omega} = \dot{\theta}\mathbf{k} \tag{b}$$

Then, it can be shown that [M17]

$$\frac{d\mathbf{u}_n}{dt} = \boldsymbol{\omega} \times \mathbf{u}_n = \dot{\theta}\mathbf{k} \times \mathbf{u}_n = -\dot{\theta}\mathbf{u}_t, \quad \frac{d\mathbf{u}_t}{dt} = \boldsymbol{\omega} \times \mathbf{u}_t = \dot{\theta}\mathbf{k} \times \mathbf{u}_t = \dot{\theta}\mathbf{u}_n \tag{c}$$

Hence, the velocity vector is simply

$$\mathbf{v} = \frac{d\mathbf{r}}{dt} = -L\dot{\mathbf{u}}_n = L\dot{\theta}\mathbf{u}_t \tag{d}$$

and the acceleration vector is

$$\mathbf{a} = \frac{d\mathbf{v}}{dt} = \frac{d^2\mathbf{r}}{dt^2} = L\ddot{\theta}\mathbf{u}_t + L\dot{\theta}\dot{\mathbf{u}}_t = L\ddot{\theta}\mathbf{u}_t + L\dot{\theta}^2\mathbf{u}_n \tag{e}$$

The two forces acting on m are the tension in the string and the weight. They have the vector form

$$\mathbf{T} = T\mathbf{u}_n, \quad \mathbf{W} = -mg(\sin\theta\,\mathbf{u}_t + \cos\theta\,\mathbf{u}_n) \tag{f}$$

Using Eqs. (1.1) and (1.2), we can write

$$\mathbf{F} = \mathbf{T} + \mathbf{W} = m\mathbf{a} \tag{g}$$

so that, inserting Eqs. (e) and (f) into Eq. (g), we obtain the equations of motion in the tangential and normal directions

$$\begin{aligned} -mg\sin\theta &= mL\ddot{\theta} \\ T - mg\cos\theta &= mL\dot{\theta}^2 \end{aligned} \tag{h}$$

The first of Eqs. (h) gives the equation of motion

$$mL\ddot{\theta} + mg\sin\theta = 0 \tag{i}$$

whereas the second of Eqs. (h) yields the tension in the string

$$T = mg\cos\theta + mL\dot{\theta}^2 \tag{j}$$

1.2 IMPULSE AND MOMENTUM

Let us multiply both sides of Eq. (1.1) by dt and integrate with respect to time between the times t_1 and t_2 to obtain

$$\int_{t_1}^{t_2} \mathbf{F}\, dt = \int_{t_1}^{t_2} \frac{d\mathbf{p}}{dt}\, dt = \mathbf{p}_2 - \mathbf{p}_1 \tag{1.3}$$

The integral $\int_{t_1}^{t_2} \mathbf{F}\, dt$ is known as the *impulse vector* and the difference $\mathbf{p}_2 - \mathbf{p}_1$ can be identified as the change $\Delta \mathbf{p}$ in the momentum vector corresponding to the time interval $\Delta t = t_2 - t_1$. Hence, *the impulse vector is equal to the change in the momentum vector.*

In the absence of external forces acting upon the particle, $\mathbf{F} = \mathbf{0}$, Eq. (1.3) yields

$$\mathbf{p} = m\mathbf{v} = \text{constant} \tag{1.4}$$

which is known as the *conservation of the linear momentum vector*. Equation (1.4) is the essence of *Newton's first law*, which is clearly a special case of the second law.

1.3 MOMENT OF A FORCE AND ANGULAR MOMENTUM

Let *XYZ* be an inertial system with the origin at 0 (Fig. 1.1). The *moment of momentum*, or *angular momentum*, of m with respect to 0 is defined as

$$\mathbf{H}_0 = \mathbf{r} \times \mathbf{p} = \mathbf{r} \times m\dot{\mathbf{r}} \tag{1.5}$$

The rate of change of the angular momentum is

$$\dot{\mathbf{H}}_0 = \dot{\mathbf{r}} \times m\dot{\mathbf{r}} + \mathbf{r} \times m\ddot{\mathbf{r}} = \mathbf{r} \times m\ddot{\mathbf{r}} \tag{1.6}$$

By Newton's second law, however, $m\ddot{\mathbf{r}}$ is the force $\mathbf{F}$. Moreover, by definition $\mathbf{r} \times \mathbf{F}$ represents the *moment of the force* $\mathbf{F}$ about 0, denoted by $\mathbf{M}_0$. Hence, Eq. (1.6) can be rewritten as

$$\mathbf{M}_0 = \dot{\mathbf{H}}_0 \tag{1.7}$$

or *the moment of a force about 0 is equal to the rate of change of the angular momentum about 0.*

By analogy with the linear momentum, we can write

$$\int_{t_1}^{t_2} \mathbf{M}_0\, dt = \Delta \mathbf{H}_0 \tag{1.8}$$

or, the *angular impulse vector is equal to the change in the angular momentum vector*. If $\mathbf{M}_0 = \mathbf{0}$, we obtain

$$\mathbf{H}_0 = \text{constant} \tag{1.9}$$

8   NEWTONIAN MECHANICS

where the latter is known as the *conservation of the angular momentum*. Note that the conservation of angular momentum does not require $\mathbf{F}$ to be zero. Indeed, $\mathbf{M}_0$ is zero when $\mathbf{F} = \mathbf{0}$, but it is zero also when $\mathbf{F}$ passes through 0.

Example 1.3 Derive the equation of motion for the simple pendulum of Fig. 1.3a by means of Eq. (1.7).

The simple pendulum was discussed in Example 1.2. In deriving the moment equation of motion, we propose to use many of the expressions derived there. Hence, inserting Eqs. (a) and (d) of Example 1.2 into Eq. (1.5), we can write the angular momentum vector about 0 in the form

$$\mathbf{H}_0 = \mathbf{r} \times m\dot{\mathbf{r}} = (-L\mathbf{u}_n) \times mL\dot{\theta}\mathbf{u}_t = mL^2\dot{\theta}\,\mathbf{k} \tag{a}$$

Moreover, using Eqs. (a) and (f) of Example 1.2, we obtain the moment of the force about 0

$$\mathbf{M}_0 = \mathbf{r} \times \mathbf{F} = \mathbf{r} \times (\mathbf{T} + \mathbf{W})$$
$$= (-L\mathbf{u}_n) \times [T\mathbf{u}_n - mg(\sin\theta\,\mathbf{u}_t + \cos\theta\,\mathbf{u}_n)] = -mgL\sin\theta\,\mathbf{k} \tag{b}$$

and we observe that both the angular momentum and the moment of the force have only one component, namely, that in the z-direction. This direction is fixed in the inertial space, so that the unit vector $\mathbf{k}$ is constant. Introducing Eqs. (a) and (b) into Eq. (1.7) and dispensing with the vector notation, we have

$$M_0 = -mgL\sin\theta = \dot{H}_0 = \frac{d}{dt}(mL^2\dot{\theta}) = mL^2\ddot{\theta} \tag{c}$$

Upon rearranging, we obtain the moment equation of motion

$$mL^2\ddot{\theta} + mgL\sin\theta = 0 \tag{d}$$

which represents Eq. (i) of Example 1.2 multiplied by L.

1.4 WORK AND ENERGY

Consider a particle of mass m moving along curve C while under the action of the force $\mathbf{F}$ (Fig. 1.4). By definition, the *increment of work* associated with the change in position $d\mathbf{r}$ is

$$\overline{dW} = \mathbf{F} \cdot d\mathbf{r} \tag{1.10}$$

where the overbar indicates that $\overline{dW}$ is not to be regarded as the true

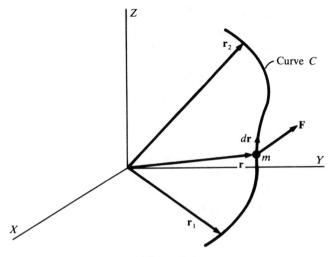

Figure 1.4

differential of a function W but simply as an infinitesimal expression. By Newton's second law, however, $\mathbf{F} = m\ddot{\mathbf{r}}$, so that Eq. (1.10) yields

$$\overline{dW} = m\ddot{\mathbf{r}} \cdot d\mathbf{r} = m\frac{d\dot{\mathbf{r}}}{dt} \cdot \dot{\mathbf{r}}\, dt = m\dot{\mathbf{r}} \cdot d\dot{\mathbf{r}} = d\left(\frac{1}{2} m\dot{\mathbf{r}} \cdot \dot{\mathbf{r}}\right) = dT \quad (1.11)$$

where

$$T = \frac{1}{2} m\dot{\mathbf{r}} \cdot \dot{\mathbf{r}} \quad (1.12)$$

is called the *kinetic energy*. Integrating Eq. (1.11) and considering Eq. (1.10), we obtain

$$\int_{\mathbf{r}_1}^{\mathbf{r}_2} \mathbf{F} \cdot d\mathbf{r} = \int_{T_1}^{T_2} dT = T_2 - T_1 \quad (1.13)$$

which implies that *the work performed by* $\mathbf{F}$ *in carrying m from* $\mathbf{r}_1$ *to* $\mathbf{r}_2$ *is equal to the change in the kinetic energy* between the two points.

In many physical problems, the force $\mathbf{F}$ depends on position alone, $\mathbf{F} = \mathbf{F}(\mathbf{r})$. In this case, the increment of work can be expressed as the perfect differential

$$dW_c = \mathbf{F} \cdot d\mathbf{r} = -dV \quad (1.14)$$

where V is the *potential energy*. Its definition is

$$V(\mathbf{r}) = \int_{\mathbf{r}}^{\mathbf{r}_0} \mathbf{F} \cdot d\mathbf{r} \quad (1.15)$$

where $\mathbf{r}_0$ is a reference position. Forces that depend on position alone are said to be *conservative*, which explains the subscript c in Eq. (1.14). In general, there are both conservative and *nonconservative forces*, so that

$$\mathbf{F} = \mathbf{F}_c + \mathbf{F}_{nc} \tag{1.16}$$

where the notation is obvious. Taking the dot product of $\mathbf{F}$ and $d\mathbf{r}$ and considering Eqs. (1.10), (1.11) and (1.14), we obtain

$$\overline{dW} = (\mathbf{F}_c + \mathbf{F}_{nc}) \cdot d\mathbf{r} = -dV + \overline{dW}_{nc} = dT \tag{1.17}$$

where $\overline{dW}_{nc}$ is the increment of work performed by the nonconservative force $\mathbf{F}_{nc}$. Equation (1.17) can be rewritten as

$$\overline{dW}_{nc} = dE \tag{1.18}$$

where

$$E = T + V \tag{1.19}$$

is known as the *total energy*. In the absence of nonconservative forces, $\overline{dW}_{nc} = 0$ and Eq. (1.18) can be integrated to yield

$$E = \text{constant} \tag{1.20}$$

which is recognized as the *principle of conservation of energy*.

Example 1.4 Write Eq. (1.18) for the system of Fig. 1.2a and demonstrate that the equation can be used to derive the equation of motion.

Using Eq. (1.12), we obtain the kinetic energy

$$T = \frac{1}{2} m\dot{x}^2 \tag{a}$$

Moreover, recognizing that the only conservative force is the force in the spring, we can take $x = 0$ as the reference position and use Eq. (1.15) to write the potential energy

$$V = \int_x^0 (-k\zeta) \, d\zeta = -\frac{1}{2} k\zeta^2 \Big|_x^0 = \frac{1}{2} kx^2 \tag{b}$$

where ζ is a mere dummy variable of integration. The remaining two forces, F and $-c\dot{x}$, are nonconservative. Hence, using Eq. (1.17), we can write the increment of work performed by the nonconservative forces

$$\overline{dW}_{nc} = F_{nc}\, dx = (F - c\dot{x})\, dx \qquad (c)$$

Inserting Eqs. (a) and (b) into Eq. (1.19), we obtain the total energy

$$E = T + V = \frac{1}{2} m\dot{x}^2 + \frac{1}{2} kx^2 \qquad (d)$$

so that Eq. (1.18) yields

$$d\left(\frac{1}{2} m\dot{x}^2 + \frac{1}{2} kx^2\right) = (F - c\dot{x})\, dx \qquad (e)$$

To derive the equation of motion, we consider

$$d\left(\frac{1}{2} m\dot{x}^2\right) = m\dot{x}\, d\dot{x} \qquad (f)$$

From kinematics, however,

$$\dot{x}\, d\dot{x} = \ddot{x}\, dx \qquad (g)$$

so that

$$d\left(\frac{1}{2} m\dot{x}^2\right) = m\ddot{x}\, dx \qquad (h)$$

Moreover,

$$d\left(\frac{1}{2} kx^2\right) = kx\, dx \qquad (i)$$

Inserting Eqs. (h) and (i) into Eq. (e), dividing through by dx and rearranging, we obtain

$$m\ddot{x} + c\dot{x} + kx = F \qquad (j)$$

which is identical to Eq. (b) of Example 1.1.

1.5 SYSTEMS OF PARTICLES

Consider the system of N particles of mass m_i ($i = 1, 2, \ldots, N$) shown in Fig. 1.5. The position of the *mass center C* of the system is defined as

$$\mathbf{r}_C = \frac{1}{m} \sum_{i=1}^{N} m_i \mathbf{r}_i \qquad (1.21)$$

12　NEWTONIAN MECHANICS

where $m = \sum_{i=1}^{N} m_i$ is the total mass of the system. Letting $\mathbf{F}_i$ be the external force acting upon a typical particle i and $\mathbf{f}_{ij}$ the internal force exerted by particle j upon particle i, Newton's second law for particle i can be written as

$$\mathbf{F}_i + \sum_{j=1}^{N} (1 - \delta_{ij})\mathbf{f}_{ij} = m_i \ddot{\mathbf{r}}_i, \quad i = 1, 2, \ldots, N \qquad (1.22)$$

where δ_{ij} is the Kronecker delta.

At times, the motion of the individual particles is of no particular interest, and the interest lies in the aggregate motion of the system. To investigate such motion, we sum over the system of particles and obtain

$$\sum_{i=1}^{N} \mathbf{F}_i + \sum_{i=1}^{N} \sum_{j=1}^{N} (1 - \delta_{ij})\mathbf{f}_{ij} = \sum_{i=1}^{N} m_i \ddot{\mathbf{r}}_i \qquad (1.23)$$

But, the internal forces are such that $\mathbf{f}_{ij} = -\mathbf{f}_{ji}$, which is an expression of Newton's third law, so that the double sum in Eq. (1.23) reduces to zero. Moreover, $\sum_{i=1}^{N} \mathbf{F}_i = \mathbf{F}$ is recognized as the resultant of all the external forces acting upon the system. Hence, using Eq. (1.21), Eq. (1.23) reduces to

$$\mathbf{F} = m \ddot{\mathbf{r}}_C \qquad (1.24)$$

Equation (1.24) indicates that the motion of the mass center is the same as the motion of a fictitious body equal in mass to the mass of the system, concentrated at the mass center and acted upon by the resultant of all the external forces.

In the absence of external forces, $\mathbf{F} = \mathbf{0}$, Eq. (1.24) can be integrated to yield $m \dot{\mathbf{r}}_C = \mathbf{p} =$ constant, which represents the *principle of conservation of linear momentum for a system of particles*.

Next, let us consider the angular momentum of the system of particles with respect to any moving point A. Its definition is

$$\mathbf{H}_A = \sum_{i=1}^{N} \mathbf{H}_{Ai} = \sum_{i=1}^{N} \mathbf{r}_{Ai} \times m_i \dot{\mathbf{r}}_i \qquad (1.25)$$

Differentiation of Eq. (1.25) with respect to time yields

$$\dot{\mathbf{H}}_A = \sum_{i=1}^{N} \dot{\mathbf{r}}_{Ai} \times m_i \dot{\mathbf{r}}_i + \sum_{i=1}^{N} \mathbf{r}_{Ai} \times m_i \ddot{\mathbf{r}}_i \qquad (1.26)$$

From Fig. 1.5, we have

$$\mathbf{r}_{Ai} = \mathbf{r}_{AC} + \boldsymbol{\rho}_i, \quad \mathbf{r}_i = \mathbf{r}_C + \boldsymbol{\rho}_i, \quad i = 1, 2, \ldots, N \qquad (1.27\text{a, b})$$

Introducing Eqs. (1.27) into Eq. (1.26) and recalling that pairs of internal

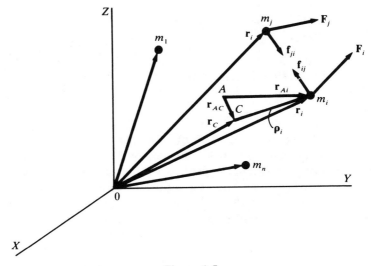

Figure 1.5

forces are not only equal and opposite in direction but also collinear, we can write

$$\dot{\mathbf{H}}_A = \sum_{i=1}^{N} (\dot{\mathbf{r}}_{AC} + \dot{\boldsymbol{\rho}}_i) \times m_i(\dot{\mathbf{r}}_C + \dot{\boldsymbol{\rho}}_i) + \sum_{i=1}^{n} \mathbf{r}_{Ai} \times \mathbf{F}_i$$
$$= \dot{\mathbf{r}}_{AC} \times m\dot{\mathbf{r}}_C + \mathbf{M}_A \tag{1.28}$$

where, from Eq. (1.21), we recognized that $\sum_{i=1}^{N} m_i \boldsymbol{\rho}_i = \mathbf{0}$. Moreover, $\mathbf{M}_A$ is recognized as the torque of the external forces about A.

If point A coincides with the fixed origin 0, then $\mathbf{r}_{AC} = \mathbf{r}_C$ and Eq. (1.28) reduces to the simple form

$$\mathbf{M}_0 = \dot{\mathbf{H}}_0 \tag{1.29}$$

On the other hand, if A coincides with the mass center C, then $\mathbf{r}_{AC} = \mathbf{0}$ and Eq. (1.28) becomes

$$\mathbf{M}_C = \dot{\mathbf{H}}_C \tag{1.30}$$

If $\mathbf{M}_0 = \mathbf{0}$, or $\mathbf{M}_C = \mathbf{0}$, then *the angular momentum about* 0, *or about* C, *is conserved.*

The kinetic energy of a system of particles is defined as

$$T = \frac{1}{2} \sum_{i=1}^{N} m_i \dot{\mathbf{r}}_i \cdot \dot{\mathbf{r}}_i \tag{1.31}$$

Introducing Eq. (1.27b) into Eq. (1.31), we can write

14 NEWTONIAN MECHANICS

$$T = \frac{1}{2} \sum_{i=1}^{N} m_i (\dot{\mathbf{r}}_C + \dot{\boldsymbol{\rho}}_i) \cdot (\dot{\mathbf{r}}_C + \dot{\boldsymbol{\rho}}_i) = \frac{1}{2} m \dot{\mathbf{r}}_C \cdot \dot{\mathbf{r}}_C + \frac{1}{2} \sum_{i=1}^{N} m_i \dot{\boldsymbol{\rho}}_i \cdot \dot{\boldsymbol{\rho}}_i \quad (1.32)$$

so that the kinetic energy of a system of particles is equal to the kinetic energy obtained by regarding the entire mass of the system as concentrated at the mass center plus the kinetic energy of motion relative to the mass center.

***Example* 1.5** Derive the equations of motion for the double pendulum shown in Fig. 1.6a by means of Newton's second law.

Because the double pendulum involves two masses, m_1 and m_2, we must use two free-body diagrams. They are shown in Figs. 1.6b and c, where the notation is consistent with that used in Example 1.2 for the simple pendulum. From Figs. 1.6b and c, we write the equations of motion in the vector form

$$\mathbf{T}_1 - \mathbf{T}_2 + \mathbf{W}_1 = m_1 \mathbf{a}_1, \quad \mathbf{F} + \mathbf{T}_2 + \mathbf{W}_2 = m_2 \mathbf{a}_2 \quad (a)$$

Using Eq. (c) of Example 1.2, the acceleration of m_1 is simply

$$\mathbf{a}_1 = L_1 \ddot{\theta}_1 \mathbf{u}_{t1} + L_1 \dot{\theta}_1^2 \mathbf{u}_{n1} \quad (b)$$

and, from kinematics, the acceleration of m_2 is

$$\mathbf{a}_2 = \mathbf{a}_1 + \mathbf{a}_{12} = \mathbf{a}_1 + L_2 \ddot{\theta}_2 \mathbf{u}_{t2} + L_2 \dot{\theta}_2^2 \mathbf{u}_{n2} \quad (c)$$

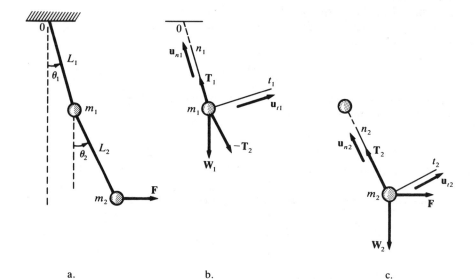

a.   b.   c.

Figure 1.6

where $\mathbf{a}_{12}$ is the acceleration of m_2 relative to m_1. Moreover, the various force vectors can be written as

$$\mathbf{T}_1 = T_1 \mathbf{u}_{n1}, \quad \mathbf{W}_1 = -m_1 g(\sin\theta_1 \mathbf{u}_{t1} + \cos\theta_1 \mathbf{u}_{n1})$$
$$\mathbf{T}_2 = T_2 \mathbf{u}_{n2}, \quad \mathbf{W}_2 = -m_2 g(\sin\theta_2 \mathbf{u}_{t2} + \cos\theta_2 \mathbf{u}_{n2}) \quad \text{(d)}$$
$$\mathbf{F} = F(\cos\theta_2 \mathbf{u}_{t2} - \sin\theta_2 \mathbf{u}_{n2})$$

Equations (a) represent four scalar equations. We propose to write them in explicit form: the equations of motion for m_1 in terms of components along t_1 and n_1, and those for m_2 in terms of components along t_2 and n_2. This necessitates coordinate transformations between the two sets of unit vectors. They can be verified to have the expressions

$$\mathbf{u}_{t1} = \cos(\theta_2 - \theta_1)\mathbf{u}_{t2} - \sin(\theta_2 - \theta_1)\mathbf{u}_{n2}$$
$$\mathbf{u}_{n1} = \sin(\theta_2 - \theta_1)\mathbf{u}_{t2} + \cos(\theta_2 - \theta_1)\mathbf{u}_{n2} \quad \text{(e)}$$
$$\mathbf{u}_{t2} = \cos(\theta_2 - \theta_1)\mathbf{u}_{t1} + \sin(\theta_2 - \theta_1)\mathbf{u}_{n1}$$
$$\mathbf{u}_{n2} = -\sin(\theta_2 - \theta_1)\mathbf{u}_{t1} + \cos(\theta_2 - \theta_1)\mathbf{u}_{n1}$$

Inserting Eqs. (b)–(e) into Eqs. (a), we obtain

$$T_2 \sin(\theta_2 - \theta_1) - m_1 g \sin\theta_1 = m_1 L_1 \ddot{\theta}_1$$
$$T_1 - T_2 \cos(\theta_2 - \theta_1) - m_1 g \cos\theta_1 = m_1 L_1 \dot{\theta}_1^2$$
$$F \cos\theta_2 - m_2 g \sin\theta_2 = m_2[L_1 \ddot{\theta}_1 \cos(\theta_2 - \theta_1) + L_1 \dot{\theta}_1^2 \sin(\theta_2 - \theta_1) + L_2 \ddot{\theta}_2] \quad \text{(f)}$$
$$-F \sin\theta_2 + T_2 - m_2 g \cos\theta_2$$
$$= m_2[-L_1 \ddot{\theta}_1 \sin(\theta_2 - \theta_1) + L_1 \dot{\theta}_1^2 \cos(\theta_2 - \theta_1) + L_2 \dot{\theta}_2^2]$$

The four equations of motion, Eqs. (f), contain the string tensions T_1 and T_2. They play the role of constraint forces and complicate the solution. Fortunately, it is not difficult to eliminate them and to obtain two equations of motion in terms of θ_1 and θ_2 and their time derivatives. In fact, the third of Eqs. (f) is already in such a form. Inserting T_2 from the fourth of Eqs. (f) into the first, using the third of Eqs. (f) and rearranging, we can write the two equations of motion in the form

$$m_1 L_1 \ddot{\theta}_1 + m_2[L_1 \ddot{\theta}_1 \sin(\theta_2 - \theta_1) - L_1 \dot{\theta}_1^2 \cos(\theta_2 - \theta_1)$$
$$- L_2 \dot{\theta}_2^2] \sin(\theta_2 - \theta_1) + m_1 g \sin\theta_1 - m_2 g \cos\theta_2 \sin(\theta_2 - \theta_1)$$
$$= F \sin\theta_2 \sin(\theta_2 - \theta_1) \quad \text{(g)}$$
$$m_2[L_1 \ddot{\theta}_1 \cos(\theta_2 - \theta_1) + L_1 \dot{\theta}_1^2 \sin(\theta_2 - \theta_1) + L_2 \ddot{\theta}_2] + m_2 g \sin\theta_2$$
$$= F \cos\theta_2$$

1.6 RIGID BODIES

A rigid body can be regarded as a system of particles in which the distance between any two particles is constant, so that the motion relative to the mass center is simply due to rigid-body rotation. Rigid bodies are ordinarily treated as continuous bodies, so that the index i is replaced by the spatial position. Moreover, summation over the system is replaced by integration in a process in which $m_i \to dm$.

The force equation remains in the form of Eq. (1.24) and the moment equation about the mass center C remains in the form of Eq. (1.30), but the angular momentum about C can be given a more specific form. Letting A coincide with C, so that $\mathbf{r}_{Ai} \to \boldsymbol{\rho}$ and $\dot{\mathbf{r}}_i \to \dot{\mathbf{r}}$, we obtain from Eq. (1.25) and Fig. 1.7

$$\mathbf{H}_C = \int_m \boldsymbol{\rho} \times \dot{\mathbf{r}} \, dm \tag{1.33}$$

But the absolute velocity of dm can be written in the form [M17]

$$\dot{\mathbf{r}} = \dot{\mathbf{r}}_C + \dot{\boldsymbol{\rho}} = \dot{\mathbf{r}}_C + \boldsymbol{\omega} \times \boldsymbol{\rho} \tag{1.34}$$

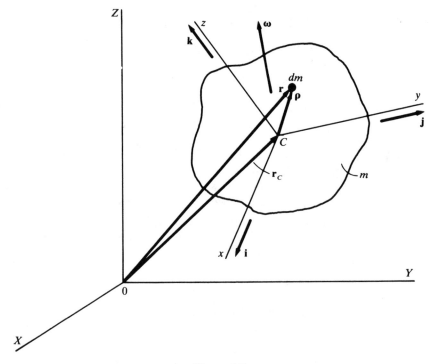

Figure 1.7

where $\boldsymbol{\rho}$ is the position of dm relative to the mass center C. Introducing Eq. (1.34) into Eq. (1.33) and recognizing that Eq. (1.21) implies that $\int_m \boldsymbol{\rho} \, dm = \mathbf{0}$, we obtain

$$\mathbf{H}_C = \int_m \boldsymbol{\rho} \times (\boldsymbol{\omega} \times \boldsymbol{\rho}) \, dm \tag{1.35}$$

The vectors $\boldsymbol{\rho}$ and $\boldsymbol{\omega}$ can be written in terms of components along the body axes xyz as follows:

$$\boldsymbol{\rho} = x\mathbf{i} + y\mathbf{j} + z\mathbf{k} \tag{1.36a}$$

$$\boldsymbol{\omega} = \omega_x \mathbf{i} + \omega_y \mathbf{j} + \omega_z \mathbf{k} \tag{1.36b}$$

where $\mathbf{i}$, $\mathbf{j}$ and $\mathbf{k}$ are unit vectors (Fig. 1.7). Introducing Eqs. (1.36) into Eq. (1.35), we can write the vector $\mathbf{H}_C$ in terms of cartesian components in the form

$$\mathbf{H}_C = H_x \mathbf{i} + H_y \mathbf{j} + H_z \mathbf{k} \tag{1.37}$$

where

$$\begin{aligned} H_x &= I_{xx}\omega_x - I_{xy}\omega_y - I_{xz}\omega_z \\ H_y &= -I_{xy}\omega_x + I_{yy}\omega_y - I_{yz}\omega_z \\ H_z &= -I_{xz}\omega_x - I_{yz}\omega_y + I_{zz}\omega_z \end{aligned} \tag{1.38}$$

in which

$$I_{xx} = \int_m (\rho^2 - x^2) \, dm, \quad I_{yy} = \int_m (\rho^2 - y^2) \, dm, \quad I_{zz} = \int_m (\rho^2 - z^2) \, dm$$

$$I_{xy} = I_{yx} = \int_m xy \, dm, \quad I_{xz} = I_{zx} = \int_m xz \, dm, \quad I_{yz} = I_{zy} = \int_m yz \, dm \tag{1.39}$$

are mass moments and products of inertia.

Because $\mathbf{H}_C$ is expressed in terms of components along moving axes, the moment equation, Eq. (1.30), has the form [M17]

$$\mathbf{M}_C = \dot{\mathbf{H}}_C = \dot{\mathbf{H}}'_C + \boldsymbol{\omega} \times \mathbf{H}_C \tag{1.40}$$

where $\dot{\mathbf{H}}'_C$ is the time derivative of $\mathbf{H}_C$ obtained by regarding axes xyz as fixed. The cartesian components of $\mathbf{M}_C$ can be obtained by inserting Eqs. (1.36b) and (1.38) into Eq. (1.40). This task is left as an exercise to the reader.

18 NEWTONIAN MECHANICS

Because $\dot{\boldsymbol{\rho}}$ is due to the rotation of the body about C, the kinetic energy, Eq. (1.32), can be written in the form

$$T = T_{tr} + T_{rot} \tag{1.41}$$

where

$$T_{tr} = \frac{1}{2} m \dot{\mathbf{r}}_C \cdot \dot{\mathbf{r}}_C \tag{1.42}$$

is the kinetic energy of translation and

$$T_{rot} = \frac{1}{2} \int_m (\boldsymbol{\omega} \times \boldsymbol{\rho}) \cdot (\boldsymbol{\omega} \times \boldsymbol{\rho}) \, dm$$

$$= \frac{1}{2} (I_{xx} \omega_x^2 + I_{yy} \omega_y^2 + I_{zz} \omega_z^2 - 2 I_{xy} \omega_x \omega_y - 2 I_{xz} \omega_x \omega_z - 2 I_{yz} \omega_y \omega_z) \tag{1.43}$$

is the kinetic energy of rotation.

Example 1.6 The system shown in Fig. 1.8a consists of two uniform rigid links of mass m and length L and a spring of stiffness k. The links are hinged at points 0, A and B and the pulley at B is massless. When the links are horizontal the spring is undeformed. Derive the equation of motion by means of Newton's second law.

Although the system consists of two rigid bodies, and it requires two free-body diagrams, it is constrained in such a fashion that its motion is fully

a.
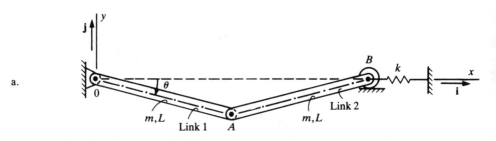

b.
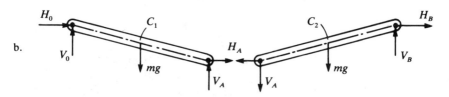

Figure 1.8

defined by a single coordinate, namely the angle $\theta(t)$. The free-body diagrams are shown in Fig. 1.8b. Because the motion is planar, there are three equations of motion for each link, two force equations for the motion of the mass center in the x- and y-directions and one moment equation in the z-direction for rotation about the mass center. To describe the translation of the two bodies, we consider the position vectors of the displaced mass centers

$$\mathbf{r}_{C1} = \frac{L}{2}(\cos\theta\,\mathbf{i} - \sin\theta\,\mathbf{j}), \quad \mathbf{r}_{C2} = \frac{L}{2}(3\cos\theta\,\mathbf{i} - \sin\theta\,\mathbf{j}) \tag{a}$$

Then, taking time derivatives, we obtain the velocity vectors

$$\mathbf{v}_{C1} = \dot{\mathbf{r}}_{C1} = -\frac{L}{2}(\dot{\theta}\sin\theta\,\mathbf{i} + \dot{\theta}\cos\theta\,\mathbf{j})$$
$$\mathbf{v}_{C2} = \dot{\mathbf{r}}_{C2} = -\frac{L}{2}(3\dot{\theta}\sin\theta\,\mathbf{i} + \dot{\theta}\cos\theta\,\mathbf{j}) \tag{b}$$

and the acceleration vectors

$$\mathbf{a}_{C1} = \ddot{\mathbf{r}}_{C1} = -\frac{L}{2}[(\ddot{\theta}\sin\theta + \dot{\theta}^2\cos\theta)\mathbf{i} + (\ddot{\theta}\cos\theta - \dot{\theta}^2\sin\theta)\mathbf{j}]$$
$$\mathbf{a}_{C2} = \ddot{\mathbf{r}}_{C2} = -\frac{L}{2}[3(\ddot{\theta}\sin\theta + \dot{\theta}^2\cos\theta)\mathbf{i} + (\ddot{\theta}\cos\theta - \dot{\theta}^2\sin\theta)\mathbf{j}] \tag{c}$$

Considering the free-body diagrams of Fig. 1.8b and using Eqs. (1.24) and (1.30) in conjunction with Eqs. (c), we can write the equations of motion by components

$$F_{1x} = m_1 \ddot{r}_{C1x} \rightarrow H_0 + H_A = m\left[-\frac{L}{2}(\ddot{\theta}\sin\theta + \dot{\theta}^2\cos\theta)\right]$$

$$F_{1y} = m_1 \ddot{r}_{C1y} \rightarrow V_0 + V_A - mg = m\left[-\frac{L}{2}(\ddot{\theta}\cos\theta - \dot{\theta}^2\sin\theta)\right]$$

$$M_{C1} = I_{C1}\dot{\omega}_1 \rightarrow (V_A - V_0)\frac{L}{2}\cos\theta + (H_A - H_0)\frac{L}{2}\sin\theta = -\frac{mL^2}{12}\ddot{\theta} \tag{d}$$

$$F_{2x} = m_2 \ddot{r}_{C2x} \rightarrow -H_A + H_B = m\left[-\frac{3L}{2}(\ddot{\theta}\sin\theta + \dot{\theta}^2\cos\theta)\right]$$

$$F_{2y} = m_2 \ddot{r}_{C2y} \rightarrow -V_A + V_B - mg = m\left[-\frac{L}{2}(\ddot{\theta}\cos\theta - \dot{\theta}^2\sin\theta)\right]$$

$$M_{C2} = I_{C2}\dot{\omega}_2 \rightarrow (V_A + V_B)\frac{L}{2}\cos\theta - (H_A + H_B)\frac{L}{2}\sin\theta = \frac{mL^2}{12}\ddot{\theta}$$

where we used $m_1 = m_2 = m$, $I_{C1} = I_{C2} = mL^2/12$ and $\omega_1 = -\omega_2 = -\dot{\theta}$. Moreover,

$$H_B = 2kL(1 - \cos \theta) \tag{e}$$

Equations (d) represent six equations and, in addition to $\theta(t)$, there are five unknowns, H_0, V_0, H_A, V_A and V_B, where we note that H_B can be removed from Eqs. (d) by using Eq. (e). It is possible to eliminate these five unknowns to obtain an equation in terms of $\theta(t)$ and its time derivatives alone. After lengthy manipulations, we obtain the desired equation of motion

$$mL^2\ddot{\theta}\left(\frac{2}{3} + 2\sin\theta\right) + 2mL^2\dot{\theta}^2 \sin\theta \cos\theta - mgL \cos\theta \\ + 4kL^2(1 - \cos\theta)\sin\theta = 0 \tag{f}$$

1.7 EULER'S MOMENT EQUATIONS

The moment equations, Eq. (1.40), reduce to a considerably simpler form when the body axes xyz coincide with the principal axes. In this case, the products of inertia vanish, so that the components of the angular momentum vector, Eq. (1.38), reduce to

$$H_x = I_{xx}\omega_x, \quad H_y = I_{yy}\omega_y, \quad H_z = I_{zz}\omega_z \tag{1.44}$$

Inserting Eq. (1.44) into Eq. (1.40), recalling Eq. (1.36b) and expressing the result in terms of cartesian components, we obtain

$$\begin{aligned} I_{xx}\dot{\omega}_x + (I_{zz} - I_{yy})\omega_y\omega_z &= M_x \\ I_{yy}\dot{\omega}_y + (I_{xx} - I_{zz})\omega_x\omega_z &= M_y \\ I_{zz}\dot{\omega}_z + (I_{yy} - I_{xx})\omega_x\omega_y &= M_z \end{aligned} \tag{1.45}$$

which are the celebrated *Euler's moment equations*. The equations are used widely in problems involving the rotational motion of rigid bodies.

CHAPTER 2

PRINCIPLES OF ANALYTICAL MECHANICS

Newtonian mechanics describes the motion in terms of physical coordinates. Equations of motion must be derived for every mass separately and constraint forces appear explicitly in the problem formulation. Newtonian mechanics works with vector quantities. By contrast, Lagrangian mechanics describes the motion in terms of more abstract generalized coordinates. For this reason, Lagrangian mechanics is commonly referred to as analytical mechanics. In analytical mechanics, the equations of motion are derived by means of scalar functions, namely, the kinetic and potential energy, and an infinitesimal expression known as the virtual work. Moreover, constraint forces do not appear explicitly. For many applications, analytical mechanics has many advantages. Methods of analytical mechanics, and in particular Hamilton's principle and Lagrange's equations, are used extensively in this text to derive equations of motion for both lumped and distributed structures.

2.1 DEGREES OF FREEDOM AND GENERALIZED COORDINATES

Let us consider a system of N particles whose positions are given by the vectors $\mathbf{r}_i = \mathbf{r}_i(x_i, y_i, z_i)$, where x_i, y_i, z_i ($i = 1, 2, \ldots, N$) are the cartesian coordinates of the ith particle. The motion of the system is defined completely by the positions of all particles i as functions of the time t, $x_i = x_i(t)$, $y_i = y_i(t)$, $z_i = z_i(t)$. The motion can be interpreted conveniently in terms of geometrical concepts by regarding x_i, y_i, z_i ($i = 1, 2, , \ldots, N$) as the coordinates of a representative point in a $3N$-dimensional space known as the *configuration space*. As time unfolds, the representative point traces a solution curve in the configuration space.

Quite frequently, it is more advantageous to express the motion in terms of a different set of coordinates, say, $q_1, q_2, \ldots, q_n$, where $n = 3N$. The relation between the old and new coordinates can be given in general form by the coordinate transformation

$$\begin{aligned} x_1 &= x_1(q_1, q_2, \ldots, q_n) \\ y_1 &= y_1(q_1, q_2, \ldots, q_n) \\ &\vdots \\ z_N &= z_N(q_1, q_2, \ldots, q_n) \end{aligned} \qquad (2.1)$$

Generally, the coordinates $q_1, q_2, \ldots, q_n$ are chosen so as to render the formulation of the problem easier. One simple example is the use of polar coordinates to describe orbital motion of planets and satellites, in which case $q_1 = r$, $q_2 = \theta$.

In the above discussion, it is implied that the particles are free. In many problems, however, the particles are not free but subject to certain kinematical conditions restricting their freedom of motion. As an example, if the distance between particles 1 and 2 is fixed, such as in the case of a dumbbell, the coordinates of these two particles must satisfy

$$(x_1 - x_2)^2 + (y_1 - y_2)^2 + (z_1 - z_2)^2 = L^2 = \text{constant} \qquad (2.2)$$

where L is the distance between the two particles. Equation (2.2) is known as a constraint equation. If a system of N particles is subject to C constraints, then there are only n independent coordinates q_k ($k = 1, 2, \ldots, n$), where

$$n = 3N - C \qquad (2.3)$$

is the number of *degrees of freedom* of the system. Hence, the number of degrees of freedom of a system coincides with the number of independent coordinates necessary to describe the motion of the system. The n coordinates $q_1, q_2, \ldots, q_n$ are known as *generalized coordinates*. We shall see later that generalized coordinates do not necessarily have physical meaning. Indeed, in some cases they can be the coefficients of a series expansion.

2.2 THE PRINCIPLE OF VIRTUAL WORK

The principle of virtual work is essentially a statement of the static equilibrium of a system. Our interest in the principle is as a means for carrying out the transition from Newtonian to Lagrangian mechanics.

THE PRINCIPLE OF VIRTUAL WORK

Let us consider a system of N particles and denote by $\mathbf{R}_i$ ($i = 1, 2, \ldots, N$) the resultant force acting on particle i. For a system in equilibrium, $\mathbf{R}_i = \mathbf{0}$ and the same can be said about the dot product $\mathbf{R}_i \cdot \delta \mathbf{r}_i$, where $\mathbf{R}_i \cdot \delta \mathbf{r}_i$ represents the *virtual work* performed by the force $\mathbf{R}_i$ through the *virtual displacement* $\delta \mathbf{r}_i$. Note that virtual displacements are not necessarily actual displacements but infinitesimal changes in the coordinates. The virtual displacements are consistent with the system constraints, but they are otherwise arbitrary. Moreover, they are imagined to take place contemporaneously. This implies that the virtual displacements are the result of imagining the system in a slightly displaced position, a process that takes no time. Because $\mathbf{R}_i \cdot \delta \mathbf{r}_i$ is zero for every particle, the virtual work for the entire system vanishes, or

$$\overline{\delta W} = \sum_{i=1}^{N} \mathbf{R}_i \cdot \delta \mathbf{r}_i = 0 \qquad (2.4)$$

If there are constraint forces acting on the system, then we can write

$$\mathbf{R}_i = +\mathbf{F}_i + \mathbf{F}'_i, \quad i = 1, 2, \ldots, N \qquad (2.5)$$

where $\mathbf{F}_i$ are applied forces and $\mathbf{F}'_i$ are constraint forces. Introducing Eq. (2.5) into Eq. (2.4), we obtain

$$\overline{\delta W} = \sum_{i=1}^{N} \mathbf{F}_i \cdot \delta \mathbf{r}_i + \sum_{i=1}^{N} \mathbf{F}'_i \cdot \delta \mathbf{r}_i = 0 \qquad (2.6)$$

But constraint forces perform no work. As an example, if a particle is constrained to move on a smooth surface, then the constraint force is normal to the surface, whereas the virtual displacement must be parallel to the surface, consistent with the constraint. Hence, the second sum in Eq. (2.6) is equal to zero, so that the equation reduces to

$$\overline{\delta W} = \sum_{i=1}^{N} \mathbf{F}_i \cdot \delta \mathbf{r}_i = 0 \qquad (2.7)$$

or, the work done by the applied forces through virtual displacements compatible with the system constraints is zero. This is the statement of the *principle of virtual work*.

The principle of virtual work can be expressed in terms of generalized coordinates in the form

$$\overline{\delta W} = \sum_{k=1}^{n} Q_k \, \delta q_k = 0 \qquad (2.8)$$

where δq_k ($k = 1, 2, \ldots, n$) are *virtual generalized displacements*. Accordingly, Q_k ($k = 1, 2, \ldots, n$) are known as *generalized forces*.

Example 2.1 Consider the system of Example 1.6 and determine the equilibrium position by means of the virtual work principle.

Figure 2.1 shows a diagram of the system with all the forces acting upon it. The forces at points C_1 and C_2 are due to gravity, and we note that, in a uniform gravitational field, the resultant force can be assumed to act at the center of mass of a body. Moreover, the force at point B, denoted by $\mathbf{F}_3$, is due to the tension in the spring. The three forces can be expressed in the vector form

$$\mathbf{F}_1 = \mathbf{F}_2 = -mg\,\mathbf{j}, \quad \mathbf{F}_3 = 2kL(1 - \cos\theta)\mathbf{i} \tag{a}$$

where $2L(1 - \cos\theta)$ is the elongation of the spring. The corresponding position vectors are

$$\mathbf{r}_1 = \frac{L}{2}(\cos\theta\,\mathbf{i} - \sin\theta\,\mathbf{j})$$

$$\mathbf{r}_2 = \frac{L}{2}(3\cos\theta\,\mathbf{i} - \sin\theta\,\mathbf{j}) \tag{b}$$

$$\mathbf{r}_3 = 2L\cos\theta\,\mathbf{i}$$

so that the virtual displacement vectors are

$$\delta\mathbf{r}_1 = -\frac{L}{2}(\sin\theta\,\mathbf{i} + \cos\theta\,\mathbf{j})\,\delta\theta$$

$$\delta\mathbf{r}_2 = -\frac{L}{2}(3\sin\theta\,\mathbf{i} + \cos\theta\,\mathbf{j})\,\delta\theta \tag{c}$$

$$\delta\mathbf{r}_3 = -2L\sin\theta\,\delta\theta\,\mathbf{i}$$

Letting $N = 3$ in Eq. (2.7) and using Eqs. (a) and (c), we obtain

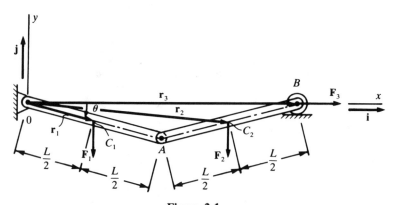

Figure 2.1

$$\overline{\delta W} = \sum_{i=1}^{3} \mathbf{F}_i \cdot \delta \mathbf{r}_i = -mg\,\mathbf{j} \cdot \left[-\frac{L}{2}(\sin\theta\,\mathbf{i} + \cos\theta\,\mathbf{j}) \right] \delta\theta$$

$$- mg\,\mathbf{j} \cdot \left[-\frac{L}{2}(3\sin\theta\,\mathbf{i} + \cos\theta\,\mathbf{j}) \right] \delta\theta$$

$$+ 2kL(1-\cos\theta)\mathbf{i} \cdot (-2L\sin\theta\,\delta\theta\,\mathbf{i})$$

$$= [mgL\cos\theta - 4kL^2(1-\cos\theta)\sin\theta]\,\delta\theta \quad \text{(d)}$$

Due to the arbitrariness of the virtual displacement $\delta\theta$, Eq. (d) can be satisfied only if the coefficient of $\delta\theta$ is zero. It follows that the equilibrium position is given by the angle θ satisfying the equation

$$mgL\cos\theta - 4kL^2(1-\cos\theta)\sin\theta = 0 \quad \text{(e)}$$

Hence, Eq. (e) represents the equilibrium equation. Note that the above equilibrium equation can also be obtained by letting $\dot{\theta} = \ddot{\theta} = 0$ in Eq. (f) of Example 1.6.

It should be pointed out that the constraint forces appearing in Example 1.6 do not appear here, which is consistent with the assumption that constraint forces perform no work.

2.3 D'ALEMBERT'S PRINCIPLE

The principle of virtual work is concerned with the static equilibrium of a system. It can be extended to dynamical systems by means of a principle attributed to d'Alembert. The principle can be used to derive the system equations of motion, as shown below in Example 2.2. However, our interest here is primarily to use the principle to effect the transition from the Newtonian mechanics to the Lagrangian mechanics.

Newton's second law for a typical particle m_i can be written in the form

$$\mathbf{F}_i + \mathbf{F}'_i - m_i \ddot{\mathbf{r}} = \mathbf{0}, \quad i = 1, 2, \ldots, N \quad (2.9)$$

Similarly, we can write

$$(\mathbf{F}_i + \mathbf{F}'_i - m_i \ddot{\mathbf{r}}_i) \cdot \delta\mathbf{r}_i = 0, \quad i = 1, 2, \ldots, N \quad (2.10)$$

Summing over the system of particles and recalling that $\sum_{i=1}^{N} \mathbf{F}'_i \cdot \delta\mathbf{r}_i = 0$, we obtain

$$\sum_{i=1}^{N} (\mathbf{F}_i - m_i \ddot{\mathbf{r}}_i) \cdot \delta\mathbf{r}_i = 0 \quad (2.11)$$

which is referred to as the *generalized principle of d'Alembert*.

Example 2.2 Derive the equation of motion for the system of Example 1.6 by means of the generalized principle of d'Alembert. Discuss the role of constraints in both approaches.

The generalized d'Alembert's principle stated by Eq. (2.11) is suitable for systems of particles and the problem at hand involves rigid bodies. But rigid bodies can be regarded as systems of particles, so that we propose to adapt Eq. (2.11) to the case at hand. Considering Fig. 2.2, we can write the generalized d'Alembert's principle in the form

$$\sum_{i=1}^{2} \int_{\text{link } i} (d\mathbf{F}_i - \ddot{\mathbf{r}}_i \, dm_i) \cdot \delta \mathbf{r}_i + \mathbf{F}_3 \cdot \delta \mathbf{r}_3 = 0 \tag{a}$$

where, from Example 1.6,

$$d\mathbf{F}_i = -g \, dm_i \, \mathbf{j}, \quad i = 1, 2; \quad \mathbf{F}_3 = 2kL(1 - \cos \theta) \, \mathbf{i} \tag{b}$$

Moreover,

$$dm_i = \frac{m}{L} d\xi_i, \quad i = 1, 2 \tag{c}$$

The position vectors corresponding to a mass element in links 1 and 2 are

$$\begin{aligned}\mathbf{r}_1 &= \xi_1(\cos \theta \, \mathbf{i} - \sin \theta \, \mathbf{j}) \\ \mathbf{r}_2 &= (L + \xi_2) \cos \theta \, \mathbf{i} - (L - \xi_2) \sin \theta \, \mathbf{j}\end{aligned} \tag{d}$$

from which we obtain the virtual displacements

$$\begin{aligned}\delta \mathbf{r}_1 &= -\xi_1(\sin \theta \, \mathbf{i} + \cos \theta \, \mathbf{j}) \, \delta \theta \\ \delta \mathbf{r}_2 &= -[(L + \xi_2) \sin \theta \, \mathbf{i} + (L - \xi_2) \cos \theta \, \mathbf{j}] \, \delta \theta\end{aligned} \tag{e}$$

In addition, from Example 1.6, we have

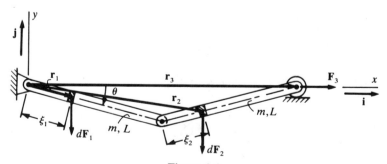

Figure 2.2

D'ALEMBERT'S PRINCIPLE

$$\delta \mathbf{r}_3 = -2L \sin \theta \, \delta\theta \, \mathbf{i} \tag{f}$$

The acceleration vectors are obtained from Eqs. (d) by differentiating twice with respect to time, with the result

$$\ddot{\mathbf{r}}_1 = -\xi_1[(\ddot\theta \sin\theta + \dot\theta^2 \cos\theta)\mathbf{i} + (\ddot\theta \cos\theta - \dot\theta^2 \sin\theta)\mathbf{j}]$$
$$\ddot{\mathbf{r}}_2 = -[(L+\xi_2)(\ddot\theta \sin\theta + \dot\theta^2 \cos\theta)\mathbf{i} + (L-\xi_2)(\ddot\theta \cos\theta - \dot\theta^2 \sin\theta)\mathbf{j}] \tag{g}$$

Inserting Eqs. (b), (c), (e), (f) and (g) into Eq. (a), we obtain

$$\int_0^L \left\{ -\frac{m}{L} g \mathbf{j} + \frac{m}{L} \xi_1 [(\ddot\theta \sin\theta + \dot\theta^2 \cos\theta)\mathbf{i} + (\ddot\theta \cos\theta - \dot\theta^2 \sin\theta)\mathbf{j}] \right\}$$
$$\cdot [-\xi_1(\sin\theta \, \mathbf{i} + \cos\theta \, \mathbf{j})\, \delta\theta] \, d\xi_1 + \int_0^L \left\{ -\frac{m}{L} g \mathbf{j} + \frac{m}{L}[(L+\xi_2)(\ddot\theta \sin\theta \right.$$
$$+ \dot\theta^2 \cos\theta)\mathbf{i} + (L-\xi_2)(\ddot\theta \cos\theta - \dot\theta^2 \sin\theta)\mathbf{j}]\bigg\} \cdot \{-[(L+\xi_2) \sin\theta \, \mathbf{i}$$
$$+ (L-\xi_2) \cos\theta \, \mathbf{j}] \, \delta\theta\} \, d\xi_2 - 4kL^2(1-\cos\theta) \sin\theta \, \delta\theta$$
$$= \left[\frac{m}{L} \int_0^L \{-\xi_1^2(\ddot\theta \sin\theta + \dot\theta^2 \cos\theta) \sin\theta + [g - \xi_1(\ddot\theta \cos\theta \right.$$
$$- \dot\theta^2 \sin\theta)]\xi_1 \cos\theta\} \, d\xi_1 + \frac{m}{L} \int_0^L \{-(L+\xi_2)^2(\ddot\theta \sin\theta + \dot\theta^2 \cos\theta) \sin\theta$$
$$+ [g - (L-\xi_2)(\ddot\theta \cos\theta - \dot\theta^2 \sin\theta)](L-\xi_2) \cos\theta\} \, d\xi_2$$
$$\left. - 4kL^2(1-\cos\theta) \sin\theta \right] \delta\theta = 0 \tag{h}$$

Invoking the arbitrariness of $\delta\theta$, we conclude that Eq. (h) can be satisfied only if the coefficient of $\delta\theta$ is identically zero. Carrying out the indicated integrations and setting the coefficient of $\delta\theta$ equal to zero, we obtain the equation of motion

$$mgL \cos\theta - \frac{2}{3} mL^2[\ddot\theta(1 + 3\sin^2\theta) + 3\dot\theta^2 \sin\theta \cos\theta]$$
$$- 4kL^2(1 - \cos\theta) \sin\theta = 0 \tag{i}$$

which is identical to Eq. (f) in Example 1.6.

We observe that in using Newton's second law in Example 1.6 the constraint forces appear explicitly in the derivation of the equation of motion. In fact, Eq. (f) in Example 1.6 was obtained only by eliminating the constraint forces from six equations, which required substantial effort. In deriving Eq. (i) in the present example the constraint forces did not appear at all, so that in this regard the generalized principle of d'Alembert has an advantage. Still, the procedure is quite clumsy and is not advocated here. This type of problems can be solved much more efficiently by using Hamilton's principle or Lagrange's equations, as demonstrated later in this chapter.

2.4 HAMILTON'S PRINCIPLE

From Eq. (2.7), we recognize that

$$\sum_{i=1}^{N} \mathbf{F}_i \cdot \delta \mathbf{r}_i = \overline{\delta W} \tag{2.12}$$

represents the virtual work performed by the applied forces. But, in contrast with the static case, the virtual work is not zero in the dynamic case.

Next, we consider the following:

$$\frac{d}{dt}(\dot{\mathbf{r}}_i \cdot \delta \mathbf{r}_i) = \ddot{\mathbf{r}}_i \cdot \delta \mathbf{r}_i + \dot{\mathbf{r}}_i \cdot \delta \dot{\mathbf{r}}_i = \ddot{\mathbf{r}}_i \cdot \delta \mathbf{r}_i + \delta\left(\frac{1}{2}\dot{\mathbf{r}}_i \cdot \dot{\mathbf{r}}_i\right)$$

so that

$$\ddot{\mathbf{r}}_i \cdot \delta \mathbf{r}_i = \frac{d}{dt}(\dot{\mathbf{r}}_i \cdot \delta \mathbf{r}_i) - \delta\left(\frac{1}{2}\dot{\mathbf{r}}_i \cdot \dot{\mathbf{r}}_i\right) \tag{2.13}$$

Multiplying Eq. (2.13) by m_i and summing up over the entire system, we obtain

$$\sum_{i=1}^{N} m_i \ddot{\mathbf{r}}_i \cdot \delta \mathbf{r}_i = \sum_{i=1}^{N} m_i \frac{d}{dt}(\dot{\mathbf{r}}_i \cdot \delta \mathbf{r}_i) - \delta T \tag{2.14}$$

where T is the system kinetic energy. Introducing Eqs. (2.12) and (2.14) into Eq. (2.11), we have

$$\delta T + \overline{\delta W} = \sum_{i=1}^{N} m_i \frac{d}{dt}(\dot{\mathbf{r}}_i \cdot \delta \mathbf{r}_i) \tag{2.15}$$

The solution $\mathbf{r}_i = \mathbf{r}_i(t)$ ($i = 1, 2, \ldots, N$) in the configuration space is known as the *true path*. By conceiving of the virtual displacements $\delta \mathbf{r}_i(t)$, we imagine that the system travels along the *varied path* $\mathbf{r}_i(t) + \delta \mathbf{r}_i(t)$. We choose a varied path that coincides with the true path at the two times $t = t_1$ and $t = t_2$, which implies that $\delta \mathbf{r}_i(t_1) = \delta \mathbf{r}_i(t_2) = \mathbf{0}$ (Fig. 2.3). Multiplying Eq. (2.15) by dt and integrating with respect to time between the times t_1 and t_2, we obtain

HAMILTON'S PRINCIPLE

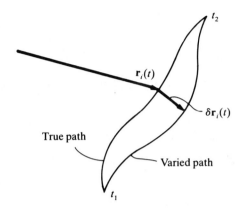

Figure 2.3

$$\int_{t_1}^{t_2} (\delta T + \overline{\delta W}) \, dt = \int_{t_1}^{t_2} \sum_{i=1}^{N} m_i \frac{d}{dt} (\dot{\mathbf{r}}_i \cdot \delta \mathbf{r}_i) \, dt = \sum_{i=1}^{N} m_i \dot{\mathbf{r}}_i \cdot \delta \mathbf{r}_i \bigg|_{t_1}^{t_2} \quad (2.16)$$

Due to the nature of the path chosen, Eq. (2.16) reduces to

$$\int_{t_1}^{t_2} (\delta T + \overline{\delta W}) \, dt = 0, \quad \delta \mathbf{r}_i(t_1) = \delta \mathbf{r}_i(t_2) = \mathbf{0}, \quad i = 1, 2, \ldots, N \quad (2.17)$$

which is known as the *extended Hamilton's principle*.

The extended Hamilton's principle can be written in a slightly more convenient form. To this end, we divide the virtual work into two parts, one due to conservative forces and the other due to nonconservative forces, or

$$\overline{\delta W} = \overline{\delta W}_c + \overline{\delta W}_{nc} \quad (2.18)$$

But, consistent with Eq. (1.14), we can write for the virtual work due to conservative forces

$$\overline{\delta W}_c = -\delta V \quad (2.19)$$

where V is the potential energy. Hence, introducing the Lagrangian

$$L = T - V \quad (2.20)$$

and considering Eqs. (2.18) and (2.19), the extended Hamilton's principle, Eq. (2.17), can be rewritten in the form

30 PRINCIPLES OF ANALYTICAL MECHANICS

$$\int_{t_1}^{t_2} (\delta L + \overline{\delta W}_{nc}) \, dt = 0, \quad \delta \mathbf{r}_i(t_1) = \delta \mathbf{r}_i(t_2) = \mathbf{0}, \quad i = 1, 2, \ldots, N \tag{2.21}$$

For conservative systems $\overline{\delta W}_{nc} = 0$, so that Eq. (2.21) reduces to

$$\int_{t_1}^{t_2} \delta L \, dt = 0, \quad \delta \mathbf{r}_i(t_1) = \delta \mathbf{r}_i(t_2) = \mathbf{0}, \quad i = 1, 2, \ldots, N \tag{2.22}$$

If the system is such that the integration and variation processes are interchangeable, then

$$\delta I = \delta \int_{t_1}^{t_2} L \, dt = 0, \quad \delta \mathbf{r}_i(t_1) = \delta \mathbf{r}_i(t_2) = \mathbf{0}, \quad i = 1, 2, \ldots, N \tag{2.23}$$

which is the mathematical statement of *Hamilton's principle*. In words, *the actual path in the configuration space renders the value of the definite integral $I = \int_{t_1}^{t_2} L \, dt$ stationary with respect to all arbitrary variations of the path between the times t_1 and t_2, provided the variations vanish at the two end points corresponding to $t = t_1$ and $t = t_2$.*

Example 2.3 Use the extended Hamilton's principle, Eq. (2.21), to derive the equation of motion for the system of Fig. 1.2.

As can be concluded from Eq. (2.21), derivation of the equation of motion by means of the extended Hamilton's principle requires the Lagrangian and the virtual work of the nonconservative forces. Inserting Eqs. (a) and (b) from Example 1.4 into Eq. (2.20), we obtain the Lagrangian

$$L = T - V = \frac{1}{2} m\dot{x}^2 - \frac{1}{2} kx^2 \tag{a}$$

The variation of the Lagrangian is simply

$$\delta L = \delta \left(\frac{1}{2} m\dot{x}^2 - \frac{1}{2} kx^2 \right) = m\dot{x} \, \delta \dot{x} - kx \, \delta x \tag{b}$$

Moreover, from Eq. (c) of Example 1.4, we can write virtual work of the nonconservative forces in the form

$$\overline{\delta W} = (F - c\dot{x}) \, \delta x \tag{c}$$

Hence, introducing Eqs. (b) and (c) into Eq. (2.21), we obtain

$$\int_{t_1}^{t_2} [m\dot{x} \, \delta \dot{x} - kx \, \delta x + (F - c\dot{x}) \, \delta x] \, dt = 0, \quad \delta x(t_1) = \delta x(t_2) = 0 \tag{d}$$

The integrand in Eq. (d) involves both δx and $\delta \dot{x}$. Before the equation can be used to derive the equation of motion, we must eliminate the explicit dependence on $\delta \dot{x}$. To this end, we consider the following integration by parts

$$\int_{t_1}^{t_2} m\dot{x}\, \delta \dot{x}\, dt = \int_{t_1}^{t_2} m\dot{x}\, \frac{d}{dt} \delta x\, dt = m\dot{x}\, \delta x \Big|_{t_1}^{t_2} - \int_{t_1}^{t_2} \frac{d}{dt}(m\dot{x})\, \delta x\, dt$$

$$= -\int_{t_1}^{t_2} m\ddot{x}\, \delta x\, dt \qquad (e)$$

where we considered the facts that $\delta x(t_1) = \delta x(t_2) = 0$ and that the mass m does not depend on time. Inserting Eq. (e) into the integral in Eq. (d), we obtain

$$\int_{t_1}^{t_2} (-m\ddot{x} - kx + F - c\dot{x})\, \delta x\, dt = 0 \qquad (f)$$

But, the virtual displacement is arbitrary, which implies that it can be assigned values at will. Because the integral must vanish for all values of δx, Eq. (f) can be satisfied if and only if the coefficient of δx is equal to zero, or

$$m\ddot{x} + c\dot{x} + kx = F \qquad (g)$$

which is the same equation of motion as that obtained in Examples 1.1 and 1.4.

Example 2.4 Derive the equations of motion for the double pendulum of Example 1.5 by means of the extended Hamilton's principle. Compare the equations with those obtained in Example 1.5 and draw conclusions.

Before using the extended Hamilton's principle, we must derive expressions for the variation of the Lagrangian, which involves the kinetic energy and the potential energy, and the virtual work due to the nonconservative force **F**. The kinetic energy requires the velocity of m_1 and m_2. Using results from Example 1.5, they can be verified to have the form

$$\mathbf{v}_1 = L_1 \dot{\theta}_1\, \mathbf{u}_{t1}$$

$$\mathbf{v}_2 = L_1 \dot{\theta}_1\, \mathbf{u}_{t1} + L_2 \dot{\theta}_2\, \mathbf{u}_{t2} = [L_1 \dot{\theta}_1 \cos(\theta_2 - \theta_1) + L_2 \dot{\theta}_2]\mathbf{u}_{t2} \qquad (a)$$

$$- L_1 \dot{\theta}_1 \sin(\theta_2 - \theta_1)\, \mathbf{u}_{n2}$$

so that the kinetic energy is

$$T = \frac{1}{2} m_1 \mathbf{v}_1 \cdot \mathbf{v}_1 + \frac{1}{2} m_2 \mathbf{v}_2 \cdot \mathbf{v}_2$$

$$= \frac{1}{2} m_1 (L_1 \dot{\theta}_1)^2 + \frac{1}{2} m_2 \{[L_1 \dot{\theta}_1 \cos(\theta_2 - \theta_1) + L_2 \dot{\theta}_2]^2 + [L_1 \dot{\theta}_1 \sin(\theta_2 - \theta_1)]^2\}$$

$$= \frac{1}{2} (m_1 + m_2) L_1^2 \dot{\theta}_1^2 + m_2 L_1 L_2 \dot{\theta}_1 \dot{\theta}_2 \cos(\theta_2 - \theta_1) + \frac{1}{2} m_2 L_2^2 \dot{\theta}_2^2 \quad \text{(b)}$$

The potential energy is due to the rise of the two weights, and can be shown to have the expression

$$V = m_1 g L_1 (1 - \cos \theta_1) + m_2 g [L_1 (1 - \cos \theta_1) + L_2 (1 - \cos \theta_2)]$$

$$= (m_1 + m_2) g L_1 (1 - \cos \theta_1) + m_2 g L_2 (1 - \cos \theta_2) \quad \text{(c)}$$

Using Eqs. (b) and (c), the variation of the Lagrangian is simply

$$\delta L = \delta T - \delta V = (m_1 + m_2) L_1^2 \dot{\theta}_1 \, \delta \dot{\theta}_1 + m_2 L_1 L_2 [\dot{\theta}_2 \cos(\theta_2 - \theta_1) \, \delta \dot{\theta}_1$$

$$+ \dot{\theta}_1 \cos(\theta_2 - \theta_1) \, \delta \dot{\theta}_2 - \dot{\theta}_1 \dot{\theta}_2 \sin(\theta_2 - \theta_1)(\delta \theta_2 - \delta \theta_1)]$$

$$+ m_2 L_2^2 \dot{\theta}_2 \, \delta \dot{\theta}_2 - (m_1 + m_2) g L_1 \sin \theta_1 \, \delta \theta_1 - m_2 g L_2 \sin \theta_2 \, \delta \theta_2$$

$$= [(m_1 + m_2) L_1^2 \dot{\theta}_1 + m_2 L_1 L_2 \dot{\theta}_2 \cos(\theta_2 - \theta_1)] \, \delta \dot{\theta}_1$$

$$+ m_2 L_2 [L_1 \dot{\theta}_1 \cos(\theta_2 - \theta_1) + L_2 \dot{\theta}_2] \, \delta \dot{\theta}_2$$

$$+ [m_2 L_1 L_2 \dot{\theta}_1 \dot{\theta}_2 \sin(\theta_2 - \theta_1) - (m_1 + m_2) g L_1 \sin \theta_1] \, \delta \theta_1$$

$$- [m_2 L_1 L_2 \dot{\theta}_1 \dot{\theta}_2 \sin(\theta_2 - \theta_1) + m_2 g L_2 \sin \theta_2] \, \delta \theta_2 \quad \text{(d)}$$

To derive the virtual work due to $\mathbf{F}$, we observe that the projection of the position vector of m_2 on a line coinciding with $\mathbf{F}$ is $L_1 \sin \theta_1 + L_2 \sin \theta_2$. Hence, the virtual work is

$$\delta W_{nc} = F \delta (L_1 \sin \theta_1 + L_2 \sin \theta_2) = F(L_1 \cos \theta_1 \, \delta \theta_1 + L_2 \cos \theta_2 \, \delta \theta_2) \quad \text{(e)}$$

At this point we have all the ingredients required by the extended Hamilton's principle.

Inserting Eqs. (d) and (e) into the extended Hamilton's principle, Eq. (2.21), and collecting terms, we obtain

$$\int_{t_1}^{t_2} \{[(m_1 + m_2) L_1^2 \dot{\theta}_1 + m_2 L_1 L_2 \dot{\theta}_2 \cos(\theta_2 - \theta_1)] \, \delta \dot{\theta}_1$$

$$+ m_2 L_2 [L_1 \dot{\theta}_1 \cos(\theta_2 - \theta_1) + L_2 \dot{\theta}_2] \, \delta \dot{\theta}_2$$

$$+ [m_2 L_1 L_2 \dot{\theta}_1 \dot{\theta}_2 \sin(\theta_2 - \theta_1) - (m_1 + m_2) g L_1 \sin \theta_1$$

$$+ F L_1 \cos \theta_1] \, \delta \theta_1 - [m_2 L_1 L_2 \dot{\theta}_1 \dot{\theta}_2 \sin(\theta_2 - \theta_1) + m_2 g L_2 \sin \theta_2$$

$$- F L_2 \cos \theta_2] \, \delta \theta_2 \} \, dt = 0, \quad \delta \theta_i(t_1) = \delta \theta_i(t_2) = 0, \quad i = 1, 2 \quad \text{(f)}$$

LAGRANGE'S EQUATIONS OF MOTION   33

The integrand in Eq. (f) involves $\delta\dot{\theta}_1$ and $\delta\dot{\theta}_2$. To eliminate them, we consider the typical integration by parts

$$\int_{t_1}^{t_2} f\,\delta\dot{\theta}\,dt = \int_{t_1}^{t_2} f\,\frac{d}{dt}\delta\theta\,dt = f\,\delta\theta \bigg|_{t_1}^{t_2} - \int_{t_1}^{t_2} \dot{f}\,\delta\theta\,dt = -\int_{t_1}^{t_2} \dot{f}\,\delta\theta\,dt \quad (g)$$

where we recognized that $\delta\theta(t_1) = \delta\theta(t_2) = 0$. In view of Eq. (g), Eq. (f) can be rewritten in the form

$$\int_{t_1}^{t_2} \bigg[\bigg\{ -\frac{d}{dt}\big[(m_1 + m_2)L_1^2\dot{\theta}_1 + m_2 L_1 L_2 \dot{\theta}_2 \cos(\theta_2 - \theta_1)\big]$$
$$+ m_2 L_1 L_2 \dot{\theta}_1 \dot{\theta}_2 \sin(\theta_2 - \theta_1) - (m_1 + m_2)gL_1 \sin\theta_1$$
$$+ FL_1 \cos\theta_1 \bigg\} \delta\theta_1 + \bigg\{ -m_2 L_2 \frac{d}{dt}[L_1\dot{\theta}_1 \cos(\theta_2 - \theta_1) + L_2\dot{\theta}_2]$$
$$- [m_2 L_1 L_2 \dot{\theta}_1 \dot{\theta}_2 \sin(\theta_2 - \theta_1) + m_2 gL_2 \sin\theta_2$$
$$- FL_2 \cos\theta \bigg\} \delta\theta_2 \bigg]\,dt = 0 \quad (h)$$

But, $\delta\theta_1$ and $\delta\theta_2$ are arbitrary. Hence, for the above integral to be zero for all values of $\delta\theta_1$ and $\delta\theta_2$, the coefficients of $\delta\theta_1$ and $\delta\theta_2$ must be identically zero. Hence, carrying out the indicated differentiations, setting the coefficients of $\delta\theta_1$ and $\delta\theta_2$ equal to zero and eliminating the terms cancelling out, we obtain the desired equations of motion

$$(m_1 + m_2)L_1\ddot{\theta}_1 + m_2 L_2 \ddot{\theta}_2 \cos(\theta_2 - \theta_1) - m_2 L_2 \dot{\theta}_2^2 \sin(\theta_2 - \theta_1)$$
$$+ (m_1 + m_2)g \sin\theta_1 = F \cos\theta_1$$
$$m_2[L_1\ddot{\theta}_1 \cos(\theta_2 - \theta_1) + L_1\dot{\theta}_1^2 \sin(\theta_2 - \theta_1) + L_2\ddot{\theta}_2] + m_2 g \sin\theta_2$$
$$= F \cos\theta_2 \quad (i)$$

Comparing Eqs. (i) with the results obtained in Example 1.5, we conclude that the use of the extended Hamilton's principle yields directly equations of motion that are free of constraining forces, as expected. We also note that, whereas the second of Eqs. (i) is identical to the second of Eqs. (g) of Example 1.5, the first equations are different. This apparent discrepancy can be easily resolved by multiplying the second of Eqs. (g) of Example 1.5 by $\cos(\theta_2 - \theta_1)$ and adding to the first. The result of these operations is indeed the first of Eqs. (i) obtained here.

2.5 LAGRANGE'S EQUATIONS OF MOTION

The extended Hamilton's principle can be used to derive the system of equations of motion in explicit form. All that is required is the Lagrangian L

34 PRINCIPLES OF ANALYTICAL MECHANICS

and the virtual work of the nonconservative forces $\overline{\delta W}_{nc}$. From Examples 2.3 and 2.4, however, we observe that certain operations must be repeated every time the extended Hamilton's principle is used. It is possible to eliminate the repetitive work by carrying out these operations for a generic system, thus obtaining the system differential equations of motion in general form, where the latter are the well-known *Lagrange's equations*.

Let us consider a system of N particles and write the kinetic energy

$$T = \frac{1}{2} \sum_{i=1}^{N} m_i \dot{\mathbf{r}}_i \cdot \dot{\mathbf{r}}_i \qquad (2.24)$$

where m_i and $\dot{\mathbf{r}}_i$ are the mass and velocity of the ith particle, respectively. We assume that the position vector $\mathbf{r}_i$ is a function of the generalized coordinates $q_1, q_2, \ldots, q_n$ and the time t, or

$$\mathbf{r}_i = \mathbf{r}_i(q_1, q_2, \ldots, q_n, t), \quad i = 1, 2, \ldots, N \qquad (2.25)$$

Differentiating Eq. (2.25) with respect to time, we obtain the velocity vector

$$\dot{\mathbf{r}}_i = \sum_{k=1}^{n} \frac{\partial \mathbf{r}_i}{\partial q_k} \dot{q}_k + \frac{\partial \mathbf{r}_i}{\partial t}, \quad i = 1, 2, \ldots, N \qquad (2.26)$$

where the partial derivatives $\partial \mathbf{r}_i / \partial q_k$ depend in general on $q_1, q_2, \ldots, q_n$ and t. Introducing Eqs. (2.26) into Eq. (2.24), we can write the kinetic energy in the general form

$$T = T(q_1, q_2, \ldots, q_n, \dot{q}_1, \dot{q}_2, \ldots, \dot{q}_n, t) \qquad (2.27)$$

Assuming that the system is subjected to conservative forces derivable from a potential energy function and nonconservative forces, we can use Eqs. (2.18) and (2.19) and write the virtual work in the form

$$\overline{\delta W} = \overline{\delta W}_c + \overline{\delta W}_{nc} = -\delta V + \overline{\delta W}_{nc} \qquad (2.28)$$

where $V = V(q_1, q_2, \ldots, q_n)$ is the potential energy and $\overline{\delta W}_{nc}$ is the virtual work performed by the nonconservative forces, where the latter include control forces. Denoting the nonconservative forces by $\mathbf{F}_i$ ($i = 1, 2, \ldots, N$), the virtual work due to nonconservative forces can be expressed as

$$\overline{\delta W}_{nc} = \sum_{i=1}^{N} \mathbf{F}_i \cdot \delta \mathbf{r}_i \qquad (2.29)$$

where, from Eqs. (2.25), the virtual displacements $\delta \mathbf{r}_i$ have the form

$$\delta \mathbf{r}_i = \sum_{k=1}^{n} \frac{\partial \mathbf{r}_i}{\partial q_k} \delta q_k, \quad i = 1, 2, \ldots, N \qquad (2.30)$$

Introducing Eqs. (2.30) into Eq. (2.29), we obtain

$$\delta W_{nc} = \sum_{k=1}^{n} Q_k \delta q_k \qquad (2.31)$$

where the generalized nonconservative forces Q_k have the expression

$$Q_k = \sum_{i=1}^{N} \mathbf{F}_i \cdot \frac{\partial \mathbf{r}_i}{\partial q_k}, \quad k = 1, 2, \ldots, n \qquad (2.32)$$

Inserting Eqs. (2.27), (2.29) and (2.31) into Eq. (2.17), the extended Hamilton principle can be rewritten as

$$\int_{t_1}^{t_2} \left(\delta L + \sum_{k=1}^{n} Q_k \delta q_k \right) dt = 0, \quad \delta q_k(t_1) = \delta q_k(t_2) = 0, \quad k = 1, 2, \ldots, n \qquad (2.33)$$

where

$$L = L(q_1, q_2, \ldots, q_n, \dot{q}_1, \dot{q}_2, \ldots, \dot{q}_n, t) \qquad (2.34)$$

is the system Lagrangian. Equation (2.33) is used in the following to derive Lagrange's equations in terms of the generalized coordinates $q_1, q_2, \ldots, q_n$.

From Eq. (2.34), we can write

$$\delta L = \sum_{k=1}^{n} \left(\frac{\partial L}{\partial q_k} \delta q_k + \frac{\partial L}{\partial \dot{q}_k} \delta \dot{q}_k \right) \qquad (2.35)$$

Introducing Eq. (2.35) into Eq. (2.33) and integrating by parts, we obtain

$$\int_{t_1}^{t_2} \left(\delta L + \sum_{k=1}^{n} Q_k \delta q_k \right) dt = -\int_{t_1}^{t_2} \sum_{k=1}^{n} \left[\frac{d}{dt} \left(\frac{\partial L}{\partial \dot{q}_k} \right) - \frac{\partial L}{\partial q_k} - Q_k \right] \delta q_k \, dt = 0 \qquad (2.36)$$

where we took into consideration the fact that $(\partial L/\partial \dot{q}_k) \delta q_k |_{t_1}^{t_2} = 0$ ($k = 1, 2, \ldots, n$). But the virtual displacements are arbitrary, which implies that their values can be changed at will. Hence, Eq. (2.36) can be satisfied if and only if the coefficient of every δq_k vanishes, or

$$\frac{d}{dt} \left(\frac{\partial L}{\partial \dot{q}_k} \right) - \frac{\partial L}{\partial q_k} = Q_k, \quad k = 1, 2, \ldots, n \qquad (2.37)$$

which represent the *Lagrange equations of motion*. They permit the derivation of the differential equations of motion from the Lagrangian L and the generalized forces Q_k ($k = 1, 2, \ldots, n$). Note that, to obtain Q_k, it is not really necessary to use expressions (2.32), as there are more direct ways of writing the virtual work in terms of generalized coordinates. This is demonstrated in the following example.

36 PRINCIPLES OF ANALYTICAL MECHANICS

Example 2.5 Derive Lagrange's equations of motion for the double pendulum of Example 2.4.

Using Eqs. (b) and (c) from Example 2.4, we obtain the Lagrangian

$$L = T - V = \frac{1}{2}(m_1 + m_2)L_1^2\dot{\theta}_1^2 + m_2 L_1 L_2 \dot{\theta}_1 \dot{\theta}_2 \cos(\theta_2 - \theta_1) + \frac{1}{2}m_2 L_2^2 \dot{\theta}_2^2$$
$$- (m_1 + m_2)gL_1(1 - \cos\theta_1) - m_2 gL_2(1 - \cos\theta_2) \quad (a)$$

Moreover, letting $\theta_1 = q_1$ and $\theta_2 = q_2$ and using Eq. (e) of Example 2.4, we can write the virtual work due to the nonconservative force **F** in the form

$$\overline{\delta W}_{nc} = \Theta_1 \delta\theta_1 + \Theta_2 \delta\theta_2 \quad (b)$$

where

$$\Theta_1 = FL_1 \cos\theta_1, \quad \Theta_2 = FL_2 \cos\theta_2 \quad (c)$$

are the generalized nonconservative forces Q_1 and Q_2, respectively.

Next, let us write

$$\frac{\partial L}{\partial \dot{\theta}_1} = (m_1 + m_2)L_1^2 \dot{\theta}_1 + m_2 L_1 L_2 \dot{\theta}_2 \cos(\theta_2 - \theta_1)$$

$$\frac{\partial L}{\partial \dot{\theta}_2} = m_2 L_1 L_2 \dot{\theta}_1 \cos(\theta_2 - \theta_1) + m_2 L_2^2 \dot{\theta}_2$$

$$\frac{\partial L}{\partial \theta_1} = m_2 L_1 L_2 \dot{\theta}_1 \dot{\theta}_2 \sin(\theta_2 - \theta_1) - (m_1 + m_2)gL_1 \sin\theta_1 \quad (d)$$

$$\frac{\partial L}{\partial \theta_2} = -m_2 L_1 L_2 \dot{\theta}_1 \dot{\theta}_2 \sin(\theta_2 - \theta_1) - m_2 gL_2 \sin\theta_2$$

Inserting Eqs. (c) and (d) into Eqs. (2.37), we obtain Lagrange's equations

$$\frac{d}{dt}[(m_1 + m_2)L_1^2\dot{\theta}_1 + m_2 L_1 L_2 \dot{\theta}_2 \cos(\theta_2 - \theta_1)] - m_2 L_1 L_2 \dot{\theta}_1 \dot{\theta}_2 \sin(\theta_2 - \theta_1)$$
$$+ (m_1 + m_2)gL_1 \sin\theta_1 = FL_1 \cos\theta_1 \quad (e)$$
$$\frac{d}{dt}[m_2 L_1 L_2 \dot{\theta}_1 \cos(\theta_2 - \theta_1) + m_2 L_2^2 \dot{\theta}_2] + m_2 L_1 L_2 \dot{\theta}_1 \dot{\theta}_2 \sin(\theta_2 - \theta_1)$$
$$+ m_2 gL_2 \sin\theta_2 = FL_2 \cos\theta_2$$

Carrying out the indicated differentiations, Eqs. (e) reduce to Eqs. (i) of Example 2.4.

LAGRANGE'S EQUATIONS OF MOTION

Example 2.6 Derive Lagrange's equation of motion for the system of Examples 1.6 and 2.2 and draw conclusions.

Recognizing that the system under consideration is conservative, we see that the Lagrange equation of motion can be obtained by letting $q_1 = \theta$ and $Q_1 = 0$ in Eqs. (2.37). To derive the Lagrangian, we must first derive the kinetic energy and potential energy. From Fig. 1.8a, we conclude that the kinetic energy for link 1 can be obtained by regarding the motion as a rotation about 0. On the other hand, recalling Eq. (1.41), we can divide the kinetic energy into one part due to the translation of the mass center and a second part due to rotation about the mass center. The kinetic energy for link 1 is simply

$$T_1 = \frac{1}{2} I_0 \dot{\theta}^2 = \frac{1}{2} \frac{mL^2}{3} \dot{\theta}^2 \tag{2}$$

Then, recalling the second of Eqs. (b) in Example 1.6, we can write the kinetic energy for link 2 in the form

$$T_2 = \frac{1}{2} m_2 \dot{\mathbf{r}}_{C2} \cdot \dot{\mathbf{r}}_{C2} + \frac{1}{2} I_{C2} \dot{\theta}^2 = \frac{1}{2} m \left(\frac{L}{2}\right)^2 [(3\dot{\theta} \sin \theta)^2 + (\dot{\theta} \cos \theta)^2]$$
$$+ \frac{1}{2} \frac{mL^2}{12} \dot{\theta}^2 = \frac{1}{2} mL^2 \left(\frac{1}{3} + 2 \sin^2 \theta \right) \dot{\theta}^2 \tag{b}$$

Hence, the total kinetic energy is

$$T = T_1 + T_2 = \frac{1}{2} mL^2 \left(\frac{2}{3} + 2 \sin^2 \theta \right) \dot{\theta}^2 \tag{c}$$

The potential energy can be divided into two parts, one due to the drop in the mass of the two links and another part due to the energy stored in the spring. From Fig. 1.8a, we see that the mass center of both links has dropped by the amount $(L/2) \sin \theta$ and the spring is stretched by the amount $2L(1 - \cos \theta)$. Hence, the potential energy is

$$V = 2mg\left(-\frac{L}{2} \sin \theta\right) + \frac{1}{2} k[2L(1 - \cos \theta)]^2$$
$$= -mgL \sin \theta + \frac{1}{2} 4kL^2(1 - \cos \theta)^2 \tag{d}$$

so that the Lagrangian is

$$L = T - V = \frac{1}{2} mL^2 \dot{\theta}^2 \left(\frac{2}{3} + 2 \sin^2 \theta\right) + mgL \sin \theta - \frac{1}{2} 4kL^2 (1 - \cos \theta)^2 \tag{e}$$

Next, we write

$$\frac{\partial L}{\partial \dot{\theta}} = mL^2 \dot{\theta}\left(\frac{2}{3} + 2\sin^2\theta\right)$$

$$\frac{d}{dt}\left(\frac{\partial L}{\partial \dot{\theta}}\right) = mL^2\left[\ddot{\theta}\left(\frac{2}{3} + 2\sin^2\theta\right) + 4\dot{\theta}^2 \sin\theta \cos\theta\right] \tag{f}$$

$$\frac{\partial L}{\partial \theta_1} = 2mL^2\dot{\theta}^2 \sin\theta \cos\theta + mgL \cos\theta - 4kL^2(1 - \cos\theta)\sin\theta$$

Inserting the above into Eqs. (2.37) with $q_1 = \theta$, we obtain Lagrange's equation of motion

$$2mL^2\left[\ddot{\theta}\left(\frac{1}{3} + \sin^2\theta\right) + \dot{\theta}^2 \sin\theta \cos\theta\right] - mgL \cos\theta + 4kL^2(1 - \cos\theta)\sin\theta = 0 \tag{g}$$

which is identical to the equation derived in Examples 1.6 and 2.2.

It is clear from the above that the Lagrangian formulation permits the derivation of the equations of motion with appreciable more ease than the Newtonian approach or d'Alembert's principle. It shares with d'Alembert's principle the advantage of not having to deal with constraint forces, but the advantages go beyond that. One clear advantage is that all the equations of motion can be derived from scalar quantities, as opposed to Newton's equations and d'Alembert's principle, which are basically vectorial approaches. Moreover, the Lagrangian approach requires velocities only instead of accelerations.

2.6 HAMILTON'S CANONICAL EQUATIONS. MOTION IN THE PHASE SPACE

Lagrange's equations, Eqs. (2.37), represent a set of n second-order differential equations for the generalized coordinates q_k ($k = 1, 2, \ldots, n$). The coordinates $q_k(t)$ can be regarded as the components of an n-dimensional vector $\mathbf{q}(t)$ defining the so-called *configuration space*. As time elapses, the tip of the vector $\mathbf{q}(t)$ traces a curve in the configuration space known as the *dynamical path*. It turns out that the geometric description of the solution in the configuration space is not satisfactory because two dynamical paths can intersect, so that a given point in the configuration space does not define the state of the system uniquely. To define the state uniquely, in addition to the generalized coordinates, it is necessary to specify the generalized velocities $\dot{q}_k(t)$ or, alternatively, the generalized momenta $p_k(t)$ ($k = 1, 2, \ldots, n$). If the generalized momenta are used as auxiliary dependent variables, then the motion can be described geometrically in a $2n$-dimensional Euclidean space defined by q_k and p_k ($k = 1, 2, \ldots, n$) and known as the *phase space*. The set of $2n$ variables q_k and p_k define the so-called *phase vector* and, as time

HAMILTON'S CANONICAL EQUATIONS. MOTION IN THE PHASE SPACE

elapses, the tip of the vector traces a *trajectory* in the phase space. The advantage of the geometric representation in the phase space is that two trajectories never intersect, so that a point in the phase space defines the state of the system uniquely. By introducing n auxiliary coordinates in the form of the generalized momenta, it is possible to transform the n second-order Lagrange's equations into $2n$ first-order equations, where the latter are known as Hamilton's canonical equations. First-order differential equations have many advantages over second-order equations, particularly in the integration of the equations.

The *generalized momenta* are defined as

$$p_k = \frac{\partial L}{\partial \dot{q}_k}, \quad k = 1, 2, \ldots, n \qquad (2.38)$$

Moreover, the *Hamiltonian* is defined in terms of the Lagrangian as follows:

$$H = \sum_{k=1}^{n} \frac{\partial L}{\partial \dot{q}_k} \dot{q}_k - L = \sum_{k=1}^{n} p_k \dot{q}_k - L \qquad (2.39)$$

Substituting the generalized momenta p_k for the generalized velocities $\dot{q}_k$, the Hamiltonian can be written in the functional form

$$H = H(q_1, q_2, \ldots, q_n, p_1, p_2, \ldots, p_n, t) \qquad (2.40)$$

Taking the variation of H in both Eqs. (2.39) and (2.40) and considering Eqs. (2.38), we obtain

$$\delta H = \sum_{k=1}^{n} \left(\dot{q}_k \delta p_k + p_k \delta \dot{q}_k - \frac{\partial L}{\partial q_k} \delta q_k - \frac{\partial L}{\partial \dot{q}_k} \delta \dot{q}_k \right) = \sum_{k=1}^{n} \left(\dot{q}_k \delta p_k - \frac{\partial L}{\partial q_k} \delta q_k \right)$$

$$= \sum_{k=1}^{n} \left(\frac{\partial H}{\partial q_k} \delta q_k + \frac{\partial H}{\partial p_k} \delta p_k \right) \qquad (2.41)$$

from which we conclude that

$$\dot{q}_k = \frac{\partial H}{\partial p_k}, \quad -\frac{\partial L}{\partial q_k} = \frac{\partial H}{\partial q_k}, \quad k = 1, 2, \ldots, n \qquad (2.42)$$

Equations (2.42) are strictly the result of the transformation from the Lagrangian to the Hamiltonian. To complete the transformation, we must invoke the laws of dynamics. To this end, we turn to Lagrange's equations, Eqs. (2.37), and write

$$\dot{p}_k = \frac{d}{dt} \left(\frac{\partial L}{\partial \dot{q}_k} \right) = \frac{\partial L}{\partial q_k} + Q_k, \quad k = 1, 2, \ldots, n \qquad (2.43)$$

so that, retaining the first half of Eqs. (2.42) and introducing the second half of Eqs. (2.42) into Eqs. (2.43), we obtain

$$\dot{q}_k = \frac{\partial H}{\partial p_k}, \quad \dot{p}_k = -\frac{\partial H}{\partial q_k} + Q_k, \quad k = 1, 2, \ldots, n \quad (2.44)$$

Equations (2.44) constitute a set of $2n$ first-order differential equations in the generalized coordinates q_k and generalized momenta p_k and they represent the *Hamilton canonical equations* for a nonconservative system.

If the generalized velocities $\dot{q}_k$ are used as auxiliary variables, then the motion of the system can be described in a $2n$-dimensional Euclidean space defined by q_k and $\dot{q}_k$ ($k = 1, 2, \ldots, n$) and known as the *state space*. Consistent with this, the $2n$-dimensional vector with components q_k and $\dot{q}_k$ ($k = 1, 2, \ldots, n$) is called the *state vector*. The state-space representation of the motion is commonly used in control theory and is discussed in Section 3.7.

Example 2.7 Derive Hamilton's canonical equations for the system of Fig. 2.4.

The kinetic energy is simply

$$T = \frac{1}{2}(m_1 \dot{q}_1^2 + m_2 \dot{q}_2^2) \quad \text{(a)}$$

and the potential energy is

$$V = \frac{1}{2}[k_1 q_1^2 + k_2 (q_2 - q_1)^2] = \frac{1}{2}[(k_1 + k_2) q_1^2 - 2 k_2 q_1 q_2 + k_2 q_2^2] \quad \text{(b)}$$

Moreover, the virtual work due to nonconservative forces has the simple form

$$\overline{\delta W}_{nc} = Q_1 \delta q_1 + Q_2 \delta q_2 \quad \text{(c)}$$

Using Eqs. (a) and (b), we obtain the Lagrangian

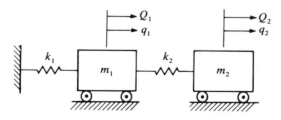

Figure 2.4

$$L = T - V = \frac{1}{2}(m_1\dot{q}_1^2 + m_2\dot{q}_2^2) - \frac{1}{2}[(k_1 + k_2)q_1^2 - 2k_2q_1q_2 + k_2q_2^2] \quad (d)$$

so that, from Eqs. (2.38), the generalized momenta are

$$p_i = \frac{\partial L}{\partial \dot{q}_i} = m_i\dot{q}_i, \quad i = 1, 2 \quad (e)$$

Then, using Eq. (2.39), we obtain the Hamiltonian

$$H = p_1\dot{q}_1 + p_2\dot{q}_2 - L = \frac{1}{2}(m_1\dot{q}_1^2 + m_2\dot{q}_2^2) + \frac{1}{2}[(k_1 + k_2)q_1^2 - 2k_2q_1q_2 + k_2q_2^2] = \frac{1}{2}\left(\frac{1}{m_1}p_1^2 + \frac{1}{m_2}p_2^2\right) + \frac{1}{2}[(k_1 + k_2)q_1^2 - 2k_2q_1q_2 + k_2q_2^2] \quad (f)$$

Finally, inserting Eq. (f) into Eqs. (2.44), we obtain the Hamilton canonical equations

$$\begin{aligned}
\dot{q}_1 &= \frac{\partial H}{\partial p_1} = \frac{1}{m_1}p_1 \\
\dot{q}_2 &= \frac{\partial H}{\partial p_2} = \frac{1}{m_2}p_2 \\
\dot{p}_1 &= -\frac{\partial H}{\partial q_1} + Q_1 = -(k_1 + k_2)q_1 + k_2q_2 + Q_1 \\
\dot{p}_2 &= -\frac{\partial H}{\partial q_2} + Q_2 = k_2q_1 - k_2q_2 + Q_2
\end{aligned} \quad (g)$$

2.7 LAGRANGE'S EQUATIONS OF MOTION IN TERMS OF QUASI-COORDINATES

The motion of a rigid body in space, such as that shown in Fig. 1.7, can be defined in terms of translations and rotations relative to an inertial space of a reference frame embedded in the body and known as body axes. With reference to Fig. 1.7, the inertial axes are denoted by XYZ and the body axes by xyz. The equations describing the motion of a rigid body in space can be obtained with ease by means of Lagrange's equations, Eq. (2.37). It is common practice to define the orientation of the body axes xyz relative to the inertial axes XYZ by means of a set of three rotations about nonorthogonal axes, such as Euler's angles [M17]. In many cases, it is better to work with angular velocity components about the orthogonal body axes than with components about nonorthogonal axes. This is certainly the case when the interest lies in the attitude control of a spacecraft, in which case the sensors measure angular motions and the actuators apply torques in terms of components about the body axes. In such cases, it is more advantageous to derive the equations of motion by means of Lagrange's equations in terms of quasi-coordinates [M17] instead of the standard Lagrange's equations. It should be pointed out here that the term quasi-coordinates refers to certain variables that cannot be strictly defined. To explain this statement, we note that the angular velocity components ω_x, ω_y, ω_z about the orthogonal body axes xyz encountered in Section 1.6 can be regarded as time derivatives of quasi-coordinates. Because these angular velocity components cannot be integrated to obtain actual angular coordinates, the associated variabes are referred to as quasi-coordinates [M17].

Lagrange's equations in terms of quasi-coordinates can be obtained from the ordinary Lagrange equations, as demonstrated in Meirovitch [M51]. The transition from one to the other is very laborious and the details are not particularly interesting. Omitting the details and assuming that point C in Fig. 1.7 is an arbitrary point, not necessarily the mass center, we can write [from M51] Lagrange's equations in terms of quasi-coordinates in the symbolic vector form

$$\frac{d}{dt}\left(\frac{\partial L}{\partial \mathbf{V}}\right) + \tilde{\omega}\frac{\partial L}{\partial \mathbf{V}} - C\frac{\partial L}{\partial \mathbf{R}} = \mathbf{F} \qquad (2.45a)$$

$$\frac{d}{dt}\left(\frac{\partial L}{\partial \boldsymbol{\omega}}\right) + \tilde{V}\frac{\partial L}{\partial \mathbf{V}} + \tilde{\omega}\frac{\partial L}{\partial \boldsymbol{\omega}} - (D^{\mathrm{T}})^{-1}\frac{\partial L}{\partial \boldsymbol{\theta}} = \mathbf{M} \qquad (2.45b)$$

where $L = L(R_X, R_Y, R_Z, \theta_1, \theta_2, \theta_3, V_x, V_y, V_z, \omega_x, \omega_y, \omega_z)$ is the Lagrangian, in which R_X, R_Y, R_Z are inertial components of the position vector $\mathbf{R}$ of the origin of the body axes, $\theta_1, \theta_2, \theta_3$ are angular displacements defining the orientation of the body axes relative to the inertial axes, and we note that $\boldsymbol{\theta}$ is not really a vector, V_x, V_y, V_z are the body axes components of the velocity vector $\mathbf{V}$ of the origin of the body axes and $\omega_x, \omega_y, \omega_z$ are the body axes components of the angular velocity vector $\boldsymbol{\omega}$ of the body. Moreover, $\tilde{V}$ and $\tilde{\omega}$ are skew symmetric matrices defined by

$$\tilde{V} = \begin{bmatrix} 0 & -V_z & V_y \\ V_z & 0 & -V_x \\ -V_y & V_x & 0 \end{bmatrix}, \quad \tilde{\omega} = \begin{bmatrix} 0 & -\omega_z & \omega_y \\ \omega_z & 0 & -\omega_x \\ -\omega_y & \omega_x & 0 \end{bmatrix} \qquad (2.46a, b)$$

$C = C(\theta_1, \theta_2, \theta_3)$ is a rotation matrix between xyz and XYZ, $D = D(\theta_1, \theta_2, \theta_3)$ is a transformation matrix defined by $\boldsymbol{\omega} = D\dot{\boldsymbol{\theta}}$ [M17] and $\mathbf{F}$ and $\mathbf{M}$ are the force vector and moment vector, respectively, in terms of components about the body axes.

As a special case, let us consider the rotational motion of a rigid body about a fixed point, in which case the Lagrangian depends on the angular velocity components alone, $L = L(\omega_x, \omega_y, \omega_z)$. Then, assuming that xyz are principal axes and considering Eq. (1.43), we can write

$$L = T = \frac{1}{2}(I_{xx}\omega_x^2 + I_{yy}\omega_y^2 + I_{zz}\omega_z^2) \tag{2.47}$$

where I_{xx}, I_{yy}, I_{zz} are principal moments of inertia. Inserting Eqs. (2.46b) and (2.47) into Eq. (2.45b) and carrying out the indicated operations, we can write the moment equations in terms of the cartesian components

$$\begin{aligned} I_{xx}\dot{\omega}_x + (I_{zz} - I_{yy})\omega_y\omega_z &= M_x \\ I_{yy}\dot{\omega}_y + (I_{xx} - I_{zz})\omega_x\omega_z &= M_y \\ I_{zz}\dot{\omega}_z + (I_{yy} - I_{xx})\omega_x\omega_y &= M_z \end{aligned} \tag{2.48}$$

which are recognized as Euler's moment equations, first derived in Section 1.7.

CHAPTER 3

CONCEPTS FROM LINEAR SYSTEM THEORY

In Chapter 2, we studied techniques for the derivation of the equations of motion. The next task is to learn as much as possible about the behavior of the system from these equations, which involves the study of the stability characteristics, as well as the derivation of the system response. The task involves subjects such as input-output relations, state equations, equilibrium positions, linearization about equilibrium, the transition matrix, the eigenvalue problem, controllability and observability and control sensitivity. These subjects are in the domain of system theory and are covered in this chapter.

3.1 CONCEPTS FROM SYSTEM ANALYSIS

A *system* is defined as an assemblage of components acting together as a whole. A system acted upon by a given *excitation* exhibits a certain *response*. *Dynamics* is the study of this cause-and-effect relation.

In order to study the dynamic behavior of a system, it is necessary to construct a *mathematical model* of the system. This amounts to identifying essential components of the system, establishing their excitation-response characteristics, perhaps experimentally, as well as considering the manner in which they work together. Throughout this process, appropriate physical laws must be invoked. The mathematical model permits the derivation of the equations governing the behavior of the system. For the most part, the equations have the form of *differential equations*, but at times they have the form of *integral equations*. In effect, we have already derived such differential equations in Chapter 1 by means of Newton's second law and in Chapter

46 CONCEPTS FROM LINEAR SYSTEM THEORY

2 by means of Hamilton's principle and Lagrange's equations. Before a mathematical model can be accepted with confidence, it must be ascertained that it is capable of predicting the observed behavior of the actual system.

In system analysis terminology, systems are often referred to as *plants*, or *processes*. Moreover, the excitation is known as the *input signal*, or simply the input, and the response as the *output signal*, or simply the output. It is convenient to represent the relation between the input and output schematically in terms of a *block diagram*, as shown in Fig. 3.1.

Quite frequently, the plant characteristics and the input are known and the object is to determine the output. This task is known as *analysis*. At times, however, the plant characteristics are not known and the object is to determine them. To this end, it is necessary to acquire information about the input and output, very likely through measurement, and to infer the system characteristics from this information. The task of determining the system properties from known inputs and outputs is referred to as *system identification*. In engineering, the functions a system must perform and the conditions under which it must operate are well defined, but the system itself does not exist yet. In such cases, one must *design* a system capable of performing satisfactorily under the expected conditions. System control generally involves all three tasks to some degree, although for the most part the plant characteristics can be regarded as known, albeit not exactly. Moreover, reference here is not to the design of the plant itself but of a mechanism guaranteeing satisfactory performance in the presence of uncertainties.

Figure 3.1 depicts a very simple input-output relation, in the sense that the response is a natural result of the excitation. The figure is typical of an *uncontrolled system* and it describes a situation arising ordinarily in the physical world. For example, in the case of structures during earthquakes, the structure represents the plant and the motion of the foundation represents the input. Of course, the output is the motion of the structure itself. In *controlled systems*, on the other hand, the object is to elicit a satisfactory response. To this end, the desired output acts as input to a *controller* and the output of the controller acts as input to the plant, as shown in the block diagram of Fig. 3.2. This desired output is preprogrammed into the controller and remains the same irrespective of the actual output, so that the output does not affect the input. The type of control depicted in Fig. 3.2 is known as *open-loop control*. An example of open-loop control is a heating system in a building, where the heating unit is set to start working at a given time independently of the temperature inside the building. Of course, in

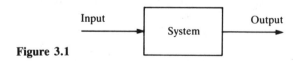

Figure 3.1

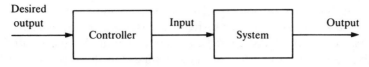

Figure 3.2

many applications the performance of open-loop control is not satisfactory. Indeed, in the case of the heating system, it makes more sense to design the system so as to actuate the heating unit when the temperature drops to a preset level and to shut it off when it reaches a higher temperature level, which requires a temperature-sensing device. This type of control is shown in Fig. 3.3 and is known as *feedback control*, or *closed-loop control*, because the block diagram represents a closed loop. This also explains the term open-loop for the control shown in the block diagram of Fig. 3.2. Modern heating systems are of the closed-loop type. Feedback control tends to be more stable than open-loop control and is capable of compensating for unexpected disturbances, uncertainties in the plant model, sensor measurements and actuator outputs. However, in addition to being more expensive, feedback control tends to be more complicated than open-loop control and is hard to justify when the disturbances and uncertainties are sufficiently small that open-loop control can perform satisfactorily.

In developing a mathematical model for a structure, it is necessary to identify the various members in the structure and ascertain their excitation-response characteristics. These characteristics are governed by given physical laws, such as laws of motion or force-deformation relations, and are described in terms of *system parameters*, such as mass and stiffness. These parameters are of two types, namely, *lumped* and *distributed*. Lumped parameters depend on the spatial position only implicitly. On the other hand, distributed parameters depend explicitly on the spatial coordinates. Consistent with this, in lumped-parameter systems the input and output are functions of time alone and in distributed-parameter systems the input and

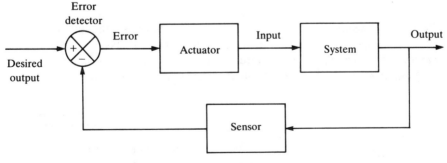

Figure 3.3

48 CONCEPTS FROM LINEAR SYSTEM THEORY

output are functions of both spatial coordinates and time. As a result, the behavior of lumped-parameter systems is governed by *ordinary differential equations* and that of distributed-parameter systems by *partial differential equations*. In both cases, the parameters appear in the form of *coefficients*. For the most part, in structures the coefficients do not depend on time, and such systems are said to be *time-invariant*. If the coefficients do depend on time, then the system is known as *time-varying*. In this text, we study both lumped-parameter and distributed-parameter time-invariant systems. Note that lumped-parameter systems are also known as *discrete systems*. Designing controls for distributed structures is quite often very difficult, so that in such cases it is necessary to construct a discrete model to represent the behavior of the distributed-parameter system. This tends to introduce uncertainties in the plant model, as no discrete model is capable of representing a distributed model exactly. The system of Fig. 2.4 shows an example of a lumped-parameter model, while a rod in axial vibration is an example of a distributed model. If the parameters are distributed nonuniformly, then the construction of a discrete model, perhaps of the type shown in Fig. 2.4, to represent the distributed system is a virtual necessity.

One property of structures with profound implications in mathematical analysis is *linearity*. To illustrate the concept, let us consider a system characterized by the single input $f(t)$ and single output $y(t)$, as shown in Fig. 3.4. Then, if the excitation $f(t)$ is the sum of two different inputs, or

$$f(t) = c_1 f_1(t) + c_2 f_2(t) \tag{3.1}$$

where c_1 and c_2 are constants, and the response can be expressed in the form

$$y(t) = c_1 y_1(t) + c_2 y_2(t) \tag{3.2}$$

where $y_1(t)$ is the response to $f_1(t)$ and $y_2(t)$ is the response to $f_2(t)$, *the system is linear*. If, on the other hand,

$$y(t) \neq c_1 y_1(t) + c_2 y_2(t) \tag{3.3}$$

the *systen is nonlinear*. Equation (3.2) represents the mathematical expression of a very fundamental principle in system analysis known as the *superposition principle* and it states that: *the response of a linear system to different excitations can be obtained separately and combined linearly*. In

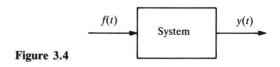

Figure 3.4

practice, it is not necessary to use the test given by Eqs. (3.1)–(3.3) to distinguish between linear and nonlinear systems. Indeed, it is possible to state whether a system is linear or nonlinear by mere inspection of the system differential equation.

Let us consider a generic system described by the ordinary differential equation

$$a_0 \frac{d^n y(t)}{dt^n} + a_1 \frac{d^{n-1} y(t)}{dt^{n-1}} + \cdots + a_{n-1} \frac{dy(t)}{dt} + a_n y(t) = f(t) \qquad (3.4)$$

where a_i ($i = 0, 1, \ldots, n$) are coefficients representing the system parameters. Because the order of the highest derivative in Eq. (3.4) is n, the differential equation, and hence the system, is said to be of *order n*. Introducing Eqs. (3.1) and (3.2) into Eq. (3.4), it is not difficult to verify that the latter equation is satisfied identically, so that we conclude that the system is linear. The reason for this can be traced to the fact that the dependent variable $y(t)$ and its time derivatives appear in Eq. (3.4) to the first power only. Note that linearity rules out not only higher powers and fractional powers of $y(t)$ and its derivatives but also mixed products thereof. Of course, some derivatives of $y(t)$, or even $y(t)$ itself, can be absent from the differential equation. The above criterion is valid whether the coefficients a_i are constant or time-dependent. The only restriction on the coefficients a_i is that they do not depend on $y(t)$ or its derivatives.

In open-loop control, the input $f(t)$ does not depend on $y(t)$ and/or its time derivatives, so that the linearity of the plant implies the linearity of the combined controller-plant system. This is not necessarily the case in closed-loop control. Indeed, in this case the input depends explicitly on $y(t)$ and/or its derivatives, so that for the controlled system to be linear it is necessary that both the plant and the input be linear in $y(t)$ and/or its time derivatives.

In many cases, whether the system is linear or nonlinear depends on the magnitude of the output $y(t)$ and its derivatives. Indeed, when $y(t)$ and the time derivatives of $y(t)$ are sufficiently small that the nonlinear terms can be ignored, the system behaves like a linear one. On the other hand, when $y(t)$ and its time derivatives are so large that some of the nonlinear terms cannot be ignored, the same system acts as a nonlinear system. We shall return to this subject later in this chapter.

Example 3.1 Consider the simple pendulum of Example 1.2 and determine whether the system is linear or nonlinear.

From Eq. (i) of Example 1.2, we obtain the differential equation of motion

$$mL\ddot{\theta}(t) + mg \sin \theta(t) = 0 \qquad \text{(a)}$$

Recognizing that

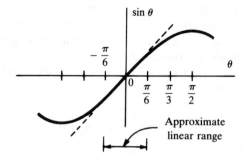

Figure 3.5

$$\sin \theta(t) = \theta(t) - \frac{1}{3!} \theta^3(t) + \frac{1}{5!} \theta^5(t) - \cdots \qquad (b)$$

is a nonlinear function of θ, we conclude that Eq. (a) is nonlinear. The plot of $\sin \theta$ versus θ shown in Fig. 3.5 reveals that in the neighborhood of $\theta = 0$ the function $\sin \theta$ is approximately equal to θ. Hence, if the motion is confined to small angles θ, then $\sin \theta \cong \theta$ and Eq. (a) can be replaced by

$$mL\ddot{\theta}(t) + mg\theta(t) = 0 \qquad (c)$$

which is linear. It follows that for small angles θ the system is linear and for large angles the system is nonlinear. Figure 3.5 indicates a range in which the simple pendulum acts approximately like a linear system. Note that θ can reach values of $\pi/6$ and the error in replacing $\sin \theta$ by θ is still less than 5%.

3.2 FREQUENCY RESPONSE

Equation (3.4) can be written in the symbolic form

$$D(t)y(t) = f(t) \qquad (3.5)$$

where

$$D(t) = a_0 \frac{d^n}{dt^n} + a_1 \frac{d^{n-1}}{dt^{n-1}} + \cdots + a_{n-1} \frac{d}{dt} + a_n \qquad (3.6)$$

represents a *linear homogeneous differential operator*. The operator $D(t)$ contains all the dynamic characteristics of the plant, i.e., it contains all the coefficients multiplying the derivatives of appropriate order or unity, where unity can be regarded as the derivative of zero order. Hence, the block diagram of Fig. 3.4 can be replaced by that of Fig. 3.6. It should be pointed out that Fig. 3.6 represents only a symbolic and not a genuine block

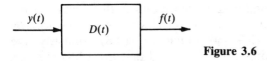

Figure 3.6

diagram. Indeed, we observe that the flow in Fig. 3.1 is from input to plant to output, which is the natural way; in Fig. 3.6 the flow is in reverse order. To direct the flow in the correct sense, we express the solution of Eq. (3.5) in the symbolic form

$$y(t) = D^{-1}(t)f(t) \tag{3.7}$$

The operation indicated by Eq. (3.7) is shown in the block diagram of Fig. 3.7. The symbol $D^{-1}(t)$ can be interpreted as an operator representing the inverse of $D(t)$, in the sense that, if $D(t)$ is a differential operator, $D^{-1}(t)$ is an integral operator. If Eq. (3.5) were to describe a first-order differential equation, then Eq. (3.7) would represent a mere integration. In general, however, the operation is considerably more involved. Hence, although the block diagram of Fig. 3.7 depicts the input-output relation in a more natural way, it still falls short of being a practical tool for producing a solution.

Equations (3.5) and (3.7) express the input-output relation in the time domain and they are symbolic in nature; by themselves they do not help in producing the system response. In the case in which $f(t)$ is harmonic, the situation is considerably better, as the inverse operation $D^{-1}(t)$ is relatively simple. To demonstrate this, let us consider the case in which the input is harmonic, or

$$f(t) = f_0 \cos \omega t \tag{3.8}$$

where f_0 is the *amplitude* and ω the *frequency* of the excitation. Note that the latter is sometimes called the *driving frequency*. Instead of working with Eq. (3.8), it will prove convenient to consider the excitation in the complex form

$$f(t) = f_0 e^{i\omega t} \tag{3.9}$$

where we recall that

$$e^{i\omega t} = \cos \omega t + i \sin \omega t \tag{3.10}$$

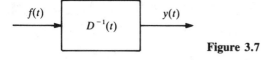

Figure 3.7

represents a complex vector of unit magnitude rotating in the complex plane with the angular velocity ω in the counterclockwise sense. Hence, by retaining the real part of Eq. (3.9), we obtain Eq. (3.8). Similarly, the imaginary part of Eq. (3.9) represents $f_0 \sin \omega t$, so that Eq. (3.9) contains the excitations $f_0 \cos \omega t$ and $f_0 \sin \omega t$ in a single expression. By working with the excitation in the complex form given by Eq. (3.9), it is possible to obtain the response to either $f_0 \cos \omega t$ or $f_0 \sin \omega t$ by simply retaining the real part or the imaginary part of the response, respectively. However, the real advantage of the complex notation is that the response to the excitation in the exponential form is easier to obtain than the response to the excitation in the trigonometric form.

Next, let us assume that the response to the excitation in the exponential form (3.9) can be written as

$$y(t) = Y(i\omega)e^{i\omega t} \qquad (3.11)$$

Then, observing that

$$\frac{d^r}{dt^r} e^{i\omega t} = (i\omega)^r e^{i\omega t}, \quad r = 1, 2, \ldots, n \qquad (3.12)$$

and using Eq. (3.6), we conclude that

$$D(t)e^{i\omega t} = Z(i\omega)e^{i\omega t} \qquad (3.13)$$

where

$$Z(i\omega) = (i\omega)^n a_0 + (i\omega)^{n-1} a_1 + \cdots + i\omega a_{n-1} + a_n \qquad (3.14)$$

is known as the *impedance function*. Hence, introducing Eqs. (3.9) and (3.11) into Eq. (3.5) and considering Eqs. (3.6), (3.13) and (3.14), we obtain

$$Z(i\omega)Y(i\omega)e^{i\omega t} = f_0 e^{i\omega t} \qquad (3.15)$$

so that

$$Y(i\omega) = \frac{f_0}{Z(i\omega)} \qquad (3.16)$$

Finally, inserting Eq. (3.16) into Eq. (3.11), we can write the response to the harmonic excitation in the form

$$y(t) = \frac{1}{Z(i\omega)} f_0 e^{i\omega t} = G(i\omega) f(t) \qquad (3.17)$$

where

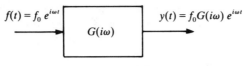

Figure 3.8

$$G(i\omega) = \frac{1}{Z(i\omega)} = \frac{1}{(i\omega)^n a_0 + (i\omega)^{n-1} a_1 + \cdots + i\omega a_{n-1} + a_n} \quad (3.18)$$

is known as the *admittance function*, or the *frequency response*.

Like $D(t)$, the frequency response $G(i\omega)$ contains all the information concerning the system characteristics. In contrast to $D(t)$, which represents a differential operator, $G(i\omega)$ represents an algebraic expression. The input-output relation for the case of harmonic excitation is displayed in the block diagram of Fig. 3.8. This time, however, the block diagram of Fig. 3.8 has practical implications, as the input is merely multiplied by an algebraic expression, namely the frequency response, to obtain the output. The frequency response plays a very important role in vibrations and control.

3.3 RESPONSE BY TRANSFORM METHODS. THE TRANSFER FUNCTION

The simple input-output relation given by Eq. (3.17) is valid only for the case of harmonic excitation. In the case of arbitrary excitation, it is advisable to leave the time domain in favor of a complex domain obtained through the use of the Laplace transformation, where the complex domain is sometimes referred to as the Laplace domain. The procedure is valid only if the coefficients $a_0, a_1, \ldots, a_n$ in Eq. (3.4) are constant.

Let us define the Laplace transform of the input $f(t)$ as

$$F(s) = \mathcal{L}f(t) = \int_0^\infty e^{-st} f(t)\, dt \quad (3.19)$$

where the symbol $\mathcal{L}$ denotes the Laplace transform of a function and s is a subsidiary variable, generally a complex quantity. The variable s defines the Laplace domain mentioned above. Similarly, the Laplace transform of the output $y(t)$ is defined as

$$Y(s) = \mathcal{L}y(t) = \int_0^\infty e^{-st} y(t)\, dt \quad (3.20)$$

Our interest is to transform differential equations of the type given by Eq. (3.4), which makes it necessary to evaluate Laplace transforms of derivatives of $y(t)$ with respect to time. Hence, let us consider the Laplace transform of $d^r y(t)/dt^r$. Integrating by parts repeatedly, it is not difficult to show that

$$\mathscr{L}\frac{d^r y(t)}{dt^r} = \int_0^\infty e^{-st}\frac{d^r y(t)}{dt^r}\,dt = s^r Y(s) - s^{r-1} y(0)$$
$$- s^{r-2}\frac{dy(0)}{dt} - \cdots - \frac{d^{r-1} y(0)}{dt^{r-1}}, \quad r = 1, 2, \ldots, n \quad (3.21)$$

where $d^k y(0)/dt^k$ is the initial value of $d^k y(t)/dt^k$, i.e., the value of $d^k y(t)/dt^k$ evaluated at $t = 0$ ($k = 0, 1, \ldots, r-1$).

For convenience, we assume that all the initial conditions are zero. So that Eqs. (3.21) reduce to

$$\mathscr{L}\frac{d^r y(t)}{dt^r} = s^r Y(s), \quad r = 1, 2, \ldots, n \quad (3.22)$$

and we note that differentiations in the time domain correspond to multiplications by s in the Laplace domain. Then, assuming that the coefficients $a_0, a_1, \ldots, a_n$ in Eq. (3.4) are all constant, transforming both sides of the equation and considering Eqs. (3.19), (3.20) and (3.22), we obtain

$$(a_0 s^n + a_1 s^{n-1} + \cdots + a_{n-1} s + a_n) Y(s) = F(s) \quad (3.23)$$

Equation (3.23) can be rewritten in the form

$$Y(s) = G(s) F(s) \quad (3.24)$$

where

$$G(s) = \frac{1}{a_0 s^n + a_1 s^{n-1} + \cdots + a_{n-1} s + a_n} \quad (3.25)$$

is an algebraic expression relating the transformed output $Y(s)$ to the transformed input $F(s)$ in the Laplace domain. The algebraic expression $G(s)$ is known as the *system transfer function*. The relation given by Eq. (3.24) is shown schematically in Fig. 3.9. The block diagram of Fig. 3.9 represents the Laplace-domain counterpart of the time-domain block diagram of Fig. 3.7. But, unlike Fig. 3.7, which expresses a mere symbolic

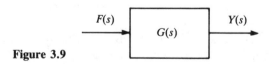

Figure 3.9

relation, Fig. 3.9 represents a genuine block diagram stating that *the transformed output Y(s) can be obtained by multiplying the transformed input F(s) by the transfer function G(s)*. Note that *the transfer function represents the ratio of the transformed output to the transformed input*. Hence, by working in the s-domain instead of the time domain, the block diagram is no longer a symbolic way of describing the input-output relation, but one with significant practical implications as far as the response of systems described by linear differential equations with constant coefficients is concerned.

The block diagram of Fig. 3.9 represents the same relation in the s-domain as the block diagram of Fig. 3.8 represents in the frequency domain. Hence, there is reason to believe that the frequency response and the transfer function are related. Indeed, an examination of Eqs. (3.18) and (3.25) permits us to conclude that *the transfer function is simply the frequency response with $i\omega$ replaced by s, and vice versa*.

Equation (3.24) gives only the transformed response. To obtain the actual response, it is necessary to carry out an inverse Laplace transformation from the Laplace domain to the time domain. The inverse transformation can be expressed in the symbolic form

$$y(t) = \mathscr{L}^{-1}Y(s) = \mathscr{L}^{-1}G(s)F(s) \qquad (3.26)$$

Strictly speaking, the inverse Laplace transformation $\mathscr{L}^{-1}$ is defined as a line integral in the complex s-plane. In most applications, however, line integrations are not really necessary, and the inverse of a function can be obtained from tables of Laplace transform pairs [M42]. In the cases in which the function is too complicated to be found in tables, it may be possible to use the method of partial fractions [M42] to decompose the function into a linear combination of simpler functions whose transforms are in tables.

3.4 SINGULARITY FUNCTIONS

There are several functions of particular importance in vibrations and control. Their importance can be traced to the fact that these functions can be used to synthesize a large variety of complicated functions. Then, involving the superposition principle, the response of linear systems to complicated excitations can be derived with relative ease. The functions are also important because they are frequently used as reference inputs to evaluate the performance of control designs. These functions are the unit impulse, the unit step function and the unit ramp function, and they belong to the class of singularity functions.

The singularity functions form the basis for the time-domain analysis of linear systems. The class of singularity functions is characterized by the fact that every singularity function and all its time derivatives are continuous functions of time, except at a given value of time. The singularity functions

56 CONCEPTS FROM LINEAR SYSTEM THEORY

are also characterized by the fact that they can be obtained from one another by successive differentiation or integration.

One of the most important and widely used singularity functions is the *unit impulse*, or the *Dirac delta function*, which is defined mathematically by

$$\delta(t - a) = 0, \quad t \neq a \tag{3.27a}$$

$$\int_{-\infty}^{\infty} \delta(t - a) \, dt = 1 \tag{3.27b}$$

The delta function is shown in Fig. 3.10 in the form of a thin rectangle of width ε and height $1/\varepsilon$ in the neighborhood of $t = a$. The function must be visualized in the context of a limiting process. In particular, as the width ε approaches zero, the height $1/\varepsilon$ approaches infinity but in a way that the area under the curve remains constant and equal to unity, which explains the term "unit impulse." The units of $\delta(t - a)$ are seconds^{-1} (s^{-1}).

Integrals involving the Dirac delta function are very easy to evaluate. Indeed, considering a continuous function $f(t)$, invoking the mean-value theorem [W7] and using Eq. (3.27b), we can write

$$\int_{-\infty}^{\infty} f(t)\delta(t - a) \, dt = f(a) \int_{-\infty}^{\infty} \delta(t - a) \, dt = f(a) \tag{3.28}$$

so that the integral of a function $f(t)$ weighted by the Dirac delta function applied at $t = a$ is simply the function $f(t)$ evaluated at $t = a$. This property of the delta function is sometimes referred to as the *sampling property*. Clearly, the above result is valid also for finite limits of integration as long as these limits bracket the point $t = a$.

Another singularity function of considerable importance is the *unit step function*, defined by

$$u(t - a) = \begin{cases} 0 & \text{for } t < a \\ 1 & \text{for } t > a \end{cases} \tag{3.29}$$

The unit step function, shown in Fig. 3.11, is characterized by a jump of unit amplitude at $t = a$. The unit step function is dimensionless.

The unit step function is very useful in describing discontinuous functions

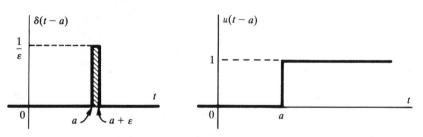

Figure 3.10 **Figure 3.11**

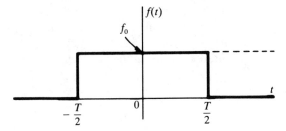

Figure 3.12

or functions defined over given time intervals. For example, the *rectangular pulse* shown in Fig. 3.12 can be expressed mathematically in the form

$$f(t) = f_0\left[u\left(t + \frac{T}{2}\right) - u\left(t - \frac{T}{2}\right)\right] \qquad (3.30)$$

Similarly the decaying exponential defined only for $t > 0$, as shown in Fig. 3.13, can be described in the compact form

$$f(t) = f_0 e^{-t/\tau} u(t) \qquad (3.31)$$

The unit step function is intimately related to the unit impulse. Indeed, it is not difficult to verify that the unit step function is the integral of the unit impulse, or

$$u(t - a) = \int_{-\infty}^{t} \delta(\zeta - a)\, d\zeta \qquad (3.32)$$

Conversely, the unit impulse is the time derivative of the unit step function, or

$$\delta(t - a) = \frac{du(t - a)}{dt} \qquad (3.33)$$

The unit impulse is not a function in an ordinary sense and can be defined rigorously as a *distribution*, or as a *generalized function*, with the ordinary functions representing special cases of distributions. The properties of the distribution are given by Eqs. (3.27a), (3.28) and (3.32). The unit impulse $\delta(t - a)$ is not defined at $t = a$, and the same can be said about the time

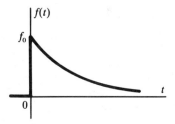

Figure 3.13

derivative of the unit step function $u(t-a)$, as can be concluded from Eq. (3.33). Consistent with this, we can refer to $du(t-a)/dt$ as a *generalized derivative*.

A third function of interest is the *unit ramp function*, defined as

$$r(t-a) = \begin{cases} t-a & \text{for } t>a \\ 0 & \text{for } t<a \end{cases} \tag{3.34}$$

Here we have the opportunity to use the unit step function to express Eq. (3.34) in the compact form

$$r(t-a) = (t-a)u(t-a) \tag{3.35}$$

The unit ramp function is shown in Fig. 3.14. Its units are seconds (s). It is easy to verify that the unit ramp function is the integral of the unit step function, or

$$r(t-a) = \int_{-\infty}^{t} u(\zeta-a)\,d\zeta \tag{3.36}$$

and, conversely, the unit step function is the time derivative of the unit ramp function

$$u(t-a) = \frac{dr(t-a)}{dt} \tag{3.37}$$

The unit ramp function can also be used to synthesize given functions, sometimes together with the unit step function. As an illustration, the sawtooth pulse displayed in Fig. 3.15 can be expressed mathematically in the compact form

$$f(t) = \frac{f_0}{T}[r(t) - r(t-T)] - f_0 u(t-T) \tag{3.38}$$

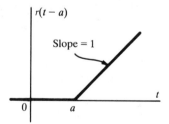

Figure 3.14

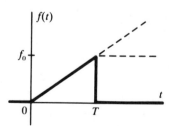

Figure 3.15

3.5 RESPONSE TO SINGULARITY FUNCTIONS

As demonstrated in Section 3.4, certain discontinuous functions can be synthesized by means of singularity function. Then, using the superposition principle, the response of linear systems to such discontinuous excitations can be expressed as a linear combination of responses to the singularity functions involved. In a different context, the response to singularity functions is of interest in the investigation of control system performance.

The *impulse response*, denoted by $g(t)$, is defined as the response of a system to a unit impulse applied at $t = 0$, with the initial conditions equal to zero. Hence, in this case the input is $f(t) = \delta(t)$ and the output is $y(t) = g(t)$. The Laplace transform of the unit impulse is

$$\Delta(s) = \mathscr{L}\delta(t) = \int_0^\infty e^{-st}\delta(t)\,dt = e^{-st}\Big|_{t=0} \int_0^\infty \delta(t)\,dt = 1 \qquad (3.39)$$

so that, using Eq. (3.26), we obtain simply

$$g(t) = \mathscr{L}^{-1}G(s)\Delta(s) = \mathscr{L}^{-1}G(s) \qquad (3.40)$$

or *the impulse response is equal to the inverse Laplace transform of the transfer function*. It follows that *the impulse response and the transfer function represent a Laplace transform pair*. Quite often in linear system theory the impulse response is denoted by $h(t)$. In view of Eq. (3.40), denoting the impulse response by $g(t)$ seems more indicated.

The *step response* $s(t)$ is defined as the response of a system to a unit step function applied at $t = 0$, with the initial conditions equal to zero. The Laplace transform of the unit step function is

$$U(s) = \mathscr{L}u(t) = \int_0^\infty e^{-st}u(t)\,dt = \int_0^\infty e^{-st}\,dt = \frac{e^{-st}}{-s}\Big|_0^\infty = \frac{1}{s} \qquad (3.41)$$

Hence, using Eq. (3.26), we obtain the step response

$$s(t) = \mathscr{L}^{-1}G(s)U(s) = \mathscr{L}^{-1}\frac{G(s)}{s} \qquad (3.42)$$

The ramp response can be defined in analogous fashion. Because its use is less frequent than the impulse response and the step response, we shall not pursue the subject here.

Example 3.2 Derive the impulse response for the second-order system

$$m\ddot{q}(t) + kq(t) = 0 \qquad \text{(a)}$$

Using Eq. (3.25), we can write the transfer function in the form

$$G(s) = \frac{1}{ms^2 + k} = \frac{1}{m} \frac{1}{s^2 + \omega_n^2} \tag{b}$$

where $\omega_n^2 = k/m$. Hence, from Eq. (3.40), the impulse response is

$$g(t) = \mathcal{L}^{-1} G(s) = \frac{1}{m} \mathcal{L}^{-1} \frac{1}{s^2 + \omega_n^2} \tag{c}$$

Using tables of Laplace transforms [M42], we obtain simply

$$g(t) = \frac{1}{m\omega_n} \sin \omega_n t \, u(t) \tag{d}$$

where we multiplied the result by $u(t)$ in recognition of the fact that $g(t) = 0$ for $t < 0$.

Example 3.3 Derive the response of the second-order system of Example 3.2 to the rectangular pulse shown in Fig. 3.12.

The rectangular pulse of Fig. 3.12 can be expressed as the linear combination of two step functions of amplitude f_0, as indicated by Eq. (3.30). Hence, invoking the superposition principle, the response can be expressed as a linear combination of the same form, or

$$q(t) = f_0 \left[s\left(t + \frac{T}{2}\right) - s\left(t - \frac{T}{2}\right) \right] \tag{a}$$

where $s(t)$ is the step response. Inserting Eq. (b) of Example 3.2 into Eq. (3.42), the step response can be written in the form of the inverse Laplace transform

$$s(t) = \mathcal{L}^{-1} \frac{G(s)}{s} = \frac{1}{m} \mathcal{L}^{-1} \frac{1}{s(s^2 + \omega_n^2)} \tag{b}$$

Before we perform the inverse Laplace transformation, we carry out the following partial fractions expansion

$$\frac{1}{s(s^2 + \omega_n^2)} = \frac{1}{\omega_n^2} \left(\frac{1}{s} - \frac{s}{s^2 + \omega_n^2} \right) \tag{c}$$

Hence, inserting Eq. (c) into Eq. (b) and using tables of Laplace transforms [M42], we obtain the step response

$$s(t) = \frac{1}{m\omega_n^2} \mathcal{L}^{-1} \left(\frac{1}{s} - \frac{s}{s^2 + \omega_n^2} \right) = \frac{1}{k} (1 - \cos \omega_n t) u(t) \tag{d}$$

Finally, introducing Eq. (d) into Eq. (a), we can write the response to the rectangular pulse in the form

$$q(t) = \frac{f_0}{k}\left\{\left[1 - \cos \omega_n\left(t + \frac{T}{2}\right)\right]u\left(t + \frac{T}{2}\right)\right.$$
$$\left. - \left[1 - \cos \omega_n\left(t - \frac{T}{2}\right)\right]u\left(t - \frac{T}{2}\right)\right\} \quad (e)$$

3.6 RESPONSE TO ARBITRARY EXCITATION. THE CONVOLUTION INTEGRAL

The impulse response can be used to derive the response to any arbitrary excitation. To show this, we consider the arbitrary excitation $f(t)$ shown in Fig. 3.16 and regard the shaded area as an impulse of magnitude $f(\tau)\Delta\tau$ applied at $t = \tau$. Hence, the function $f(t)$ can be regarded as a superposition of impulses, or

$$f(t) = \sum f(\tau)\Delta\tau \, \delta(t - \tau) \quad (3.43)$$

But, the response of a linear system to the impulse $f(\tau)\Delta\tau \, \delta(t - \tau)$ is simply

$$\Delta y(t, \tau) = f(\tau)\Delta\tau \, g(t - \tau) \quad (3.44)$$

where $g(t - \tau)$ is the impulse response shifted by the amount $t = \tau$. Invoking the superposition principle, the response of a linear combination of impulses of given magnitudes can be expressed as a linear combination of impulse responses, each multiplied by the corresponding magnitude, or

$$y(t) = \sum f(\tau)\Delta\tau \, g(t - \tau) \quad (3.45)$$

In the limit, as $\Delta\tau \to 0$, we obtain

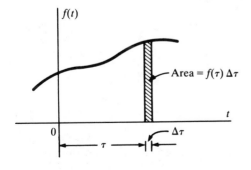

Figure 3.16

$$y(t) = \int_{-\infty}^{t} f(\tau)g(t-\tau)\, d\tau \tag{3.46}$$

The right side of Eq. (3.46) is known as the *convolution integral*, or the *superposition integral*. Because $g(t-\tau)=0$ for $\tau>t$, the upper limit in the integral can be replaced by ∞. Moreover, using a change of variables, the integral can be shown to be symmetric in $f(t)$ and $g(t)$, or

$$y(t) = \int_{-\infty}^{\infty} f(\tau)g(t-\tau)\, d\tau = \int_{-\infty}^{\infty} f(t-\tau)g(\tau)\, d\tau \tag{3.47}$$

In the case in which the input function $f(t)$ is defined only for $t>0$, the convolution integral can be written in the most commonly encountered form

$$y(t) = \int_{0}^{t} f(\tau)g(t-\tau)\, d\tau = \int_{0}^{t} f(t-\tau)g(\tau)\, d\tau \tag{3.48}$$

The question can be raised whether to shift the input function or the impulse response. As a rule of thumb, it is advisable to shift the simpler of the two functions.

It should be pointed out that Eq. (3.48) could have been obtained directly from Eq. (3.26) by using Borel's theorem [M44]. Indeed, if the Laplace transform of a function is the product of two Laplace transforms, then the function can be obtained by inverting each of the transforms in the product separately and evaluating the convolution integral involving the two inverses. In this case, the two functions can be any arbitrary functions and need not represent the input and the impulse response.

Example 3.4 Use the convolution integral to obtain the step response of the second-order system of Example 3.2.

The second-order system in question is given by Eq. (a) of Example 3.2 and the impulse response is given by Eq. (d). Hence, using Eq. (3.48) and shifting the unit step function, we can write the step response as follows:

$$s(t) = \int_{0}^{t} u(t-\tau)g(\tau)\, d\tau = \frac{1}{m\omega_n} \int_{0}^{t} u(t-\tau)\sin \omega_n \tau \, d\tau \tag{a}$$

Because $u(t-\tau)$ is equal to 1 for $\tau<t$ and equal to 0 for $\tau>t$, Eq. (a) yields

$$s(t) = \frac{1}{m\omega_n} \int_{0}^{t} \sin \omega_n \tau \, d\tau = \frac{1}{m\omega_n} \left(-\frac{\cos \omega_n \tau}{\omega_n} \right) \Big|_{0}^{t}$$

$$= -\frac{1}{m\omega_n^2}(\cos \omega_n t - 1) = \frac{1}{k}(1 - \cos \omega_n t), \quad t>0 \tag{b}$$

Of course, $s(t)=0$ for $t<0$, so that the result is the same as that obtained in Example 3.3, in the form of Eq. (d).

3.7 STATE EQUATIONS. LINEARIZATION ABOUT EQUILIBRIUM

In Section 2.6, we presented Hamilton's canonical equations, a set of first-order equations in which the generalized momenta p_k play the role of auxiliary variables. As indicated in Section 2.6, a satisfactory description of the motion can be obtained by using generalized velocities as auxiliary variables instead of generalized momenta. If the generalized velocities $\dot{q}_k(t)$ are used as auxiliary variables, then the motion can be described geometrically in a $2n$-dimensional Euclidean space defined by q_k and $\dot{q}_k$ ($k = 1, 2, \ldots, n$) and known as the *state space*; the $2n$-dimensional vector with components $q_k(t)$ and $\dot{q}_k(t)$ ($k = 1, 2, \ldots, n$) is called the *state vector*. Topologically, there is no difference between the phase-space and the state-space representations of the motion. In particular, the two representations possess the same uniqueness properties. The counterpart of Hamilton's canonical equations are the state equations. Our choice is to work with the state equations instead of Hamilton's equations. It is the formulation used exclusively in control theory.

To derive the state equations, we return to Lagrange's equations, Eqs. (2.37), and write them in the general form

$$\ddot{q}_k(t) = f_k(q_1(t), q_2(t), \ldots, q_n(t), \dot{q}_1(t), \dot{q}_2(t), \ldots, \dot{q}_n(t),$$
$$u_1(t), u_2(t), \ldots, u_r(t)), \quad k = 1, 2, \ldots, n \quad (3.49)$$

where $u_1(t), u_2(t), \ldots, u_r(t)$ are impressed forces, including control forces. Note that in general f_k are nonlinear functions of $q_1, q_2, \ldots, q_n$, $\dot{q}_1, \dot{q}_2, \ldots, \dot{q}_n$ but linear functions of $u_1, u_2, \ldots, u_r$. Equations (3.49) can be expressed in the vector form

$$\ddot{\mathbf{q}}(t) = \mathbf{f}(\mathbf{q}(t), \dot{\mathbf{q}}(t), \mathbf{u}(t)) \quad (3.50)$$

where $\mathbf{f}$ and $\mathbf{q}$ are n-dimensional vectors and $\mathbf{u}$ is an r-dimensional vector. Note that $\mathbf{q}$ is known as the *configuration vector* and $\mathbf{u}$ as the *control vector*. Equation (3.50) can be supplemented by the identity $\dot{\mathbf{q}}(t) = \dot{\mathbf{q}}(t)$. Then, introducing the *state vector*

$$\mathbf{x}(t) = \left[\begin{array}{c} \mathbf{q}(t) \\ \hline \dot{\mathbf{q}}(t) \end{array} \right] \quad (3.51)$$

as well as the vector

$$\mathbf{a}(t) = \left[\begin{array}{c} \dot{\mathbf{q}}(t) \\ \hline \mathbf{f}(\mathbf{q}(t), \dot{\mathbf{q}}(t), \mathbf{u}(t)) \end{array} \right] \quad (3.52)$$

we can write the *state equations* in the general vector form

$$\dot{\mathbf{x}}(t) = \mathbf{a}(\mathbf{x}(t), \mathbf{u}(t)) \tag{3.53}$$

Equation (3.53) defines a dynamical system in the sense that if $\mathbf{x}(t)$ is given at $t = t_0$ and $\mathbf{u}(t)$ is given over the time interval $t_0 < t < t_0 + T$, then Eq. (3.53) can be integrated to yield $\mathbf{x}(t)$ at the end of the time interval T. Hence, a dynamical system is a system such that if the initial state vector and the control vector are given, then the state vector can be computed for any subsequent time.

Equation (3.53) represents a set of $2n$ nonlinear first-order ordinary differential equations. In general, no closed-form solution of the equations is possible. A solution can be obtained by some numerical integration procedure, such as one of the Runge–Kutta methods [M42], but such a solution is confined to a given set of parameters and may not shed much light on the system characteristics. More often than not, the motion characteristics in the neighborhood of special solutions are of particular significance. The special solutions are known as equilibrium points and they are much easier to determine than general solutions. Then, considerable information concerning the system behavior can be gained by studying the linearized state equations about the equilibrium points.

An *equilibrium point* is defined as a constant solution in the state space, $\mathbf{x} = \mathbf{x}_e = \text{constant}$. All other points are said to be *ordinary*. The concept of equilibrium point implies the absence of externally impressed forces, so that we conclude from Eq. (3.53) that the equilibrium points are solutions of the algebraic vector equation

$$\mathbf{a}(\mathbf{x}_e, \mathbf{0}) = \mathbf{0} \tag{3.54}$$

If the components of $\mathbf{a}$ are nonlinear functions of the state there can be more than one equilibrium point and if they are linear functions there can be only one equilibrium point. From the definition of the state vector, Eq. (3.51), we conclude that $\mathbf{x}_e = \text{constant}$ implies that $\mathbf{q}_e = \text{constant}$ and $\dot{\mathbf{q}}_e = \mathbf{0}$, where the latter implies in turn that $\ddot{\mathbf{q}}_e = \mathbf{0}$, which explains the term equilibrium point. Because $\dot{\mathbf{q}}_e = \mathbf{0}$, it follows that all the equilibrium points lie in the *configuration space*, an n-dimensional space defined by the components of the configuration vector $\mathbf{q}(t)$. Hence, to determine the vector $\mathbf{x}_e$, it is only necessary to determine the components of $\mathbf{q}_e$.

To derive the linearized equations of motion, we write Eq. (3.53) in the scalar form

$$\dot{x}_i(t) = a_i(x_1, x_2, \ldots, x_{2n}, u_1, u_2, \ldots, u_r), \quad i = 1, 2, \ldots, 2n \tag{3.55}$$

and expand Taylor's series for a_i about the equilibrium with the result

$$a_i(x_1, x_2, \ldots, x_{2n}, u_1, u_2, \ldots, u_r) = a_i(x_{1e}, x_{2e}, \ldots, x_{2ne}, 0, 0, \ldots, 0)$$
$$+ \sum_{j=1}^{2n} \left.\frac{\partial a_i}{\partial x_j}\right|_{\mathbf{x}=\mathbf{x}_e} x_j + \sum_{j=1}^{r} \left.\frac{\partial a_i}{\partial u_j}\right|_{\mathbf{x}=\mathbf{x}_e} u_j + O_i(\mathbf{x}^2), \quad i = 1, 2, \ldots, 2n \tag{3.56}$$

STATE EQUATIONS. LINEARIZATION ABOUT EQUILIBRIUM

Recognizing that the first term on the right side of Eqs. (3.56) is zero by virtue of Eq. (3.54), introducing the notation

$$a_{ij} = \frac{\partial a_i}{\partial x_j}\bigg|_{\mathbf{x}=\mathbf{x}_e}, \quad i, j = 1, 2, \ldots, 2n \qquad (3.57a)$$

$$b_{ij} = \frac{\partial a_i}{\partial u_j}\bigg|_{\mathbf{x}=\mathbf{x}_e}, \quad i = 1, 2, \ldots, 2n; \quad j = 1, 2, \ldots, r \qquad (3.57b)$$

and ignoring the higher-order terms $O_i(\mathbf{x}^2)$, we can write the linear approximation

$$a_i(x_1, x_2, \ldots, x_{2n}, u_1, u_2, \ldots, u_r) \cong \sum_{j=1}^{2n} a_{ij} x_j + \sum_{j=1}^{r} b_{ij} u_j,$$
$$i = 1, 2, \ldots, 2n \qquad (3.58)$$

Hence, inserting Eqs. (3.58) into Eqs. (3.55), we obtain the *linearized state equations* about the equilibrium $\mathbf{x}_e$ as follows:

$$\dot{x}_i = \sum_{j=1}^{2n} a_{ij} x_j + \sum_{j=1}^{r} b_{ij} u_j \qquad (3.59)$$

which can be written in the matrix form

$$\dot{\mathbf{x}}(t) = A\mathbf{x}(t) + B\mathbf{u}(t) \qquad (3.60)$$

where A and B are constant matrices of coefficients with entries given by Eqs. (3.57).

It should be pointed out that in certain cases the interest lies in equations linearized not about an equilibrium point but about a known time-dependent solution of the state equations. In such cases the matrix A depends on time, although the matrix B remains constant. Equation (3.60) is used widely in the control of linear systems.

Equation (3.60) was derived on the basis of Lagrange equations, resulting in a $2n$-dimensional state vector, where n is the number of degrees of freedom of the system. There are other applications in controls leading to the same equation, but the matrices of coefficients are different and the dimension of the state may not necessarily be an even number. To show this, we consider Eq. (3.4) and let

$$y(t) = x_1(t) \qquad (3.61)$$

$$\frac{d^k y(t)}{dt^k} = x_{k+1}(t), \quad k = 1, 2, \ldots, n$$

as well as

66 CONCEPTS FROM LINEAR SYSTEM THEORY

$$f(t) = \sum_{j=1}^{r} b_j u_j(t) \tag{3.62}$$

Then, we can rewrite Eq. (3.4) in the form

$$\frac{d^n y(t)}{dt^n} = -\sum_{j=1}^{n-1} \frac{a_{n-j}}{a_0} \frac{d^j y(t)}{dt^j} - \frac{a_n}{a_0} y(t) + \sum_{j=1}^{r} \frac{b_j}{a_0} u_j(t) \tag{3.63}$$

so that, considering Eqs. (3.61), we can replace Eq. (3.4) by

$$\begin{aligned} \dot{x}_i(t) &= x_{i+1}(t), \quad i = 1, 2, \ldots, n-1 \\ \dot{x}_n &= -\sum_{j=0}^{n-1} \frac{a_{n-j}}{a_0} x_{j+1}(t) + \sum_{j=1}^{r} \frac{b_j}{a_0} u_j(t) \end{aligned} \tag{3.64}$$

Equations (3.64) can be cast in the form (3.60), where now $\mathbf{x}(t)$ is an n-dimensional state vector, $\mathbf{u}(t)$ is an r-dimensional control vector and

$$A = \begin{bmatrix} 0 & 1 & 0 & 0 & \cdots & 0 & 0 \\ 0 & 0 & 1 & 0 & \cdots & 0 & 0 \\ 0 & 0 & 0 & 1 & \cdots & 0 & 0 \\ \vdots & \vdots & \vdots & \vdots & & \vdots & \vdots \\ 0 & 0 & 0 & 0 & \cdots & 0 & 1 \\ -\frac{a_n}{a_0} & -\frac{a_{n-1}}{a_0} & -\frac{a_{n-2}}{a_0} & -\frac{a_{n-3}}{a_0} & \cdots & -\frac{a_2}{a_0} & -\frac{a_1}{a_0} \end{bmatrix} \tag{3.65a}$$

$$B = \begin{bmatrix} 0 & 0 & \cdots & 0 \\ 0 & 0 & \cdots & 0 \\ \vdots & \vdots & & \vdots \\ \frac{b_1}{a_0} & \frac{b_2}{a_0} & \cdots & \frac{b_r}{a_0} \end{bmatrix} \tag{3.65b}$$

are $n \times n$ and $n \times r$ coefficient matrices, respectively.

Example 3.5 Consider the simple pendulum of Example 3.1 and derive the state equations of motion. Then, identify the equilibrium positions and derive the linearized state equations about each of the equilibrium positions.

The equation of motion of the simple pendulum has the form

$$\ddot{\theta}(t) + \omega_n^2 \sin \theta(t) = 0, \quad \omega_n = \sqrt{g/L} \tag{a}$$

Then, introducing the state vector

$$\mathbf{x}(t) = \begin{bmatrix} x_1(t) \\ x_2(t) \end{bmatrix} = \begin{bmatrix} \theta(t) \\ \dot{\theta}(t) \end{bmatrix} \tag{b}$$

STATE EQUATIONS. LINEARIZATION ABOUT EQUILIBRIUM   67

and considering the identity $x_2(t) = \dot{x}_1(t)$, the state equations can be written in the form (3.64) in which

$$a_1(t) = x_2(t)$$
$$a_2(t) = -\omega_n^2 \sin x_1(t) \qquad \text{(c)}$$

To identify the equilibrium positions, we use Eq. (3.54), which in our case yields

$$x_{2e} = 0, \quad -\omega_n^2 \sin x_{1e} = 0 \qquad \text{(d)}$$

Mathematically, Eqs. (d) have an infinity of solutions. Physically, there are only two distinct solutions

$$e_1: \quad x_{1e_1} = 0, \quad x_{2e_1} = 0$$
$$e_2: \quad x_{1e_2} = \pi, \quad x_{2e_2} = 0 \qquad \text{(e)}$$

so that in both cases the angular velocity is zero in the equilibrium position, as expected.

To derive the linearized state equations, we use Eqs. (3.57a) to determine the elements of the matrix A. Of course, in this case the matrix B is zero. For the equilibrium position e_1, we have

$$a_{11} = \frac{\partial a_1}{\partial x_1}\bigg|_{x_{e_1}} = 0, \quad a_{12} = \frac{\partial a_1}{\partial x_2}\bigg|_{x_{e_1}} = 1$$
$$a_{21} = \frac{\partial a_2}{\partial x_1}\bigg|_{x_{e_1}} = -\omega_n^2 \cos x_{1e_1} = -\omega_n^2, \quad a_{22} = \frac{\partial a_2}{\partial x_2}\bigg|_{x_{e_1}} = 0 \qquad \text{(f)}$$

so that the coefficient matrix is

$$A = \begin{bmatrix} 0 & 1 \\ -\omega_n^2 & 0 \end{bmatrix} \qquad \text{(g)}$$

On the other hand, for the equilibrium position e_2, we obtain the coefficients

$$a_{11} = \frac{\partial a_1}{\partial x_1}\bigg|_{x_{e_2}} = 0, \quad a_{12} = \frac{\partial a_1}{\partial x_2}\bigg|_{x_{e_2}} = 1$$
$$a_{21} = \frac{\partial a_2}{\partial x_1}\bigg|_{x_{e_2}} = -\omega_n^2 \cos x_{1e_2} = \omega_n^2, \quad a_{22} = \frac{\partial a_2}{\partial x_2}\bigg|_{x_{e_2}} = 0 \qquad \text{(h)}$$

so that the coefficient matrix is

$$A = \begin{bmatrix} 0 & 1 \\ \omega_n^2 & 0 \end{bmatrix} \qquad \text{(i)}$$

Because the matrix B is zero, the matrices A define the linearized state equations about the corresponding equilibrium positions fully.

3.8 STABILITY OF EQUILIBRIUM POINTS

Let us consider the case in which the control vector $\mathbf{u}(t)$ is zero, in which case the state equations, Eq. (3.53), reduce to

$$\dot{\mathbf{x}}(t) = \mathbf{a}(\mathbf{x}(t)) \tag{3.66}$$

Our interest lies in motions in the neighborhood of equilibrium points, both qualitatively and quantitatively. A qualitative discussion implies a geometric interpretation of the motion. To this end, we envision the tip of the state vector $\mathbf{x}(t)$, representing a solution of Eq. (3.66), as tracing a curve in the state space known as a *trajectory*. In general, there is an infinity of trajectories, one for every set of initial conditions $\mathbf{x}(0) = \mathbf{x}_0$. These trajectories do not intersect at ordinary points of the state space. When $\mathbf{x}_e = \mathbf{0}$, the equilibrium is said to be *trivial*. But, through a simple coordinate transformation, it is always possible to translate the origin of the state space so as to coincide with an equilibrium point. A very important question in dynamics is the *stability of motion* in the neighborhood of equilibrium points. Without loss of generality, we assume that the equilibrium point coincides with the origin of the state space, so that the interest lies in the stability of the trivial solution. There are various definitions of stability. We concentrate here on the definitions due to Liapunov.

To introduce certain concepts in stability, it is advisable to consider second-order problems, in which case the state space reduces to the state plane defined by $x_1 = q$ and $x_2 = \dot{q}$. The magnitude of the state vector $\mathbf{x}$ can be defined as the *Euclidean length*, or *norm*

$$\|\mathbf{x}\| = (x_1^2 + x_2^2)^{1/2} \tag{3.67}$$

and it gives the distance of a point in the state plane from the origin. A circle of radius r with the center at the origin is given by

$$\|\mathbf{x}\| = r \tag{3.68}$$

The inequality

$$\|\mathbf{x}\| < r \tag{3.69}$$

defines an *open region* in the state plane. If the solution stays in the above circular region for all times, then it is said to be *bounded*. This definition is too broad and cannot be used to define stability.

At this point, we give Liapunov's definitions of stability. To this end, we consider two circles $\|\mathbf{x}\| = \delta$ and $\|\mathbf{x}\| = \varepsilon$, where $\delta < \varepsilon$. Then:

1. The trivial solution $\mathbf{x}(t)$ is said to be *stable* if for

$$\|\mathbf{x}(0)\| = \|\mathbf{x}_0\| < \delta \tag{3.70}$$

the inequality

$$\|\mathbf{x}(t)\| < \varepsilon, \quad t > 0 \tag{3.71}$$

is satisfied.

2. The solution $\mathbf{x}(t)$ is *asymptotically stable* if it is stable and in addition

$$\lim_{t \to \infty} \|\mathbf{x}(t)\| = 0 \tag{3.72}$$

3. The solution $\mathbf{x}(t)$ is *unstable* if for

$$\|\mathbf{x}(0)\| = \|\mathbf{x}_0\| = \delta \tag{3.73}$$

we have

$$\|\mathbf{x}(t_1)\| = \varepsilon \tag{3.74}$$

for some finite time t_1.

A geometric interpretation of the above definitions can be found in Fig. 3.17, where the trajectories labeled, I, II and III correspond to stable, asymptotically stable and unstable trivial solutions.

In the above, we examined the stability of equilibrium points in a qualitative manner. At this point, we turn out attention to more quantitative ways of ascertaining stability, which requires linearization of the state equations about an equilibrium point. Following the pattern of Section 3.7, it is not difficult to show that upon linearization Eq. (3.66) reduces to

$$\dot{\mathbf{x}}(t) = A\mathbf{x}(t) \tag{3.75}$$

For generality, we assume that Eq. (3.75) represents an nth-order system, where n is not necessarily even. To investigate the motion in the neighbor-

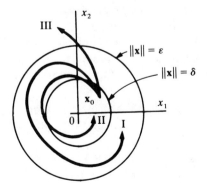

Figure 3.17

hood of equilibrium, we assume a solution of Eq. (3.75) in the exponential form

$$\mathbf{x}(t) = \mathbf{u}e^{\lambda t} \qquad (3.76)$$

where $\mathbf{u}$ is a constant vector and λ a constant scalar. Introducing Eq. (3.76) into Eq. (3.75) and dividing through by $e^{\lambda t}$, we obtain

$$[A - \lambda I]\mathbf{u} = \mathbf{0} \qquad (3.77)$$

Equation (3.77) represents a set of n homogeneous algebraic equations comprising the so-called *algebraic eigenvalue problem*. The eigenvalue problem will be discussed in detail later in this chapter. At this point, we simply state that the solution consists of n eigenvalues λ_r and associated *eigenvectors* $\mathbf{u}_r$ $(r = 1, 2, \ldots, n)$. Then, in view of Eq. (3.76), the general solution of Eq. (3.75) is

$$\mathbf{x}(t) = \sum_{r=1}^{n} \mathbf{u}_r e^{\lambda_r t} \qquad (3.78)$$

Clearly, the nature of the solution depends on the eigenvalues $\lambda_1, \lambda_2, \ldots, \lambda_n$. To obtain the eigenvalues, we recognize that a set of homogeneous algebraic equations have a solution only if the determinant of the coefficients is equal to zero, or

$$\det[A - \lambda I] = a_0 \lambda^n + a_1 \lambda^{n-1} + \cdots + a_{n-1}\lambda + a_n = 0 \qquad (3.79)$$

Equation (3.79) is known as the *characteristic equation* and the polynomial is known as the *characteristic polynomial*. The roots of the characteristic equation are the eigenvalues.

Stability is determined by the real part of the eigenvalues. Real eigenvalues can be treated as complex with zero imaginary part. Hence, we can treat all the eigenvalues as if they were complex and confine the discussion to the real part of the eigenvalues. There are basically three cases:

i. *The eigenvalues have nonpositive real part*, i.e., some real parts are zero and some are negative. Repeated roots must have negative real part. Because the solution must be real, complex eigenvalues appear in pairs of complex conjugates. The pure imaginary roots combine to give an oscillatory solution about the equilibrium, so that the linearized solution is *stable*.
ii. *All the eigenvalues have negative real part*. In this case, the solution approaches the equilibrium point as time increases, so that the solution is *asymptotically stable*.
iii. *At least one eigenvalue is real and positive or complex with positive real part*. In this case, the solution increases exponentially with time, so that the solution is *unstable*.

In Case i, the linearized system is said to possess *critical behavior*, and in Cases ii and iii the linearized system possesses *significant behavior*. If the system possesses significant behavior, then the nature of the motion in the neighborhood of equilibrium of the complete nonlinear system, Eq. (3.66), is the same as that of the linearized system, Eq. (3.75). If the system possesses critical behavior, the analysis is inconclusive and the stability characteristics are determined by the nonlinear terms $O(x^2)$.

The nature of the equilibrium can be visualized conveniently by plotting the eigenvalues in the complex λ-plane, as shown in Fig. 3.18. Then, the imaginary axis represents the region of mere stability, the left half of the plane the region of asymptotic stability, and the right half of the plane the region of instability.

Quite often, it is desirable to be able to make a statement concerning the system stability without actually solving the characteristic equation. Because the imaginary parts do not affect stability, only the information concerning the real parts is needed, and in particular the sign of the real parts. There are two conditions that must be satisfied if none of the roots, $\lambda_1, \lambda_2, \ldots, \lambda_n$ is to have positive real part. The conditions are:

1. All the coefficients $a_0, a_1, \ldots, a_n$ of the characteristic polynomial must have the same sign.
2. None of the coefficients can vanish.

Assuming that $a_0 > 0$, the conditions imply that all the coefficients must be positive.

The above conditions are only necessary, and can be used to identify unstable systems by inspection. Necessary and sufficient conditions for

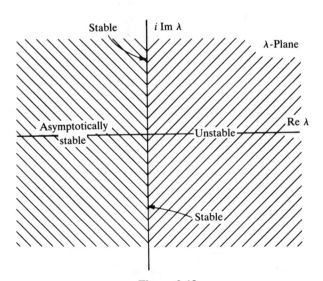

Figure 3.18

asymptotic stability are provided by the *Routh–Hurwitz criterion*. Application of the criterion calls for the use of the coefficients $a_0, a_1, \ldots, a_n$ of the characteristic polynomials to form the determinants

$$\Delta_1 = a_1$$

$$\Delta_2 = \begin{vmatrix} a_1 & a_0 \\ a_3 & a_2 \end{vmatrix}$$

$$\Delta_3 = \begin{vmatrix} a_1 & a_0 & 0 \\ a_3 & a_2 & a_1 \\ a_5 & a_4 & a_3 \end{vmatrix}$$

$$\vdots \qquad\qquad\qquad\qquad\qquad\qquad (3.80)$$

$$\Delta_n = \begin{vmatrix} a_1 & a_0 & 0 & \cdots & 0 \\ a_3 & a_2 & a_1 & \cdots & 0 \\ a_5 & a_4 & a_3 & \cdots & 0 \\ \vdots & \vdots & \vdots & & \\ a_{2n-1} & a_{2n-2} & a_{2n-3} & \cdots & a_n \end{vmatrix}$$

All the entries in the determinants $\Delta_1, \Delta_2, \ldots, \Delta_n$ corresponding to subscripts r such that $r > n$ or $r < 0$ are to be replaced by zero. Then, assuming that $a_0 > 0$, the Routh–Hurwitz criterion states that: *The necessary and sufficient conditions for all the roots $\lambda_1, \lambda_2, \ldots, \lambda_n$ of the characteristic equation to have negative real parts is that all the determinants $\Delta_1, \Delta_2, \ldots, \Delta_n$ be positive.*

The computation of large-order determinants can be time-consuming, and can be circumvented by introducing the *Routh array*

$$\begin{array}{c|cccccc} \lambda^n & a_0 & a_2 & a_4 & a_6 & \cdots \\ \lambda^{n-1} & a_1 & a_3 & a_5 & a_7 & \cdots \\ \lambda^{n-2} & c_1 & c_2 & c_3 & c_4 & \cdots \\ \lambda^{n-3} & d_1 & d_2 & d_3 & d_4 & \cdots \\ \vdots & \vdots & \vdots & \vdots & \vdots & \\ \lambda^1 & m_1 & 0 & 0 & 0 & \cdots \\ \lambda^2 & n_1 & 0 & 0 & 0 & \cdots \end{array}$$

where $a_0, a_1, \ldots, a_n$ are the coefficients of the characteristic polynomial and

$$c_1 = -\frac{1}{a_1} \begin{vmatrix} a_0 & a_2 \\ a_1 & a_3 \end{vmatrix}$$

$$c_2 = -\frac{1}{a_1} \begin{vmatrix} a_0 & a_4 \\ a_1 & a_5 \end{vmatrix} \qquad (3.81a)$$

$$c_3 = -\frac{1}{a_1} \begin{vmatrix} a_0 & a_6 \\ a_1 & a_7 \end{vmatrix}$$

$$\vdots$$

are the entries in the row corresponding to λ^{n-2},

$$d_1 = -\frac{1}{c_1} \begin{vmatrix} a_1 & a_3 \\ c_1 & c_2 \end{vmatrix}$$

$$d_2 = -\frac{1}{c_1} \begin{vmatrix} a_1 & a_5 \\ c_1 & c_3 \end{vmatrix} \qquad (3.81b)$$

$$d_3 = -\frac{1}{c_1} \begin{vmatrix} a_1 & a_7 \\ c_1 & c_4 \end{vmatrix}$$

$$\vdots$$

are the entries in the row corresponding to λ^{n-3}, etc. Then, the Routh–Hurwitz criterion can be stated in terms of the Routh array as follows: *All the roots $\lambda_1, \lambda_2, \ldots, \lambda_n$ of the characteristic equation have negative real parts if all the entries in the first column of the Routh array have the same sign.*

For large-order systems, the derivation of the characteristic polynomial can prove burdensome. In such cases, it may be more expeditious to compute the eigenvalues $\lambda_1, \lambda_2, \ldots, \lambda_n$ by solving the eigenvalue problem associated with the coefficient matrix A.

Example 3.6 Determine the nature of the two equilibrium points for the simple pendulum of Example 3.5 by computing and inspecting the eigenvalues in each case.

To obtain the eigenvalues, we turn to the characteristic equation. To this end, we use the coefficient matrices derived in Example 3.5. For e_1, we obtain

$$\det \begin{bmatrix} -\lambda & 1 \\ -\omega_n^2 & -\lambda \end{bmatrix} = \lambda^2 + \omega_n^2 = 0 \qquad (a)$$

which yields the eigenvalues

$$\begin{matrix} \lambda_1 \\ \lambda_2 \end{matrix} = \pm i\omega_n \qquad (b)$$

This represents Case i, so that the equilibrium is stable. Although this represents critical behavior, the nonlinear terms will confirm that the equilibrium point $\theta = 0$ is stable. On the other hand, for e_2 we obtain

$$\det \begin{bmatrix} -\lambda & 1 \\ \omega_m^2 & -\lambda \end{bmatrix} = \lambda^2 - \omega_n^2 = 0 \qquad (c)$$

so that the eigenvalues are

$$\begin{matrix} \lambda_1 \\ \lambda_2 \end{matrix} = \pm \omega_n \qquad (d)$$

This represents Case iii, so that the equilibrium position $\theta = \pi$ is unstable. The conclusions for both equilibrium positions conform to expectation.

3.9 RESPONSE BY THE TRANSITION MATRIX

In Section 3.8, we considered the stability characteristics of the motion in the neighborhood of equilibrium points. In this and the following two sections, we consider more quantitative results, namely the solution of the linearized state equations.

According to Eq. (3.60), the linearized state equations have the matrix form

$$\dot{\mathbf{x}}(t) = A\mathbf{x}(t) + B\mathbf{u}(t) \tag{3.82}$$

where $\mathbf{x}(t)$ is an n-dimensional state vector, $\mathbf{u}(t)$ is an r-dimensional vector of impressed forces, including control forces, and A and B are $n \times n$ and $n \times r$ constant coefficient matrices, respectively. Note that, for generality, here n denotes the order of the system, in contrast with Eq. (3.60) where the order was denoted by $2n$. This way we can accommodate systems for which the order is not an even number. Equation (3.82) is subject to the initial condition $\mathbf{x}(0)$. Before considering the complete solution, we consider the homogeneous solution, i.e., the solution of the homogeneous equation

$$\dot{\mathbf{x}}(t) = A\mathbf{x}(t) \tag{3.83}$$

It is easy to verify by substitution that the solution of Eq. (3.83) is

$$\mathbf{x}(t) = e^{At}\mathbf{x}(0) = \Phi(t)\mathbf{x}(0) \tag{3.84}$$

where $\Phi(t) = e^{At}$ is a matrix having the form of the series

$$\Phi(t) = e^{At} = I + tA + \frac{t^2}{2!}A^2 + \frac{t^3}{3!}A^3 + \cdots \tag{3.85}$$

The matrix $\Phi(t)$ is known as the *transition matrix*. Questions concerning the convergence of the series in Eq. (3.85) are discussed by Kwakernaak and Sivan [K16].

Next, we consider a particular solution of Eq. (3.82) in the form

$$\mathbf{x}(t) = \Phi(t)\mathbf{x}'(t) \tag{3.86}$$

where $\Phi(t)$ is the transition matrix just introduced and $\mathbf{x}'(t)$ is an n-vector. Differentiating Eq. (3.86) with respect to time and recalling Eq. (3.82), we obtain

$$\dot{\Phi}(t)\mathbf{x}'(t) + \Phi(t)\dot{\mathbf{x}}'(t) = A\Phi(t)\mathbf{x}'(t) + B\mathbf{u}(t) \tag{3.87}$$

It can easily be verified, however, that the transition matrix satisfies the equation

$$\dot{\Phi}(t) = A\Phi(t) \tag{3.88}$$

so that Eq. (3.87) reduces to

$$\Phi(t)\dot{\mathbf{x}}'(t) = B\mathbf{u}(t) \tag{3.89}$$

Moreover, from Eq. (3.85), we recognize that

$$\Phi^{-1}(t) = \Phi(-t) \tag{3.90}$$

so that, multiplying Eq. (3.89) on the left by $\Phi^{-1}(t)$ and integrating, we obtain

$$\mathbf{x}'(t) = \int_0^t \Phi(-\tau)B\mathbf{u}(\tau)\, d\tau \tag{3.91}$$

Hence, using Eq. (3.86), the particular solution becomes

$$\mathbf{x}(t) = \Phi(t)\int_0^t \Phi(-\tau)B\mathbf{u}(\tau)\, d\tau = \int_0^t \Phi(t-\tau)B\mathbf{u}(\tau)\, d\tau \tag{3.92}$$

which is true by virtue of the fact that the transition matrix satisfies

$$\Phi(t)\Phi(-\tau) = \Phi(t-\tau) \tag{3.93}$$

The complete solution is the sum of the homogeneous solution and the particular solution, or

$$\begin{aligned}\mathbf{x}(t) &= \Phi(t)\mathbf{x}(0) + \int_0^t \Phi(t-\tau)B\mathbf{u}(\tau)\, d\tau \\ &= e^{At}\mathbf{x}(0) + \int_0^t e^{A(t-\tau)}B\mathbf{u}(\tau)\, d\tau \end{aligned} \tag{3.94}$$

Example 3.7 Derive the transition matrix for the second-order system of Example 3.2. Then, use the approach based on the transition matrix to obtain the response of the system to the unit step function $u(t)$.

The state equations for the case in which the second-order system of Example 3.2 is subjected to a unit step function can be written in the vector form

$$\dot{\mathbf{x}}(t) = A\mathbf{x}(t) + \mathbf{b}u(t) \tag{a}$$

where $\mathbf{x}(t) = [q(t)\ \dot{q}(t)]^T$ is the state vector and

$$A = \begin{bmatrix} 0 & 1 \\ -\omega_n^2 & 0 \end{bmatrix},\quad \mathbf{b} = \begin{bmatrix} 0 \\ 1/m \end{bmatrix} \tag{b}$$

are coefficient matrices, where the second one is really a vector. Inserting the first of Eqs. (b) into Eq. (3.85), we obtain the transition matrix

$$\Phi(t) = e^{At} = I + tA + \frac{1}{2!}t^2A^2 + \frac{1}{3!}t^3A^3 + \cdots$$

$$= \begin{bmatrix} 1 & 0 \\ 0 & 1 \end{bmatrix} + t\begin{bmatrix} 0 & 1 \\ -\omega_n^2 & 0 \end{bmatrix} - \frac{1}{2!}(\omega_n t)^2 \begin{bmatrix} 1 & 0 \\ 0 & 1 \end{bmatrix}$$

$$- \frac{1}{3!}\omega_n^2 t^3 \begin{bmatrix} 0 & 1 \\ -\omega_n^2 & 0 \end{bmatrix} + \cdots$$

$$= \begin{bmatrix} 1 - \frac{1}{2!}(\omega_n t)^2 + \frac{1}{4!}(\omega_n t)^4 - \cdots & \frac{1}{\omega_n}\left[\omega_n t - \frac{1}{3!}(\omega_n t)^3 + \frac{1}{5!}(\omega_n t)^5 - \cdots\right] \\ -\omega_n\left[\omega_n t - \frac{1}{3!}(\omega_n t)^3 + \frac{1}{5!}(\omega_n t)^5 - \cdots\right] & 1 - \frac{1}{2!}(\omega_n t)^2 + \frac{1}{4!}(\omega_n t)^4 \cdots \end{bmatrix}$$

$$= \begin{bmatrix} \cos \omega_n t & \frac{1}{\omega_n} \sin \omega_n t \\ -\omega_n \sin \omega_n t & \cos \omega_n t \end{bmatrix} \quad (c)$$

To obtain the response, we recognize that $x(0) = 0$. Moreover, through the change of variables $t - \tau = \sigma$, it can be shown that it does not matter which term is shifted in the convolution integral in Eq. (3.94). Hence, we choose to shift the input, so that using the second of Eqs. (b) and Eq. (c) the desired response can be calculated as follows:

$$\mathbf{x}(t) = \int_0^t e^{A\tau} \mathbf{b} u(t - \tau) \, d\tau = \frac{1}{m} \int_0^t \begin{bmatrix} \frac{1}{\omega_n} \sin \omega_n \tau \\ \cos \omega_n \tau \end{bmatrix} u(t - \tau) \, d\tau$$

$$= \frac{1}{m\omega_n^2} \begin{bmatrix} -\cos \omega_n \tau \\ \omega_n \sin \omega_n \tau \end{bmatrix} \Bigg|_0^t = \frac{1}{k}\begin{bmatrix} 1 - \cos \omega_n t \\ \omega_n \sin \omega_n t \end{bmatrix}, \quad t > 0 \quad (d)$$

where $k = m\omega_n^2$. From Example 3.3, we conclude that the top component of Eq. (d) represents the step response $s(t)$ of the system, as expected.

3.10 COMPUTATION OF THE TRANSITION MATRIX

Equation (3.94) gives the system state in the form of a convolution integral involving the transition matrix $\Phi(t) = e^{At}$, so that the interest lies in ways of determining this matrix. For matrices A of small dimensions and simple

COMPUTATION OF THE TRANSITION MATRIX 77

structure, it is possible at times to calculate the transition matrix in closed form, as demonstrated in Example 3.7 above.

The transition matrix can also be obtained by the Laplace transformation method. Taking the Laplace transformation of Eq. (3.83), we have

$$s\mathbf{X}(s) - \mathbf{x}(0) = A\mathbf{X}(s) \qquad (3.95)$$

where $\mathbf{X}(s) = \mathscr{L}\mathbf{x}(t)$ is the Laplace transform of $\mathbf{x}(t)$. Solving for $\mathbf{X}(s)$, we obtain

$$\mathbf{X}(s) = [sI - A]^{-1}\mathbf{x}(0) \qquad (3.96)$$

so that the solution of Eq. (3.83) is simply

$$\mathbf{x}(t) = \mathscr{L}^{-1}\{[sI - A]^{-1}\}\mathbf{x}(0) \qquad (3.97)$$

where the symbol $\mathscr{L}^{-1}$ denotes the inverse Laplace transform of the quantity inside the braces. Comparing Eqs. (3.84) and (3.97), we conclude that the transition matrix can be obtained by evaluating the inverse Laplace transform

$$\Phi(t) = \mathscr{L}^{-1}\{[sI - A]^{-1}\} \qquad (3.98)$$

The procedure described by Eq. (3.98) involves the inverse of the matrix $[sI - A]$, where the matrix is a function of the subsidiary variable s. The matrix function $[sI - A]^{-1}$ is known as the *resolvent* of A and can be calculated by means of the Leverrier algorithm (see Section 5.12). The approach becomes impractical as the order of the system increases.

Another approach is to compute the transition matrix by means of series (3.85). The series converges for all finite values of At, but for large dimensions of the matrix A the computation of the transition matrix involves a large number of additions and multiplications. The process is best carried out numerically by means of a digital computer, which requires that the infinite series be truncated. If we retain terms through the mth power, then we have the approximation

$$\Phi(t) \cong \Phi_m(t) = I + tA + \frac{t^2}{2!}A^2 + \cdots + \frac{t^m}{m!}A^m \qquad (3.99)$$

The computation can be carried out in "nested" form by writing [V6]

$$\Phi_m(t) = I + tA\left(I + \frac{tA}{2}\left(I + \frac{tA}{3}\left(I + \cdots + \frac{tA}{m-1}\left(I + \frac{tA}{m}\right)\cdots\right)\right)\right) \qquad (3.100)$$

where Φ_m is actually computed by means of the recursive relations

$$\psi_1 = I + \frac{tA}{m}$$

$$\psi_2 = I + \frac{tA}{m-1}\psi_1$$

$$\psi_3 = I + \frac{tA}{m-2}\psi_2 \qquad (3.101)$$

$$\vdots$$

$$\Phi_m = I + tA\psi_{m-1}$$

There remains the question of the number of terms to be included in the series. This number depends on the time interval t as well as on the eigenvalue of the matrix A of largest modulus, as shown later in this chapter. As far as the time interval is concerned, some relief can be obtained by dividing the interval into smaller subintervals. To this end, we consider the interval $(t, \tau) = t - \tau$ and divide it into the subintervals $(t, t_1), (t_1, t_2), \ldots, (t_{k-1}, t_k)$ and (t_k, τ), where $(t, \tau) = (t, t_1) + (t_1, t_2) + \cdots + (t_{k-1}, t_k) + (t_k, \tau)$. Then, it is not difficult to verify that the transition matrix satisfies

$$\Phi(t, \tau) = \Phi(t, t_1)\Phi(t_1, t_2)\cdots\Phi(t_{k-1}, t_k)\Phi(t_k, \tau) \qquad (3.102)$$

where (3.102) describes the so-called *group property*. In fact, Eqs. (3.90) and (3.93) are special cases of Eq. (3.102).

Example 3.8 Derive the transition matrix for the second-order system of Example 3.2 by means of Eq. (3.98).

The coefficient matrix A is only 2×2, so that use of the Leverrier algorithm is not really necessary. Using the matrix A from Example 3.7, we calculate first

$$[sI - A]^{-1} = \begin{bmatrix} s & -1 \\ \omega_n^2 & s \end{bmatrix}^{-1} = \frac{1}{s^2 + \omega_n^2}\begin{bmatrix} s & 1 \\ -\omega_n^2 & s \end{bmatrix} \qquad (a)$$

so that, inserting Eq. (a) into Eq. (3.98) and using tables of Laplace transforms [M42], we obtain the transition matrix

$$\Phi(t) = \mathscr{L}^{-1}\{[sI - A]^{-1}\} = \mathscr{L}^{-1}\frac{1}{s^2 + \omega_n^2}\begin{bmatrix} s & 1 \\ -\omega_n^2 & s \end{bmatrix}$$

$$= \begin{bmatrix} \cos \omega_n t & \frac{1}{\omega_n}\sin \omega_n t \\ -\omega_n \sin \omega_n t & \cos \omega_n t \end{bmatrix} \qquad (b)$$

which is the same as that obtained in Example 3.7.

3.11 THE EIGENVALUE PROBLEM. RESPONSE BY MODAL ANALYSIS

In Section 3.10, we made the statement that the number of terms to be included in the series approximation (3.99) depends on the eigenvalue of A of largest modulus. This statement becomes obvious when we consider an alternative procedure for the computation of the transition matrix, one based on the eigenvalues and eigenvectors of A.

Let us assume a solution of Eq. (3.83) in the form

$$\mathbf{x}(t) = e^{\lambda t}\mathbf{u} \qquad (3.103)$$

where λ and $\mathbf{u}$ are constant scalar and vector, respectively. Introducing Eq. (3.103) into Eq. (3.83) and dividing through by $e^{\lambda t}$, we obtain

$$A\mathbf{u} = \lambda \mathbf{u} \qquad (3.104)$$

Equation (3.104) represents a set of homogeneous algebraic equations and is known as the *algebraic eigenvalue problem*. The problem can be stated as follows: *determine the values of the parameter λ for which Eq. (3.104) has a nontrivial solution*. Recalling that A is an $n \times n$ matrix, the eigenvalue problem can be satisfied in n different ways, namely,

$$A\mathbf{u}_i = \lambda_i \mathbf{u}_i, \quad i = 1, 2, \ldots, n \qquad (3.105)$$

where λ_i and $\mathbf{u}_i$ ($i = 1, 2, \ldots, n$) are the *eigenvalues* and *eigenvectors* of A, respectively, both complex in general. Another eigenvalue problem of interest is associated with the matrix A^T and is known as the *adjoint eigenvalue problem*; it is defined by

$$A^T \mathbf{v}_j = \lambda_j \mathbf{v}_j, \quad j = 1, 2, \ldots, n \qquad (3.106)$$

Because $\det A = \det A^T$, the eigenvalues of A^T are the same as the eigenvalues of A. On the other hand, the eigenvectors of A^T are different from the eigenvectors of A. The set of eigenvectors $\mathbf{v}_j$ ($j = 1, 2, \ldots, n$) is known as the *adjoint* of the set of eigenvectors $\mathbf{u}_i$ ($i = 1, 2, \ldots, n$). Equations (3.106) can also be written in the form

$$\mathbf{v}_j^T A = \lambda_j \mathbf{v}_j^T, \quad j = 1, 2, \ldots, n \qquad (3.107)$$

Because of their position relative to A, $\mathbf{u}_i$ are called *right eigenvectors* of A and $\mathbf{v}_j$ are known as *left eigenvectors* of A.

If all the eigenvalues are distinct, then the right and left eigenvectors corresponding to different eigenvalues are *biorthogonal*. It is convenient to normalize the eigenvectors so as to satisfy $\mathbf{v}_i^T \mathbf{u}_i = 1$ ($i = 1, 2, \ldots, n$), in which case the eigenvectors are *biorthonormal*, as expressed by

$$\mathbf{v}_j^T \mathbf{u}_i = \delta_{ij}, \quad \mathbf{v}_j^T A \mathbf{u}_i = \lambda_i \delta_{ij}, \quad i, j = 1, 2, \ldots, n \quad (3.108\text{a, b})$$

where δ_{ij} is the Kronecker delta.

Next, let us introduce the $n \times n$ matrices of right and left eigenvectors

$$U = [\mathbf{u}_1 \quad \mathbf{u}_2 \quad \cdots \quad \mathbf{u}_n], \quad V = [\mathbf{v}_1 \quad \mathbf{v}_2 \quad \cdots \quad \mathbf{v}_n] \quad (3.109)$$

as well as the $n \times n$ matrix of eigenvalues

$$\Lambda = \text{diag}[\lambda_1 \quad \lambda_2 \quad \cdots \quad \lambda_n] \quad (3.110)$$

Then, Eqs. (3.108) can be written in the compact matrix form

$$V^T U = I, \quad V^T A U = \Lambda \quad (3.111\text{a, b})$$

Equations (3.111) can be used to express the transition matrix in a computationally attractive form. To this end, we use Eq. (3.111a) and write

$$V^T = U^{-1}, \quad U = (V^T)^{-1} \quad (3.112\text{a, b})$$

As a matter of interest, if we introduce Eq. (3.112a) into Eq. (3.111b), we obtain

$$U^{-1} A U = \Lambda \quad (1.113)$$

which indicates that matrices A and Λ are related by a *similarity transformation*, so that A and Λ are *similar* matrices. In the general case, in which the eigenvalues are not distinct, the matrix A is similar to a triangular matrix, where the triangular matrix is known as the *Jordan form* [M26]. We do not pursue this subject here.

Multiplying Eqs. (3.111) on the left by U and on the right by V^T and considering Eqs. (3.112), we obtain

$$UV^T = I, \quad U\Lambda V^T = A \quad (3.114\text{a, b})$$

Equations (3.114) can be used to express the transition matrix in the desired form. Inserting Eqs. (3.114) into Eq. (3.85), we obtain

$$\begin{aligned}
\Phi(t) &= e^{At} \\
&= UV^T + tU\Lambda V^T + \frac{t^2}{2!} U\Lambda V^T U\Lambda V^T + \frac{t^3}{3!} U\Lambda V^T U\Lambda V^T U\Lambda V^T + \cdots \\
&= UV^T + tU\Lambda V^T + \frac{t^2}{2!} U\Lambda^2 V^T + \frac{t^3}{3!} U\Lambda^3 V^T + \cdots \\
&= U\left(I + t\Lambda + \frac{t^2}{2!} \Lambda^2 + \frac{t^3}{3!} \Lambda^3 + \cdots\right) V^T = U e^{\Lambda t} V^T \quad (3.115)
\end{aligned}$$

where we recognized that

$$I + t\Lambda + \frac{t^2}{2!}\Lambda^2 + \frac{t^3}{3!}\Lambda^3 + \cdots = e^{\Lambda t} \tag{3.116}$$

Because Λ is diagonal, the computation of $e^{\Lambda t}$ by means of Eq. (3.116) is relatively easy. Indeed, we can write

$$e^{\Lambda t} = \text{diag}\,[e^{\lambda_i t}] = \text{diag}\left[1 + t\lambda_i + \frac{(t\lambda_i)}{2!} + \frac{(t\lambda_i)^3}{3!} + \cdots\right] \tag{3.117}$$

so that we no longer have matrix multiplications. On the other hand, before we can use Eqs. (3.115) and (3.117) to compute the transition matrix, we must solve the eigenvalue problems associated with A and A^T.

From Eqs. (3.115) and (3.117), it is clear that the convergence of the transition matrix depends on the convergence of all $e^{\lambda_i t}$ $(i = 1, 2, \ldots, n)$. Hence, the convergence depends on $\max |\lambda_i| t$, where $\max |\lambda_i|$ denotes the eigenvalue of A of largest modulus. Of course, for smaller $\max |\lambda_i| t$, fewer terms are required for convergence of the series in Eq. (3.117) to $e^{\lambda_i t}$. In this regard, we recall that the time interval can be divided into smaller time subintervals by virtue of the group property.

The transition matrix in the form (3.115) can also be used to derive the system response. Indeed, inserting Eq. (3.115) into Eq. (3.94), we can write simply

$$\mathbf{x}(t) = Ue^{\Lambda t}V^T\mathbf{x}(0) + \int_0^t Ue^{\Lambda(t-\tau)}V^T B\mathbf{u}(\tau)\,d\tau \tag{3.118}$$

The procedure described by Eq. (3.118) can be regarded as representing a modal analysis for the response of general linear dynamic systems. Later in this text, we specialize Eq. (3.118) to certain cases of particular interest in the control of structures.

3.12 STATE CONTROLLABILITY

The concept of controllability ties together the input and the state of a system. To examine the concept, we return to Eq. (3.82), namely, to

$$\dot{\mathbf{x}}(t) = A\mathbf{x}(t) + B\mathbf{u}(t) \tag{3.119}$$

Then, we can define controllability by the following theorem: The system (3.119) is said to be *state controllable* at $t = t_0$ if there exists a piecewise continuous input $\mathbf{u}(t)$ that will drive the initial state $\mathbf{x}(t_0)$ to any final state $\mathbf{x}(t_f)$ at $t = t_f$ within a finite time interval $t_f - t_0$. If this is true for all initial times t_0 and all initial states $\mathbf{x}(t_0)$, the system is said to be *completely state controllable*.

The above definition is valid for linear time-varying systems, i.e., for systems in which the coefficient matrices are functions of time, $A = A(t)$ and $B = B(t)$. When A and B are constant, it is possible to derive a quantitative test. To this end, we let $t_0 = 0$ and $\mathbf{x}(t_f) = \mathbf{0}$, without loss of generality, and rewrite Eq. (3.94) in the form

$$\mathbf{x}(t_f) = \mathbf{0} = e^{At_f}\mathbf{x}(0) + \int_0^{t_f} e^{A(t_f - \tau)} B\mathbf{u}(\tau)\, d\tau \qquad (3.120)$$

which can be reduced to

$$-\mathbf{x}(0) = \int_0^{t_f} e^{-A\tau} B\mathbf{u}(\tau)\, d\tau \qquad (3.121)$$

But, from the Cayley–Hamilton theorem [B37]

$$A^k = \sum_{j=0}^{n-1} \alpha_{kj} A^j \qquad (3.122)$$

for any k. Then, using Eqs. (3.85) and (3.122), we obtain

$$e^{-A\tau} = \sum_{k=0}^{\infty} \frac{(-1)^k \tau^k A^k}{k!} = \sum_{k=0}^{\infty} \frac{(-1)^k \tau^k}{k!} \sum_{j=0}^{n-1} \alpha_{kj} A^j$$

$$= \sum_{j=0}^{n-1} A^j \sum_{k=0}^{\infty} (-1)^k \alpha_{kj} \frac{\tau^k}{k!} \qquad (3.123)$$

so that, introducing the notation

$$\sum_{k=0}^{\infty} (-1)^k \alpha_{kj} \frac{\tau^k}{k!} = \alpha_j(\tau) \qquad (3.124)$$

we can write Eq. (3.123) in the form

$$e^{-A\tau} = \sum_{j=0}^{n-1} \alpha_j(\tau) A^j \qquad (3.125)$$

Hence, inserting Eq. (3.125) into Eq. (3.121), we obtain

$$-\mathbf{x}(0) = \sum_{j=0}^{n-1} A^j B \int_0^{t_f} \alpha_j(\tau) \mathbf{u}(\tau)\, d\tau \qquad (3.126)$$

Each integral in Eq. (3.126) is a constant r-vector of the type

$$\mathbf{c}_j = \int_0^{t_f} \alpha_j(\tau) \mathbf{u}(\tau)\, d\tau \qquad (3.127)$$

so that Eq. (3.126) can be rewritten in the form

$$-\mathbf{x}(0) = [B \mid AB \mid A^2B \mid \cdots \mid A^{n-1}B] \begin{bmatrix} \mathbf{c}_0 \\ \mathbf{c}_1 \\ \vdots \\ \mathbf{c}_{n-1} \end{bmatrix} \quad (3.128)$$

Equation (3.128) represents a set of n equations and nr unknowns. The equations have a solution for any $\mathbf{x}(0)$ provided the $n \times nr$ matrix

$$\mathbb{C} = [B \mid AB \mid A^2B \mid \cdots \mid A^{n-1}B] \quad (3.129)$$

known as the *controllability matrix*, has n independent columns. This implies that *the system is completely state controllable if $\mathbb{C}$ has rank n*.

3.13 OUTPUT EQUATIONS. OBSERVABILITY

In feedback control, the control forces depend on the system behavior. This behavior can be inferred from measurements, where the physical quantities measured are known as *outputs* and are denoted by $y_1(t), y_2(t), \ldots, y_q(t)$, with the implication that there are q outputs. The outputs are related to the state variables and the inputs by the *output equations*

$$y_k(t) = g_k(x_1(t), x_2(t), \ldots, x_n(t), u_1(t), u_2(t), \ldots, u_r(t)), \quad k = 1, 2, \ldots, q \quad (3.130)$$

which can be expressed in the vector form

$$\mathbf{y}(t) = \mathbf{g}(\mathbf{x}(t), \mathbf{u}(t)) \quad (3.131)$$

where $\mathbf{y}(t)$ is the q-dimensional *output vector*. The state equations, Eq. (3.82), and the output equations, Eq. (3.131), together form a set of equations known as the *dynamical equations* of the system.

More often than not, the output depends linearly on the state and on the input explicitly, in which case Eq. (3.131) is replaced by

$$\mathbf{y}(t) = C\mathbf{x}(t) + D\mathbf{u}(t) \quad (3.132)$$

where C and D are $q \times n$ and $q \times r$ matrices, respectively. In general, C and D can be functions of time, but quite frequently they are constant matrices. In fact, most of the time the output does not depend on the input explicitly, so that Eq. (3.132) reduces to

$$\mathbf{y}(t) = C\mathbf{x}(t) \quad (3.133)$$

The concept of observability parallels that of controllability. However, whereas controllability ties together the input and the state of a system, observability ties together the output and the state of a system. Observability is defined by the following theorem: A linear system is said to be *observable* at t_0 if the state $\mathbf{x}(t_0)$ can be determined from the output $\mathbf{y}(t)$, $t_0 \le t < t_f$. If the system is observable for all t_0, then the system is said to be *completely observable*.

Observability does not really depend on the input $\mathbf{u}(t)$. Hence, letting $\mathbf{u}(\tau) = \mathbf{0}$ in Eq. (3.94) and inserting the result in Eq. (3.133), we obtain

$$\mathbf{y}(t) = Ce^{At}\mathbf{x}(0) \qquad (3.134)$$

But, by analogy with Eqs. (3.124) and (3.125), we can write

$$e^{At} = \sum_{j=0}^{n-1} \beta_j(t) A^j \qquad (3.135)$$

where

$$\beta_j(t) = \sum_{k=0}^{\infty} \beta_{kj} \frac{t^k}{k!} \qquad (3.136)$$

Introducing Eq. (3.135) into Eq. (3.134), we arrive at

$$\mathbf{y}(t) = \sum_{j=0}^{n-1} \beta_j(t) C A^j \mathbf{x}(0) \qquad (3.137)$$

which can be rewritten in the form

$$\mathbf{y}(t) = [\beta_1(t)I \mid \beta_2(t)I \mid \beta_3(t)I \mid \cdots \mid \beta_{n-1}(t)I] \begin{bmatrix} C \\ \hline CA \\ \hline CA^2 \\ \hline \vdots \\ \hline CA^{n-1} \end{bmatrix} \mathbf{x}(0) \qquad (3.138)$$

Equation (3.138) can be regarded as representing a set of q equations and n unknowns. It has a solution for any $\mathbf{x}(0)$ provided the $n \times nq$ matrix

$$\mathbb{O} = [C^T \mid A^T C^T \mid (A^T)^2 C^T \mid \cdots \mid (A^T)^{n-1} C^T] \qquad (3.139)$$

known as the *observability matrix*, has n independent columns. This implies that the *system is completely observable if* $\mathbb{O}$ *has rank* n.

3.14 SENSITIVITY OF THE EIGENSOLUTION TO CHANGES IN THE SYSTEM PARAMETERS

Sensitivity theory is concerned with the determination of the sensitivity of the eigenvalues λ_i, the right eigenvectors $\mathbf{u}_i$ and the left eigenvectors $\mathbf{v}_i$ of A

SENSITIVITY OF THE EIGENSOLUTION TO CHANGES IN THE SYSTEM PARAMETERS 85

($i = 1, 2, \ldots, n$) to changes in the system parameters, where the parameters enter into the coefficient matrix A. One approach is to examine how λ_i, $\mathbf{u}_i$ and $\mathbf{v}_i$ change as one of the elements of A changes by determining first derivatives of these quantities with respect to the element in question. Another approach, perhaps simpler, is based on the perturbation theory. We choose this second approach.

Let us consider the $n \times n$ matrix A_0 and denote its eigenvalues by λ_{0i}, its right eigenvectors by $\mathbf{u}_{0i}$ and its left eigenvectors by $\mathbf{v}_{0i}$, where the eigenvalues and eigenvectors satisfy

$$A_0 \mathbf{u}_{0i} = \lambda_{0i} \mathbf{u}_{0i}, \quad i = 1, 2, \ldots, n \tag{3.140a}$$

$$\mathbf{v}_{0i}^T A_0 = \lambda_{0i} \mathbf{v}_{0i}^T, \quad i = 1, 2, \ldots, n \tag{3.140b}$$

The matrix A_0 is assumed to be real, but otherwise is arbitrary. It is also assumed that all the eigenvalues are distinct. In general, the eigenvalues and eigenvectors are complex quantities. As pointed out in Section 3.11, the right and left eigenvectors possess the biorthogonality property and can be normalized so as to satisfy

$$\mathbf{v}_{0j}^T \mathbf{u}_{0i} = \delta_{ij}, \quad \mathbf{v}_{0j}^T A_0 \mathbf{u}_{0i} = \lambda_{0i} \delta_{ij}, \quad i, j = 1, 2, \ldots, n \tag{3.141}$$

Next, we consider the coefficient matrix

$$A = A_0 + A_1 \tag{3.142}$$

where A_0 is the matrix considered above and A_1 is an $n \times n$ matrix reflecting small changes in the system parameters. We refer to A as the perturbed matrix, to A_0 as the unperturbed matrix and to A_1 as the perturbation matrix. The perturbed eigenvalue problems can be written in the form

$$A \mathbf{u}_i = \lambda_i \mathbf{u}_i, \quad i = 1, 2, \ldots, n \tag{3.143a}$$

$$\mathbf{v}_i^T A = \lambda_i \mathbf{v}_i^T, \quad i = 1, 2, \ldots, n \tag{3.143b}$$

where λ_i are the perturbed eigenvalues, $\mathbf{u}_i$ the perturbed right eigenvectors and $\mathbf{v}_i$ the perturbed left eigenvectors. The eigenvectors are biorthogonal and can be normalized so as to satisfy the biorthonormality relations

$$\mathbf{v}_j^T \mathbf{u}_i = \delta_{ij}, \quad \mathbf{v}_j^T A \mathbf{u}_i = \lambda_i \delta_{ij}, \quad i, j = 1, 2, \ldots, n \tag{3.144}$$

A first-order perturbation eigensolution can be written in the general form

$$\lambda_i = \lambda_{0i} + \lambda_{1i}, \quad \mathbf{u}_i = \mathbf{u}_{0i} + \mathbf{u}_{1i}, \quad \mathbf{v}_i = \mathbf{v}_{0i} + \mathbf{v}_{1i}, \quad i = 1, 2, \ldots, n \tag{3.145a, b, c}$$

where $\lambda_{1i}, \mathbf{u}_{1i}$ and $\mathbf{v}_{1i}$ are first-order perturbations. The unperturbed eigenvectors $\mathbf{u}_{0i}$ $(i = 1, 2, \ldots, n)$ can be used as a basis for an n-dimensional vector space, so that any n-vector can be expressed as a linear combination of $\mathbf{u}_{0i}$. Hence, we can write

$$\mathbf{u}_{1i} = \sum_{k=1}^{n} \varepsilon_{ik} \mathbf{u}_{0k}, \quad \varepsilon_{ii} = 0, \quad i = 1, 2, \ldots, n \tag{3.146}$$

where ε_{ik} are small coefficients, and we note that by taking $\varepsilon_{ii} = 0$ the coefficient of $\mathbf{u}_{0i}$ in $\mathbf{u}_i$, Eq. (3.145b), remains equal to 1. It follows that the problem has been reduced to the determination of λ_{1i} and ε_{ik} $(i \neq k)$.

We seek a perturbation solution of the eigenvalue problem accurate to the first order. Hence, second-order terms must be ignored. Inserting Eqs. (3.142), (3.145) and (3.146) into Eq. (3.143a), multiplying on the left by $\mathbf{v}_{0j}^T$, ignoring second-order terms and considering Eqs. (3.141), we obtain [M26]

$$\lambda_{1i} = \mathbf{v}_{0i}^T A_1 \mathbf{u}_{0i}, \quad i = 1, 2, \ldots, n \tag{3.147}$$

and

$$\varepsilon_{ik} = \frac{\mathbf{v}_{0k}^T A_1 \mathbf{u}_{0i}}{\lambda_{0i} - \lambda_{0k}}, \quad i, k = 1, 2, \ldots, n; \quad i \neq k \tag{3.148}$$

Similarly, the perturbation of the left eigenvectors can be expressed as

$$\mathbf{v}_{1j} = \sum_{k=1}^{n} \gamma_{jk} \mathbf{v}_{0k}, \quad \gamma_{jj} = 0, \quad j = 1, 2, \ldots, n \tag{3.149}$$

where

$$\gamma_{jk} = \frac{\mathbf{u}_{0k}^T A_1 \mathbf{v}_{0j}}{\lambda_{0j} - \lambda_{0k}} \quad j, k = 1, 2, \ldots, n; \quad j \neq k \tag{3.150}$$

Equations (3.147), (3.148) and (3.150) show in an explicit manner how changes in the system parameters, reflected in the matrix A_1, affect the system eigenvalues and eigenvectors. Changes in the eigenvalues are of particular importance, as the eigenvalues determine the system stability. The interest lies in control design characterized by eigenvalues that are largely insensitive to changes in the system parameters.

3.15 DISCRETE-TIME SYSTEMS

In our discussions of dynamical systems, the independent variable, i.e., the time t, was regarded as a continuous quantity. For this reason, systems such as described by Eq. (3.82) are said to be *continuous-time systems*. There are processes, however, in which the time appears as a discrete variable. An example of this is the balance in a savings account in which the interest is compounded daily. In this case, the balance appears as a sequence of numbers, one number corresponding to the balance in any given day. Another example is the digital computer, which receives and gives out discrete data only. Such systems are known as *discrete-time systems*. There are yet other systems that are continuous in time, but for which the information is processed on a discrete basis. An example of this is a control system in which measurement of the state is performed at given times t_k ($k = 1, 2, \ldots$). More common is the case in which the response of a continuous-time system is processed on a digital computer. In this section, we propose to develop the mathematical formulation for this latter class of systems.

Let us assume that the interest lies in the response at the sampling times $t_k = kT$ ($k = 1, 2, \ldots$), where T is known as the *sampling period*. Letting $t = kT$ in Eq. (3.94), we obtain

$$\mathbf{x}(kT) = e^{kTA}\mathbf{x}(0) + \int_0^{kT} e^{(kT-\tau)A}\mathbf{Bu}(\tau)\, d\tau \tag{3.151}$$

At the next sampling time, the solution is

$$\mathbf{x}(kT + T) = e^{(kT+T)A}\mathbf{x}(0) + \int_0^{kT+T} e^{(kT+T-\tau)A}\mathbf{Bu}(\tau)\, d\tau \tag{3.152}$$

which can be rewritten as

$$\mathbf{x}(kT + T) = e^{TA}\left[e^{kTA}\mathbf{x}(0) + \int_0^{kT} e^{(kT-\tau)}\mathbf{Bu}(\tau)\, d\tau\right]$$
$$+ \int_{kT}^{kT+T} e^{(kT+T-\tau)A}\mathbf{Bu}(\tau)\, d\tau \tag{3.153}$$

We choose the sampling period sufficiently small that $\mathbf{u}(\tau)$ can be approximated by a so-called "zero-order hold" approximation, which is described by

$$\mathbf{u}(\tau) \cong \mathbf{u}(kT) = \text{constant}, \quad kT \leq \tau \leq kT + T \tag{3.154}$$

Hence, the last term in Eq. (3.153) can be approximated as follows:

$$\int_{kT}^{kT+T} e^{(kT+T-\tau)A}\mathbf{Bu}(\tau)\, d\tau \cong \left[\int_{kT}^{kT+T} e^{(kT+T-\tau)A}\, d\tau\right]\mathbf{Bu}(kT)$$
$$= \left[\int_T^0 e^{\sigma A}(-d\sigma)\right]\mathbf{Bu}(kT)$$
$$= \left(\int_0^T e^{\sigma A}\, d\sigma\right)\mathbf{Bu}(kT) \tag{3.155}$$

88 CONCEPTS FROM LINEAR SYSTEM THEORY

Then, recognizing from Eq. (3.151) that the expression inside brackets in Eq. (3.153) is simply $\mathbf{x}(kT)$ and omitting T from the argument, we can rewrite Eq. (3.153) in the form

$$\mathbf{x}(k+1) = \Phi \mathbf{x}(k) + \Gamma \mathbf{u}(k), \quad k = 0, 1, 2, \ldots \quad (3.156)$$

where

$$\Phi = e^{TA}, \quad \Gamma = \left(\int_0^T e^{\sigma A} \, d\sigma\right) B \quad (3.157)$$

The first of Eqs. (3.157) represents the transition matrix for the discrete-time model and it can be computed by means of the algorithm described by Eqs. (3.101) by simply replacing t by T. Equations (3.156) represent a sequence of *difference equations*, and they are the counterpart of the differential equation (3.82) for continuous systems.

Equations (3.156) represent a set of recursive relations which can be written in the expanded form

$$\begin{aligned}
\mathbf{x}(1) &= \Phi \mathbf{x}(0) + \Gamma \mathbf{u}(0) \\
\mathbf{x}(2) &= \Phi \mathbf{x}(1) + \Gamma \mathbf{u}(1) \\
&\vdots \\
\mathbf{x}(n+1) &= \Phi \mathbf{x}(n) + \Gamma \mathbf{u}(n)
\end{aligned} \quad (3.158)$$

Equations (3.158) permit the computation of the state sequentially. Indeed, for any initial state $\mathbf{x}(0)$ and for the excitation vector $\mathbf{u}(0)$ at $t = 0$, the first of Eqs. (3.158) yields $\mathbf{x}(1)$. Then, the just-computed $\mathbf{x}(1)$ together with the excitation vector $\mathbf{u}(1)$ at $t = T$ can be inserted in the second of Eqs. (3.158) to compute $\mathbf{x}(2)$. The process amounts to updating old states as soon as new ones have been computed.

Equations (3.158) can be used to derive an interesting result. To this end, we insert the first equation into the second, the resulting equation into the third, etc. The resulting equations are

$$\begin{aligned}
\mathbf{x}(1) &= \Phi \mathbf{x}(0) + \Gamma \mathbf{u}(0) \\
\mathbf{x}(2) &= \Phi[\Phi \mathbf{x}(0) + \Gamma \mathbf{u}(0)] + \Gamma \mathbf{u}(1) = \Phi^2 \mathbf{x}(0) + \Phi \Gamma \mathbf{u}(0) + \Gamma \mathbf{u}(1) \\
&= \Phi^2 \mathbf{x}(0) + \sum_{k=0}^{1} \Phi^{1-k} \Gamma \mathbf{u}(k) \\
&\vdots \\
\mathbf{x}(n+1) &= \Phi[\Phi \mathbf{x}(n-1) + \Gamma \mathbf{u}(n-1)] + \Gamma \mathbf{u}(n) \\
&= \Phi^2[\Phi \mathbf{x}(n-2) + \Gamma \mathbf{u}(n-2)] + \Phi \Gamma \mathbf{u}(n-1) + \Gamma \mathbf{u}(n) \\
&= \cdots = \Phi^{n+1} \mathbf{x}(0) + \sum_{k=0}^{n} \Phi^{n-k} \Gamma \mathbf{u}(k)
\end{aligned} \quad (3.159)$$

The last of Eqs. (3.159) represents the general solution $x(n+1)$ for the discrete-time model, and we observe that the sum is a *convolution sum*, which is the discrete-time counterpart of the convolution integral in Eq. (3.94). Whereas this is an interesting result, on a computer we actually use Eqs. (3.158).

Finally, let us consider the problem of stability of discrete-time systems. To this end, we confine ourselves to the homogeneous problem

$$x(k+1) = \Phi x(k), \quad k = 0, 1, 2, \ldots \tag{3.160}$$

and consider the eigenvalue problems

$$\Phi u_i^* = \lambda_i^* u_i^*, \quad \Phi^T v_i^* = \lambda_i^* v_i^*, \quad i = 1, 2, \ldots, n \tag{3.161}$$

where λ_i^* are the eigenvalues, u_i^* the right eigenvectors and v_i^* the left eigenvectors of Φ ($i = 1, 2, \ldots, n$). Then, arranging the eigenvectors in the square matrices $U^* = [u_1^* \ u_2^* \ \cdots \ u_n^*]$ and $V^* = [v_1^* \ v_2^* \ \cdots \ v_n^*]$, where the eigenvectors have been normalized according to $(v_j^*)^T u_i^* = \delta_{ij}$ ($i, j = 1, 2, \ldots, n$), we can write the biorthonormality conditions in the compact form

$$(V^*)^T U^* = I, \quad (V^*)^T \Phi U^* = \Lambda^* \tag{3.162}$$

where Λ^* is the matrix of eigenvalues.

Equations (3.162) can be used to obtain

$$U^*(V^*)^T = I, \quad \Phi = U^* \Lambda^* (V^*)^T \tag{3.163}$$

so that Eqs. (3.160) can be rewritten in the form

$$x(k+1) = U^* \Lambda^* (V^*)^T x(k), \quad k = 0, 1, 2, \ldots \tag{3.164}$$

Making repeated use of the first of Eqs. (3.163), Eqs. (3.164) yields

$$x(k+1) = U^*(\Lambda^*)^{k+1}(V^*)^T x(0), \quad k = 0, 1, 2, \ldots \tag{3.165}$$

It is clear from Eqs. (3.165) that the behavior of the system depends on the magnitudes $|\lambda_i^*|$ of the eigenvalues λ_i^* ($i = 1, 2, \ldots, n$) of Φ. Indeed, the system is merely *stable* if some of the eigenvalues are such that $|\lambda_i^*| = 1$ and the remaining ones are such that $|\lambda_i^*| < 1$, it is *asymptotically stable* if $|\lambda_i^*| < 1$ for all i and *unstable* if at least one eigenvalue is such that $|\lambda_i^*| > 1$. Hence, for stability, the eigenvalues must lie in the closed circular region $|\lambda_i^*| \leq 1$. The various types of behavior are displayed in the λ^*-plane of Fig. 3.19.

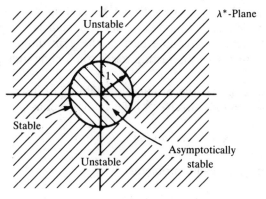

Figure 3.19

Example 3.9 Solve the problem of Example 3.7 in discrete time.

The basis for our solution consists of Eqs. (3.156) and (3.157), except that in our case Γ is a vector and $\mathbf{u}(k)$ is a scalar. Hence, Eq. (3.156) must be replaced by

$$\mathbf{x}(k+1) = \Phi\mathbf{x}(k) + \Gamma u(k), \quad \mathbf{x}(0) = \mathbf{0} \tag{a}$$

where the input is the unit step function, which has the discrete-time form

$$u(k) = 1, \quad k = 0, 1, 2, \ldots \tag{b}$$

Moreover, using results from Example 3.7,

$$\Phi = e^{TA} = \begin{bmatrix} \cos \omega_n T & \dfrac{1}{\omega_n} \sin \omega_n T \\ -\omega_n \sin \omega_n T & \cos \omega_n T \end{bmatrix} \tag{c}$$

$$\Gamma = \left(\int_0^T e^{tA}\, dt\right)\mathbf{b} = \frac{1}{m}\int_0^T \begin{bmatrix} \dfrac{1}{\omega_n}\sin \omega_n t \\ \cos \omega_n t \end{bmatrix} dt = \frac{1}{k}\begin{bmatrix} 1-\cos \omega_n T \\ \omega_n \sin \omega_n T \end{bmatrix} \tag{d}$$

Hence, inserting Eqs. (b) and (c) into Eq. (a), we obtain the sequence

$$\mathbf{x}(1) = \Phi\mathbf{x}(0) + \Gamma u(0) = \frac{1}{k}\begin{bmatrix} 1-\cos \omega_n T \\ \omega_n \sin \omega_n T \end{bmatrix}$$

$$\mathbf{x}(2) = \Phi\mathbf{x}(1) + \Gamma u(1) = \frac{1}{k}\begin{bmatrix} \cos \omega_n T & \dfrac{1}{\omega_n}\sin \omega_n T \\ -\omega_n \sin \omega_n T & \cos \omega_n T \end{bmatrix}\begin{bmatrix} 1-\cos \omega_n T \\ \omega_n \sin \omega_n T \end{bmatrix}$$

$$+ \frac{1}{k}\begin{bmatrix} 1-\cos \omega_n T \\ \omega_n \sin \omega_n T \end{bmatrix} = \frac{1}{k}\begin{bmatrix} 1-\cos 2\omega_n T \\ \omega_n \sin 2\omega_n T \end{bmatrix} \tag{e}$$

$$\mathbf{x}(3) = \Phi x(2) + \Gamma u(2) = \frac{1}{k} \begin{bmatrix} \cos\omega_n T & \frac{1}{\omega_n}\sin\omega_n T \\ -\omega_n \sin\omega_n T & \cos\omega_n T \end{bmatrix} \begin{bmatrix} 1 - \cos 2\omega_n T \\ \omega_n \sin 2\omega_n T \end{bmatrix}$$
$$+ \frac{1}{k} \begin{bmatrix} 1 - \cos\omega_n T \\ \omega_n \sin\omega_n T \end{bmatrix} = \frac{1}{k} \begin{bmatrix} 1 - \cos 3\omega_n T \\ \omega_n \sin 3\omega_n T \end{bmatrix}$$

$\vdots$

We observe that the sequence given by Eqs. (e) is simply the response to the unit step function obtained in Example 3.7 evaluated at $t_k = kT$ ($k = 1, 2, \ldots$), so that the result is excellent. The explanation for this lies in the fact that Eq. (b) reproduces the continuous-time unit step function exactly.

CHAPTER 4

LUMPED-PARAMETER STRUCTURES

Lumped-parameter structures are characterized by parameters that do not depend on the spatial variables. Such structures are described by ordinary differential equations. If the motion is confined to a small neighborhood of an equilibrium point, then the equations of motion are linear. Of particular importance for linear structures are the natural modes of vibration, as the response of such systems can be expressed as a linear combination of the natural modes.

4.1 EQUATIONS OF MOTION FOR LUMPED-PARAMETER STRUCTURES

We consider the case in which the kinetic energy can be written in the form

$$T = T_2 + T_1 + T_0 \tag{4.1}$$

where

$$T_2 = \frac{1}{2} \sum_{i=1}^{n} \sum_{j=1}^{n} m_{ij} \dot{q}_i \dot{q}_j \tag{4.2}$$

is a homogeneous quadratic function of the generalized velocities,

$$T_1 = \sum_{j=1}^{n} f_j \dot{q}_j \tag{4.3}$$

is linear in the generalized velocities and T_0 contains no generalized velocities. In general, the coefficients m_{ij} and f_j and the function T_0 depend on the generalized coordinates q_i $(i = 1, 2, \ldots, n)$. Introducing Eq. (4.1) into Eq. (2.20), we obtain the system Lagrangian

$$L = T_2 + T_1 - U \tag{4.4}$$

where

$$U = U(q_1, q_2, \ldots, q_n) = V - T_0 \tag{4.5}$$

is known as the *dynamic potential*.

As mentioned in Section 2.6, the coordinates $q_i(t)$ $(i = 1, 2, \ldots, n)$ define an n-dimensional vector $\mathbf{q}(t)$ in the configuration space. As time unfolds, the tip of the vector $\mathbf{q}(t)$ traces a curve in the configuration space known as the dynamical path. A given point in the configuration space does not define the state of the system uniquely. Indeed, the point can only specify the coordinates $q_i(t)$ and, to define the state uniquely, it is necessary to specify in addition the generalized velocities $\dot{q}_i(t)$ or, alternatively, the generalized momenta $p_i(t) = \partial L/\partial \dot{q}_i$. If the generalized velocities are used as auxiliary variables, then the motion can be described geometrically in a $2n$-dimensional Euclidean space defined by q_i and $\dot{q}_i$ and known as the *state space*; the set of $2n$ variables q_i and $\dot{q}_i$ define a vector referred to as the *state vector*. Note that, as demonstrated in Section 3.7, in control theory there exists a broader definition of the state space, or the state vector, and in fact the state space and state vector need not be defined by an even number of variables.

A constant solution in the state space, $q_i(t) = q_{ie}$ = constant, $\dot{q}_i(t) = \dot{q}_{ie} = 0$, defines a so-called *equilibrium point*. At an equilibrium point all the generalized velocities and accelerations are zero, which explains the terminology. If $q_{ie} \neq 0$, then the equilibrium point is *nontrivial* and if $q_{ie} = 0$, the equilibrium point is *trivial*. The point representing the trivial equilibrium is simply the origin of the state space. Note that the existence of equilibrium points implies that the generalized forces are either zero or constant.

We are concerned with the case in which the system admits the constant solution q_{ie} = constant, $\dot{q}_{ie} = 0$ $(i = 1, 2, \ldots, n)$. Setting $Q_k = 0$ in Eqs (2.37) and using Eqs. (4.2)–(4.5), we conclude that the constants q_{ie} must satisfy the equations

$$\frac{\partial U}{\partial q_i} = 0, \quad i = 1, 2, \ldots, n \tag{4.6}$$

If U contains terms of degree higher than two in q_i, then there can be more than one equilibrium point.

There is considerable interest in *small motions about equilibrium points*, i.e., motions confined to small neighborhoods of the equilibrium points. A simple coordinate transformation, however, can translate the origin of the state space so as to make it coincide with an equilibrium point. Hence, without loss of generality, we consider the motion in the neighborhood of the trivial solution $q_{ie} = \dot{q}_{ie} = 0$ $(i = 1, 2, \ldots, n)$.

The assumption of small motions implies linearization of the equations of motion about equilibrium, which further implies that only quadratic terms in the generalized coordinates and velocities are to be retained in the Lagrangian. As a result of linearization, the coefficients m_{ij} $(i, j = 1, 2, \ldots, n)$ in Eq. (4.2) become constant. Clearly, they are symmetric, $m_{ji} = m_{ij}$. Moreover, the coefficients f_j in Eq. (4.3) become linear in the generalized coordinates, or

$$f_j = \sum_{i=1}^{n} f_{ij} q_i, \quad j = 1, 2, \ldots, n \tag{4.7}$$

where f_{ij} $(i, j = 1, 2, \ldots, n)$ are constant coefficients, so that

$$T_1 = \sum_{i=1}^{n} \sum_{j=1}^{n} f_{ij} q_i \dot{q}_j \tag{4.8}$$

Finally, expanding the dynamic potential in a Taylor's series about the origin, we can write

$$U(q_1, q_2, \ldots, q_n) \cong \frac{1}{2} \sum_{i=1}^{n} \sum_{j=1}^{n} k_{ij} q_i q_j \tag{4.9}$$

where the term $U(0, 0, \ldots, 0)$ was ignored as a constant that has no effect on the equations of motion and the linear terms in q_i are zero by virtue of Eqs. (4.6). The coefficients

$$k_{ij} = k_{ji} = \frac{\partial^2 U}{\partial q_i \partial q_j}\bigg|_{\mathbf{q}=0} = \frac{\partial^2 U}{\partial q_j \partial q_i}\bigg|_{\mathbf{q}=0} \tag{4.10}$$

are constant and symmetric.

Introducing Eqs. (4.2), (4.8) and (4.9) into the Lagrangian, Eq. (4.4), and using Eqs. (2.37), we obtain the *linearized Lagrange's equations* of motion

$$\sum_{j=1}^{n} (m_{ij} \ddot{q}_j + g_{ij} \dot{q}_j + k_{ij} q_j) = Q_i, \quad i = 1, 2, \ldots, n \tag{4.11}$$

where m_{ij} and k_{ij} are known as *mass* and *stiffness coefficients*, respectively, and

$$g_{ij} = f_{ji} - f_{ij} = -g_{ji}, \quad i, j = 1, 2, \ldots, n \tag{4.12}$$

are referred to as *gyroscopic coefficients*. Note that g_{ij} are skew symmetric. It should be pointed out that when the potential energy is due to elastic effects the part of k_{ij} arising from V defines the *elastic stiffness coefficients* and the part of k_{ij} arising from T_0 defines the *geometric stiffness coefficients*, where the latter can be attributed to centrifugal effects in rotating structures.

There are two types of nonconservative forces not appearing explicity in Eq. (4.11), although they can be regarded as being implicit in the generalized forces Q_i. These are *viscous damping forces* and *circulatory forces*. These forces can be taken into account explicitly by writing Lagrange's equations in the form

$$\frac{d}{dt}\left(\frac{\partial L}{\partial \dot{q}_k}\right) - \frac{\partial L}{\partial q_k} + \frac{\partial \mathscr{F}}{\partial \dot{q}_k} = Q_k, \quad k = 1, 2, \ldots, n \quad (4.13)$$

in which $\mathscr{F}$ is a dissipation function given by

$$\mathscr{F} = \frac{1}{2}\sum_{i=1}^{n}\sum_{j=1}^{n} c_{ij}\dot{q}_i\dot{q}_j + \sum_{i=1}^{n}\sum_{j=1}^{n} h_{ij}\dot{q}_i q_j \quad (4.14)$$

where $c_{ij} = c_{ji}$ are symmetric *viscous damping coefficients* and $h_{ij} = -h_{ji}$ are skew symmetric *circulatory coefficients* ($i, j = 1, 2, \ldots, n$). Consistent with this, $-\sum_{j=1}^{n} c_{ij}\dot{q}_j$ are the viscous forces and $-\sum_{j=1}^{n} h_{ij}q_j$ are the circulatory forces ($i = 1, 2, \ldots, n$). In the presence of viscous and circulatory forces, the equations of motion become

$$\sum_{j=1}^{n}[m_{ij}\ddot{q}_j + (c_{ij} + g_{ij})\dot{q}_j + (k_{ij} + h_{ij})q_j] = Q_i, \quad i = 1, 2, \ldots, n \quad (4.15)$$

Equations (4.15) can be written in the matrix form

$$M\ddot{\mathbf{q}} + (C + G)\dot{\mathbf{q}} + (K + H)\mathbf{q} = \mathbf{Q} \quad (4.16)$$

where M is the symmetric *mass matrix*, C is the symmetric *damping matrix*, G is the skew symmetric *gyroscopic matrix*, K is the symmetric *stiffness matrix* and H is the skew symmetric *circulatory matrix*. Note that gyroscopic, centrifugal and circulatory terms appear in structures that are either spinning or possess spining parts.

Example 4.1 Derive the Lagrange equations of motion for the four-degree-of-freedom system shown in Fig. 4.1.

The kinetic energy has the simple form

$$T = \frac{1}{2}[2m\dot{q}_1^2(t) + 2m\dot{q}_2^2(t) + m\dot{q}_3^2(t) + m\dot{q}_4^2(t)] = \frac{1}{2}\dot{\mathbf{q}}^T(t)M\dot{\mathbf{q}}(t) \quad \text{(a)}$$

where $\dot{\mathbf{q}} = [\dot{q}_1 \quad \dot{q}_2 \quad \dot{q}_3 \quad \dot{q}_4]^T$ is the velocity vector and

EQUATIONS OF MOTION FOR LUMPED-PARAMETER STRUCTURES   97

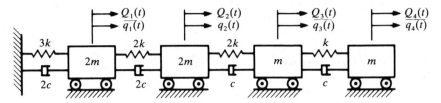

Figure 4.1

$$M = m \begin{bmatrix} 2 & 0 & 0 & 0 \\ 0 & 2 & 0 & 0 \\ 0 & 0 & 1 & 0 \\ 0 & 0 & 0 & 1 \end{bmatrix} \quad \text{(b)}$$

is the diagonal mass matrix. The effect of the viscous forces in the dampers can be taken into account through Rayleigh's dissipation function

$$\mathcal{F} = \frac{1}{2} \{2c\dot{q}_1^2(t) + 2c[\dot{q}_2(t) - \dot{q}_1(t)]^2 + c[\dot{q}_3(t) - \dot{q}_2(t)]^2$$

$$+ c[\dot{q}_4(t) - \dot{q}_3(t)]^2\} = \frac{1}{2} \dot{\mathbf{q}}^T(t) C \dot{\mathbf{q}}(t) \quad \text{(c)}$$

where

$$C = c \begin{bmatrix} 4 & -2 & 0 & 0 \\ -2 & 3 & -1 & 0 \\ 0 & -1 & 2 & -1 \\ 0 & 0 & -1 & 1 \end{bmatrix} \quad \text{(d)}$$

is the damping matrix. The potential energy is due to the energy stored in the springs and can be written as

$$V = \frac{1}{2} \{3kq_1^2(t) + 2k[q_2(t) - q_1(t)]^2 + 2k[q_3(t) - q_2(t)]^2$$

$$+ k[q_4(t) - q_3(t)]^2\} = \frac{1}{2} \mathbf{q}^T(t) K \mathbf{q}(t) \quad \text{(e)}$$

where $\mathbf{q} = [q_1 \ q_2 \ q_3 \ q_4]^T$ is the displacement vector and

$$K = k \begin{bmatrix} 5 & -2 & 0 & 0 \\ -2 & 4 & -2 & 0 \\ 0 & -2 & 3 & -1 \\ 0 & 0 & -1 & 1 \end{bmatrix} \quad \text{(f)}$$

is the stiffness matrix. Finally, the virtual work due to the external forces has the form

$$\overline{\delta W} = \sum_{i=1}^{4} Q_i(t)\delta q_i(t) = \mathbf{Q}^T(t)\delta \mathbf{q}(t) \tag{g}$$

where $\mathbf{Q} = [Q_1 \ Q_2 \ Q_3 \ Q_4]^T$ is the force vector and $\delta \mathbf{q} = [\delta q_1 \ \delta q_2 \ \delta q_3 \ \delta q_4]^T$ is the virtual displacement vector.

Lagrange's equations for the system can be obtained by introducing Eqs. (a), (c) and (e) into Eqs. (4.13), recalling that the Lagrangian has the expression $L = T - V$ and carrying out the indicated operations. The result can be expressed in the matrix form

$$M\ddot{\mathbf{q}}(t) + C\dot{\mathbf{q}}(t) + K\mathbf{q}(t) = \mathbf{Q}(t) \tag{h}$$

where M, C and K are given by Eqs. (b), (d) and (f), respectively.

We observe at this point that the task of deriving Lagrange's equations for a linear system of the type shown in Fig. 4.1 is virtually completed as soon as the expressions for the kinetic energy, Rayleigh's dissipation function, potential energy and virtual work have been derived.

4.2 ENERGY CONSIDERATIONS

In Section 4.1, the subject of nonconservative forces was mentioned. In this section, we wish to bring this subject into sharper focus. To this end, we multiply Eq. (4.16) on the left by $\dot{\mathbf{q}}^T$ and obtain

$$\dot{\mathbf{q}}^T M\ddot{\mathbf{q}} + \dot{\mathbf{q}}^T(C + G)\dot{\mathbf{q}} + \dot{\mathbf{q}}^T(K + H)\mathbf{q} = \dot{\mathbf{q}}^T \mathbf{Q} \tag{4.17}$$

Then, we observe the following:

$$\dot{\mathbf{q}}^T M\ddot{\mathbf{q}} = \frac{d}{dt}\left(\frac{1}{2}\dot{\mathbf{q}}^T M\dot{\mathbf{q}}\right) = \frac{dT_2}{dt} \tag{4.18a}$$

$$\dot{\mathbf{q}}^T G\dot{\mathbf{q}} = 0 \tag{4.18b}$$

$$\dot{\mathbf{q}}^T K\mathbf{q} = \frac{d}{dt}\left(\frac{1}{2}\mathbf{q}^T K\mathbf{q}\right) = \frac{dU}{dt} \tag{4.18c}$$

so that, introducing Eqs. (4.18) into Eq. (4.17) and rearranging, we obtain

$$\frac{dH}{dt} = \dot{\mathbf{q}}^T(\mathbf{Q} - C\dot{\mathbf{q}} - H\mathbf{q}) \tag{4.19}$$

where

$$H = T_2 + U = T_2 + V - T_0 \tag{4.20}$$

is recognized as the Hamiltonian[†] for this linear gyroscopic system. The

---

[†]No confusion should arise from the use of the same notation for the Hamiltonian and the circulatory matrix.

vectors $\mathbf{Q}$, $-C\dot{\mathbf{q}}$ and $-H\mathbf{q}$ can be identified as the vectors of external forces, viscous damping forces and circulatory forces, respectively. Hence, the time rate of change of the Hamiltonian is equal to the power due to the external forces, the viscous damping forces and the circulatory forces, where power is defined as the rate of change of work.

In the absence of external forces, viscous damping forces and circulatory forces, Eq. (4.19) yields

$$H = \text{constant} \qquad (4.21)$$

Equation (4.21) states that, when there are no external forces, viscous damping forces and circulatory forces, the Hamiltonian is conserved. Because the forces $\mathbf{Q}$, $-C\dot{\mathbf{q}}$ and $-H\mathbf{q}$ destroy the conservation of the Hamiltonian, they represent nonconservative forces. It should be noted that the gyroscopic force vector $-G\dot{\mathbf{q}}$ does not affect the conservation of the Hamiltonian, so that gyroscopic forces are conservative.

In the case in which $T_1 = T_0 = 0$, so that $T = T_2$, the Hamiltonian reduces to

$$H = T + V = E \qquad (4.22)$$

where E represents the total energy. In this case the system is known as a *natural system*. Hence, for a natural system the conservation of the Hamiltonian reduces to the conservation of energy. Note that the class of natural systems is very broad and it includes all nonrotating structures.

4.3 THE ALGEBRAIC EIGENVALUE PROBLEM. FREE RESPONSE

The vector $\mathbf{Q}$ in Eq. (4.16) represents impressed forces, such as control forces. If the vector depends on time alone, $\mathbf{Q} = \mathbf{Q}(t)$, then the control is open-loop. If the vector depends on displacements and velocities, $\mathbf{Q} = \mathbf{Q}(\mathbf{q}, \dot{\mathbf{q}})$, then the control is closed-loop. The latter is commonly known as feedback control. Control forces are normally designed to modify the system characteristics so as to produce a desired response. More often than not the object is to cause the response to certain disturbances to approach zero asymptotically. The behavior of a linear dynamical system is governed by its eigenvalues. In particular, the response approaches zero asymptotically if the real part of all the system eigenvalues are negative. If at least one eigenvalue has positive real part, the response diverges. The object of feedback control is to modify the open-loop system so that all the eigenvalues of the closed-loop system possess negative real part. This is equivalent to controlling the motion of a structure by controlling its modes. Hence, knowledge of the open-loop eigenvalues of a system is particularly important in control theory. To compute the eigenvalues, one must solve a so-called eigenvalue problem, which in the case of a discrete system of the type (4.16) is an algebraic problem.

Many control algorithms are based on the open-loop eigensolution. To obtain the open-loop eigenvalue problem, we consider the open-loop system obtained by setting $\mathbf{Q} = \mathbf{0}$ in Eq. (4.16). The result is the homogeneous matrix equation

$$M\ddot{\mathbf{q}} + (C + G)\dot{\mathbf{q}} + (K + H)\mathbf{q} = \mathbf{0} \quad (4.23)$$

As pointed out in Chapter 3, in control theory it is common practice to work with the state equations rather than the configuration equations. Hence, consistent with the approach of Section 3.7, we introduce the $2n$-dimensional state vector

$$\mathbf{x}(t) = \begin{bmatrix} \mathbf{q}(t) \\ \hline \dot{\mathbf{q}}(t) \end{bmatrix} \quad (4.24)$$

Then, premultiplying Eq. (4.23) by M^{-1} and adjoining the identity $\dot{\mathbf{q}} = \dot{\mathbf{q}}$, we can write the state equations in the compact form

$$\dot{\mathbf{x}}(t) = A\mathbf{x}(t) \quad (4.25)$$

where

$$A = \begin{bmatrix} 0 & | & I \\ \hline -M^{-1}(K + H) & | & -M^{-1}(C + G) \end{bmatrix} \quad (4.26)$$

is the $2n \times 2n$ coefficient matrix. The solution of Eq. (4.25) has the exponential form

$$\mathbf{x}(t) = e^{\lambda t}\mathbf{u} \quad (4.27)$$

where λ is a constant scalar and $\mathbf{u}$ a constant vector. Introducing Eq. (4.27) into Eq. (4.25) and dividing through by $e^{\lambda t}$, we obtain the eigenvalue problem

$$A\mathbf{u} = \lambda \mathbf{u} \quad (4.28)$$

This is the same eigenvalue problem as that introduced in Section 3.8. In general, the solution of Eq. (4.28) is complex. There are certain cases, however, in which the solution can be obtained in terms of real quantities alone. In the sequel, we examine these cases.

4.3.1 Conservative Nongyroscopic Systems

In the absence of gyroscopic, viscous damping and circulatory forces, $G = C = H = 0$, the coefficient matrix reduces to

$$A = \left[\begin{array}{c|c} 0 & I \\ \hline -M^{-1}K & 0 \end{array}\right] \tag{4.29}$$

In this particular case, it will prove more convenient to recast the eigenvalue problem in the configuration space instead of the state space. To this end, we let the configuration vector have the form

$$\mathbf{q}(t) = e^{st}\mathbf{q} \tag{4.30}$$

in which case the state vector becomes

$$\mathbf{x}(t) = e^{st}\left[\begin{array}{c} \mathbf{q} \\ \hline s\mathbf{q} \end{array}\right] \tag{4.31}$$

Introducing Eq. (4.31) into Eq. (4.25), with A in the form of Eq. (4.29), we can write the eigenvalue problem in the configuration space as follows:

$$A'\mathbf{q} = \lambda\mathbf{q} \tag{4.32}$$

where

$$A' = M^{-1}K, \quad \lambda = -s^2 \tag{4.33a, b}$$

Actually, we could have derived Eqs. (4.32) and (4.33) in a more direct fashion. Indeed, inserting Eq. (4.30) into Eq. (4.23) with $G = C = H = 0$ and dividing through by e^{st}, we have

$$K\mathbf{q} = \lambda M\mathbf{q}, \quad \lambda = -s^2 \tag{4.34a, b}$$

Then, premultiplying both sides of Eq. (4.34a) by M^{-1}, we obtain Eq. (4.32) in which A' and λ are given by Eqs. (4.33).

The matrix A' in Eq. (4.32) is nonsymmetric. Moreover, the equivalent eigenvalue problem given by Eqs. (4.34) is in terms of two real symmetric matrices. But, the most efficient computational algorithms for the solution of the eigenvalue problem are in terms of a single real symmetric matrix. However, because both M and K are real and symmetric, and furthermore M is positive definite, it is possible to reduce the eigenvalue problem to one in terms of a single real symmetric matrix. To this end, we can use the *Cholesky decomposition* [M26] and express the matrix M in the form

$$M = LL^T \tag{4.35}$$

where L is a lower triangular nonsingular real matrix. Then, introducing the linear transformation

$$q = L^{-T}p \tag{4.36}$$

where $L^{-T} = (L^T)^{-1} = (L^{-1})^T$, Eq. (4.34a) can be reduced to

$$A^*p = \lambda p \tag{4.37}$$

where

$$A^* = L^{-1}KL^{-T} = A^{*T} \tag{4.38}$$

is a real symmetric matrix. The eigenvalues obtained by solving Eq. (4.37) are the same as those of the system (4.34a). On the other hand, the eigenvectors are different, but they are related linearly according to Eq. (4.36). We discuss computational aspects of the eigensolutions later in this chapter. At this point, we confine ourselves to a qualitative discussion of the eigensolution.

Because A^* is a real symmetric matrix, *all the eigenvalues of A^* are real* [M26]. As a corollary, *all the eigenvectors of A^* are real*. Moreover, *the eigenvectors of A^* are mutually orthogonal, as well as orthogonal with respect to A^** [M26], or

$$p_r^T p_s = 0, \quad p_r^T A^* p_s = 0, \quad \lambda_r \neq \lambda_s \tag{4.39}$$

If the eigenvectors are normalized so that $p_r^T p_r = 1$ $(r = 1, 2, \ldots, n)$, then

$$p_r^T p_s = \delta_{rs}, \quad p_r^T A^* p_s = \lambda_r \delta_{rs}, \quad r, s = 1, 2, \ldots, n \tag{4.40}$$

The eigenvectors can be arranged in the square matrix $P = [p_1 \quad p_2 \quad \cdots \quad p_n]$. Then, Eqs. (4.40) can be combined into

$$P^T P = I, \quad P^T A^* P = \Lambda \tag{4.41}$$

where I is the identity matrix and

$$\Lambda = \text{diag}\,[\lambda_r] \tag{4.42}$$

is the diagonal matrix of the eigenvalues. Introducing the *modal matrix* $Q = [q_1 \quad q_2 \quad \cdots \quad q_n]$, where q_r $(r = 1, 2, \ldots, n)$ are the eigenvectors associated with the eigenvalue problem given by Eq. (4.34a), we can use Eq. (4.36) and write the relation between Q and P in the form

$$Q = L^{-T}P \tag{4.43}$$

Inserting Eq. (4.43) into Eqs. (4.41), and recalling Eqs. (4.34a) and (4.35), we obtain

THE ALGEBRAIC EIGENVALUE PROBLEM. FREE RESPONSE   103

$$Q^T M Q = I, \quad Q^T K Q = \Lambda \qquad (4.44)$$

so that *the modal matrix Q is orthonormal with respect to the mass matrix, as well as orthogonal with respect to the stiffness matrix.*

The sign of the eigenvalues λ_r depends on the nature of the stiffness matrix. Indeed, *if K is positive definite*, then A^* is positive definite, and *all the eigenvalues $\lambda_r (r = 1, 2, \ldots, n)$ are real and positive* [M26]. *If K is positive semidefinite*, then A^* is positive semidefinite, and *the eigenvalues are nonnegative*, i.e., some of the eigenvalues are zero and the rest are positive. *If K is sign-variable*, then A^* is sign-variable, and *the eigenvalues are of both signs*. In the above statements the word "positive" can be replaced by "negative" everywhere without affecting their validity.

The eigenvalues λ_r control the system behavior, as they are related to the exponent s in Eq. (4.30). Indeed, from Eq. (4.34b), we conclude that the exponent must satisfy $s = \pm\sqrt{-\lambda}$. Hence, if the eigenvalues λ_r are all positive, then we have

$$s_r = \pm i\omega_r, \quad r = 1, 2, \ldots, n, \quad i = \sqrt{-1} \qquad (4.45)$$

where $\omega_r = \sqrt{\lambda_r}$ $(r = 1, 2, \ldots, n)$ are known as the system *natural frequencies*. In this case, the general solution is

$$\mathbf{q}(t) = \sum_{r=1}^{n} (c_r e^{i\omega_r t} + \bar{c}_r e^{-i\omega_r t}) \mathbf{q}_r \qquad (4.46)$$

where $\bar{c}_r$ is the complex conjugate of c_r. Equation (4.46) indicates that the motion consists of a superposition of harmonic oscillations with frquencies equal to the natural frequencies. The motion is stable, but not asymptotically stable. If some eigenvalues are zero, then some rigid-body motions corresponding to zero natural frequencies are possible. Rigid-body motion must be regarded as unstable. If the system possesses negative eigenvalues, then the corresponding exponents s_r are real and of opposite signs. In such a case, the motion is unstable. Instability is not possible in the case in which the stiffness matrix K is due entirely to elastic restoring forces.

It should be pointed out that the nature of the eigensolution and the stability characteristics are properties of the system, and it is immaterial whether the coefficient matrix has the form (4.33a) or the real symmetric form (4.38). *The existence of these properties is predicated only on the fact that the eigenvalue problem (4.34a) is reducible to one in terms of a single real symmetric matrix.* Nevertheless, the real symmetric form is to be preferred because the Cholesky decomposition is a much more desirable operation than matrix inversion, particularly for high-order systems, and because computational algorithms for solving eigenvalue problems in terms of real symmetric matrices are by far the most efficient ones. Whereas the use of Cholesky decomposition also involves inversions, the inversion of a triangular matrix is a relatively simple process. It amounts to solving a set of simultaneous algebraic equations with the coefficients in the form of a triangular matrix by the method of back substitution [S50].

The solution to the $n \times n$ eigenvalue problem (4.37), with A^* in the form of Eq. (4.38), can be used to construct the solution to the $2n \times 2n$ eigenvalue problem (4.28), with A in the form of Eq. (4.29). Indeed, due to the nature of the eigenvectors, as reflected in Eqs. (4.24) and (4.27), and because the eigenvalues of A, Eq. (4.28), are pure imaginary having the values $\lambda_r = i\omega_r$, the right and left eigenvectors can be verified to be

$$\mathbf{u}_r = [\mathbf{q}_r^T \mid i\omega_r \mathbf{q}_r^T]^T, \quad \mathbf{v}_r = \frac{1}{2}[(M\mathbf{q}_r)^T \mid -i\omega_r^{-1}(M\mathbf{q})^T]^T, \quad r = 1, 2, \ldots, n \tag{4.47}$$

where $\omega_r = \sqrt{\lambda_r}$ and $\mathbf{q}_r$ ($r = 1, 2, \ldots, n$) are obtained by solving the eigenvalue problem (4.37) and using Eq. (4.36). Of course, the other n eigenvalues and right and left eigenvectors are the complex conjugates of the ones above.

The above discussion brings into sharp focus one clear advantage of conservative nongyroscopic systems in that they permit the solution of an $n \times n$ real symmetric eigenvalue problem, known to possess real solutions, instead of a $2n \times 2n$ real nonsymmetric eigenvalue problem, which in general possesses complex solutions. Computationally, the first eigenvalue problem is much more desirable, not only because of the lower dimensions but also because the algorithms for solving real symmetric eigenvalues problems are by far the most efficient.

Example 4.2 Derive the eigenvalue problem for the four-degree-of-freedom system of Fig. 4.2.

The system shown in Fig. 4.2 is the same as that of Fig. 4.1, except that the damping forces and external forces are absent. Hence, from Example 4.1, we can write the equations of motion in the matrix form

$$M\ddot{\mathbf{q}}(t) = K\mathbf{q}(t) = \mathbf{0} \tag{a}$$

where

$$M = m\begin{bmatrix} 2 & 0 & 0 & 0 \\ 0 & 2 & 0 & 0 \\ 0 & 0 & 1 & 0 \\ 0 & 0 & 0 & 1 \end{bmatrix}, \quad K = k\begin{bmatrix} 5 & -2 & 0 & 0 \\ -2 & 4 & -2 & 0 \\ 0 & -2 & 3 & -1 \\ 0 & 0 & -1 & 1 \end{bmatrix} \tag{b}$$

are the mass and stiffness matrices.

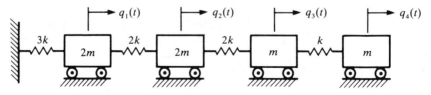

Figure 4.2

Because the system is conservative, the solution of Eq. (a) can be written as

$$\mathbf{q}(t) = e^{i\omega t}\mathbf{q} \tag{c}$$

where $\mathbf{q}$ is a constant vector. Introducing Eq. (c) into Eq. (a), dividing through by $e^{i\omega t}$ and rearranging, we obtain the eigenvalue problem

$$K\mathbf{q} = \omega^2 M\mathbf{q} \tag{d}$$

which is in terms of two real symmetric matrices. Moreover, both matrices are positive definite. Hence, the eigenvalue problem can be reduced to one in terms of a single real symmetric matrix. Indeed, in this case, the matrix L entering into the Cholesky decomposition, Eq. (4.35), has the simple form

$$L = m^{1/2}\begin{bmatrix} \sqrt{2} & 0 & 0 & 0 \\ 0 & \sqrt{2} & 0 & 0 \\ 0 & 0 & 1 & 0 \\ 0 & 0 & 0 & 1 \end{bmatrix} \tag{e}$$

The eigenvalue problem can be written as

$$A^*\mathbf{p} = \lambda\mathbf{p}, \quad \lambda = \frac{\omega^2 m}{k} \tag{f}$$

where

$$\mathbf{p} = [p_1 \quad p_2 \quad p_3 \quad p_4]^T = m^{1/2}[\sqrt{2}q_1 \quad \sqrt{2}q_2 \quad q_3 \quad q_4]^T \tag{g}$$

and

$$A^* = \begin{bmatrix} 1/\sqrt{2} & 0 & 0 & 0 \\ 0 & 1/\sqrt{2} & 0 & 0 \\ 0 & 0 & 1 & 0 \\ 0 & 0 & 0 & 1 \end{bmatrix}\begin{bmatrix} 5 & -2 & 0 & 0 \\ -2 & 4 & -2 & 0 \\ 0 & -2 & 3 & -1 \\ 0 & 0 & -1 & 1 \end{bmatrix}\begin{bmatrix} 1/\sqrt{2} & 0 & 0 & 0 \\ 0 & 1/\sqrt{2} & 0 & 0 \\ 0 & 0 & 1 & 0 \\ 0 & 0 & 0 & 1 \end{bmatrix}$$

$$= \begin{bmatrix} 2.5 & -1 & 0 & 0 \\ -1 & 2 & -\sqrt{2} & 0 \\ 0 & -\sqrt{2} & 3 & -1 \\ 0 & 0 & -1 & 1 \end{bmatrix} \tag{h}$$

4.3.2 Conservative Gyroscopic Systems

In this case, the viscous damping and the circulatory forces are zero, $C = H = 0$, and Eq. (4.23) reduces to

$$M\ddot{\mathbf{q}} + G\dot{\mathbf{q}} + K\mathbf{q} = \mathbf{0} \tag{4.48}$$

where M and K are real and symmetric and G is real and skew symmetric. Equation (4.48) can be cast in the state form (4.25), where the state vector is given by Eq. (4.24) and the coefficient matrix reduces to

$$A = \left[\begin{array}{c|c} 0 & I \\ \hline -M^{-1}K & -M^{-1}G \end{array}\right] \tag{4.49}$$

The solution of the eigenvalue problem associated with the matrix A in the form (4.49) is complex. When both M and K are positive definite, the eigensolution can be produced by working with real quantities alone. To this end, we introduce the $2n \times 2n$ matrices

$$M^* = \left[\begin{array}{c|c} K & 0 \\ \hline 0 & M \end{array}\right], \quad G^* = \left[\begin{array}{c|c} 0 & -K \\ \hline K & G \end{array}\right] \tag{4.50}$$

where M^* is real symmetric and positive definite and G^* is real skew symmetric and nonsingular. Then, adjoining the identity $K\dot{q}(t) - K\dot{q}(t) = 0$, we can replace Eq. (4.48) by the state equation

$$M^*\dot{x}(t) + G^*x(t) = 0 \tag{4.51}$$

The solution of Eq. (4.51) can be written in the form

$$x(t) = e^{st}u \tag{4.52}$$

where s is a complex scalar and u a complex vector. Introducing solution (4.52) into Eq. (4.51), we obtain the eigenvalue problem

$$sM^*u + G^*u = 0 \tag{4.53}$$

It is well known that the solution of the eigenvalue problem (4.53) consists of n pairs of pure imaginary complex conjugates, $s_r = i\omega_r$, $\bar{s}_r = -i\omega_r$, and associated eigenvectors $u_r = y_r + iz_r$, $\bar{u}_r = y_r - iz_r$ $(r = 1, 2, \ldots, n)$.

The complex eigenvalue problem (4.53) can be reduced to a real symmetric one [M18]. Indeed, introducing $s = i\omega$ and $u = y + iz$ in Eq. (4.53) and separating the real and imaginary parts, we obtain two companion equations in terms of both y and z. Then, eliminating z from the first and y from the second, we obtain

$$K^*y = \lambda M^*y, \quad K^*z = \lambda M^*z, \quad \lambda = \omega^2 \tag{4.54}$$

where

$$K^* = G^{*T}M^{*-1}G^* \tag{4.55}$$

is a real symmetric positive definite matrix. Note that the eigenvalue problem (4.54) is of the same type as eigenvalue problem (4.34a). Moreover, because M^* is positive definite, one can use once again the Cholesky decomposition to reduce the eigenvalue problem (4.54) to one in terms of a single real symmetric positive definite matrix.

Because the eigenvalue problems (4.34a) and (4.54) have the same form, and because both can be reduced to one in terms of a single real symmetric matrix, the same computational algorithms can be used to obtain their solutions. We observe, however, that every λ_r $(r = 1, 2, \ldots, n)$ has multiplicity two, as to every λ_r belong two eigenvectors, $\mathbf{y}_r$ and $\mathbf{z}_r$. But, because the problem is positive definite, $\mathbf{y}_r$ and $\mathbf{z}_r$ are independent and they can be rendered orthogonal. Of course, they are orthogonal to the remaining pairs of eigenvectors.

Because all λ_r are real and positive, all $\omega_r = \sqrt{\lambda_r}$ are real. They can be identified as the natural frequencies of the system. The response of the system can be written in the general form

$$\mathbf{x}(t) = \sum_{r=1}^{n} (c_r e^{i\omega_r t} \mathbf{u}_r + \bar{c}_r e^{-i\omega_r t} \bar{\mathbf{u}}_r) \tag{4.56}$$

where we recall that $\mathbf{x}(t)$ represents the state vector, so that the upper half represents the displacement vector and the lower half the velocity vector.

4.3.3 General Dynamical Systems

In the case of viscously damped nongyroscopic systems, $G = H = 0$, it is possible to formulate the problem in the state space in terms of two real symmetric matrices, so that the eigenvectors can be shown to be orthogonal. A procedure using this approach is described by Meirovitch [M16]. Because the eigensolution is complex, however, there is some question whether this approach has much advantage over the approach used for general dynamical systems. Hence, we do not pursue this approach here.

In the general case, the eigenvalue problem is given by Eq. (4.28), where A is a real arbitrary matrix. In view of the fact that M is real symmetric and positive definite, some computational advantage may be derived by using the Cholesky decomposition, Eq. (4.35), and replacing M^{-1} in Eq. (4.26) by

$$M^{-1} = L^{-T} L^{-1} \tag{4.57}$$

As mentioned earlier, the inverse of a triangular matrix is relatively easy to obtain. The solution of the general eigenvalue problem, Eq. (4.28), and the system response were discussed in Section 3.11.

Example 4.3 Derive the eigenvalue problem for the four-degree-of-freedom system of Example 4.1.

In the absence of external forces, the equations of motion have the matrix form

$$M\ddot{q}(t) + C\dot{q}(t) + Kq(t) = 0 \qquad (a)$$

where

$$M = m\begin{bmatrix} 2 & 0 & 0 & 0 \\ 0 & 2 & 0 & 0 \\ 0 & 0 & 1 & 0 \\ 0 & 0 & 0 & 1 \end{bmatrix}, \quad C = c\begin{bmatrix} 4 & -2 & 0 & 0 \\ -2 & 3 & -1 & 0 \\ 0 & -1 & 2 & -1 \\ 0 & 0 & -1 & 1 \end{bmatrix},$$

$$K = k\begin{bmatrix} 5 & -2 & 0 & 0 \\ -2 & 4 & -2 & 0 \\ 0 & -2 & 3 & -1 \\ 0 & 0 & -1 & 1 \end{bmatrix} \qquad (b)$$

Introducing the state vector $\mathbf{x}(t) = [\mathbf{q}^T(t) \mid \dot{\mathbf{q}}^T(t)]^T$, we can write Eq. (a) in the state form (4.25) in which

$$A = \left[\begin{array}{c|c} 0 & I \\ \hline -M^{-1}K & -M^{-1}C \end{array}\right] \qquad (c)$$

The eigenvalue problem is given by Eq. (4.28), or

$$A\mathbf{u} = \lambda \mathbf{u} \qquad (d)$$

where, inserting Eqs. (b) into Eq. (c),

$$A = \left[\begin{array}{c|c} 0 & I \\ \hline -\dfrac{k}{m}\begin{bmatrix} 2.5 & -1 & 0 & 0 \\ -1 & 2 & -1 & 0 \\ 0 & -2 & 3 & -1 \\ 0 & 0 & -1 & 1 \end{bmatrix} & -\dfrac{c}{m}\begin{bmatrix} 2 & -1 & 0 & 0 \\ -1 & 1.5 & -0.5 & 0 \\ 0 & -1 & 2 & -1 \\ 0 & 0 & -1 & 1 \end{bmatrix} \end{array}\right] \qquad (e)$$

We observe that the presence of damping in the system not only causes the order of the eigenvalue to double, but in addition the matrix A loses its symmetry. Moreover, the solution of the nonsymmetric eigenvalue problem is likely to be complex and in general more difficult to obtain.

4.4 QUALITATIVE BEHAVIOR OF THE EIGENSOLUTION

The algebraic eigenvalue problem is basically a numerical problem. Indeed, before an eigensolution can be produced, it is necessary to assign numerical values to the entries of the matrix A. Nevertheless, there exist certain principles capable of shedding a great deal of light on the nature of the eigensolution without actually carrying out the computations. These principles do not apply to general matrices, but they do apply to real symmetric matrices, which are representative of a large class of structures.

Let us consider the eigenvalue problem

$$A\mathbf{u}_i = \lambda_i \mathbf{u}_i, \quad i = 1, 2, \ldots, n \tag{4.58}$$

where A is a *real symmetric positive definite matrix*. Hence, all the eigenvalues are real and positive and the eigenvectors are real and mutually orthogonal. For convenience, we assume that $\lambda_1 \le \lambda_2 \le \cdots \le \lambda_n$, which implies that the eigenvalues are proportional to the natural frequencies squared. Moreover, we assume that the eigenvectors have been normalized, so that $\mathbf{u}_j^T \mathbf{u}_i = \delta_{ij}$ ($i, j = 1, 2, \ldots, n$). Because the vectors $\mathbf{u}_i$ are orthogonal, they can be used as a basis for an n-dimensional vector space L^n. Hence, any n-dimensional vector $\mathbf{u}$ can be expressed as a linear superposition of the system eigenvectors, or

$$\mathbf{u} = \sum_{i=1}^{n} c_i \mathbf{u}_i = U\mathbf{c} \tag{4.59}$$

where $\mathbf{c} = [c_1 \; c_2 \; \cdots \; c_n]^T$ is a vector of coefficients and $U = [u_1 \; u_2 \; \cdots \; u_n]$ is the matrix of eigenvectors. Using the orthonormality of U, we obtain simply

$$\mathbf{c} = U^T \mathbf{u} \tag{4.60}$$

If $\mathbf{u}$ is a unit vector, so that the Euclidean norm $\|\mathbf{u}\|$ is equal to one, then $\mathbf{c}$ is also a unit vector. Equations (4.59) and (4.60) constitute the so-called *expansion theorem* for a real orthogonal vector space.

Next, let us premultiply Eq. (4.58) by $\mathbf{u}_i^T$, divide through by $\mathbf{u}_i^T \mathbf{u}_i = \|\mathbf{u}_i\|^2$, where $\|\mathbf{u}_i\|$ is the Euclidean norm of $\mathbf{u}_i$, and obtain

$$\lambda_i = \frac{\mathbf{u}_i^T A \mathbf{u}_i}{\|\mathbf{u}_i\|^2}, \quad i = 1, 2, \ldots, n \tag{4.61}$$

so that every eigenvalue λ_i can be obtained as the ratio of two quadratic forms. The question arises as to what happens to the ratio if we use an arbitrary vector $\mathbf{u}$, instead of the eigenvector $\mathbf{u}_i$, and let $\mathbf{u}$ range over the space L^n. Hence, let us consider

$$\lambda(\mathbf{u}) = \frac{\mathbf{u}^T A \mathbf{u}}{\|\mathbf{u}\|^2} \tag{4.62}$$

which is known as *Rayleigh's quotient*. Using the expansion theorem, Eq. (4.59), it is possible to prove that *Rayleigh's quotient has a stationary value in the neighborhood of an eigenvector. In particular, Rayleigh's quotient is an upper bound for the lowest eigenvalue* [M26]

$$\lambda(\mathbf{u}) \geq \lambda_1 \tag{4.63}$$

and a lower bound for the highest eigenvalue

$$\lambda(\mathbf{u}) \leq \lambda_n \tag{4.64}$$

Equation (4.63) implies that

$$\lambda_1 = \min \lambda(\mathbf{u}) \tag{4.65}$$

which is often referred to as *Rayleigh's principle*.

Rayleigh's quotient can be used to provide estimates of the eigenvalues, particularly of the lowest eigenvalue. Indeed, by using a trial vector $\mathbf{u}$ resembling the lowest eigenvector, one can obtain an estimate of the lowest eigenvalue. Due to the stationarity of Rayleigh's quotient, if $\mathbf{u}$ differs from $\mathbf{u}_1$ by a quantity of order ε, then $\lambda(\mathbf{u})$ differs from λ_1 by a quantity of order ε^2, so that the estimate of λ_1 is one order of magnitude better than the guess of $\mathbf{u}_1$. Quite often, the static deformation vector of a structure under loads proportional to the masses can be used as a guess for $\mathbf{u}_1$ with remarkably good results.

To obtain estimates of higher eigenvalues, one can constrain the trial vector $\mathbf{u}$ to a space orthogonal to the lower eigenvectors. For example, to estimate λ_{r+1}, the trial vector $\mathbf{u}$ should satisfy $\mathbf{u}^T\mathbf{u}_i = 0$ $(i = 1, 2, \ldots, r)$. The difficulty with this approach is that the eigenvectors $\mathbf{u}_i$ $(i = 1, 2, \ldots, r)$ are not readily available, so that a characterization of the higher eigenvalues that is independent of the lower eigenvectors is desirable. This is the object of a theorem due to Courant and Fischer, which can be stated as follows: *The eigenvalue λ_{r+1} of a real symmetric matrix is the maximum value which can be given to $\min \mathbf{u}^T A\mathbf{u}/\|\mathbf{u}\|^2$ by the imposition of the r constraints $\mathbf{u}^T\mathbf{w}_i = 0$ $(i = 1, 2, \ldots, r)$, where $\mathbf{w}_i$ are arbitrary vectors* [M26]. We refer to the preceding statement as the *maximum-minimum theorem*.

The maximum-minimum theorem has limited appeal as an end in itself, but it has important implications in truncation problems. Quite often, in vibration problems the question is how truncation of the mathematical model affects the system eigenvalues. To answer this question, we consider a system described by an $n \times n$ real symmetric matrix A whose eigenvalues satisfy $\lambda_1 \leq \lambda_2 \leq \cdots \leq \lambda_n$, as well as another system described by an $(n-1) \times (n-1)$ real symmetric matrix B with eigenvalues satisfying $\gamma_1 \leq \gamma_2 \leq \cdots \leq \gamma_{n-1}$, where B is obtained from A by deleting the last row and column. In the first place, we observe that the quadratic form $\mathbf{v}^T B \mathbf{v}$ is identical to the quadratic form $\mathbf{u}^T A \mathbf{u}$, provided $u_i = v_i$ $(i = 1, 2, \ldots, n-1)$ and $u_n = 0$. The equation $u_n = 0$ can be regarded as a constraint on $\mathbf{x}$. Then, using the maximum-minimum theorem, it can be shown that [M26]

QUALITATIVE BEHAVIOR OF THE EIGENSOLUTION   111

$$\lambda_1 \le \gamma_1 \le \lambda_2 \le \gamma_2 \le \cdots \le \lambda_{n-1} \le \gamma_{n-1} \le \lambda_n \qquad (4.66)$$

Inequalities (4.66) constitute the so-called *inclusion principle*. The implication of the principle is that truncation of the mathematical model is likely to raise the estimated natural frequencies of the model. Another implication is that the estimated natural frequencies become less accurate as the mathematical model is increasingly truncated. We return to this subject when we discuss discretization (in space) of distributed systems.

The inclusion principle can be used to derive a criterion for the positive definiteness of a real symmetric matrix. It is called *Sylvester's criterion* and it reads as follows: *A real symmetric matrix A is positive definite if all its principal minor determinants are positive*, det $A_i > 0$ ($i = 1, 2, \ldots, n$) [M26]. Note that the positive definiteness of A can always be checked by examining its eigenvalues. Application of Sylvester's criterion, however, is likely to be more expedient for low-order matrices.

Example 4.4 Consider the system of Fig. 4.2 and use Rayleigh's quotient to obtain an estimate of the lowest ntural frequency.

With reference to Eq. (4.62), we can write the Rayleigh quotient in the form

$$R(\mathbf{p}) = \lambda(\mathbf{p}) = \omega^2(\mathbf{p}) = \frac{\mathbf{p}^T A^* \mathbf{p}}{\|\mathbf{p}\|^2} \qquad (a)$$

where, from Eqs. (4.36) and (4.38),

$$\mathbf{p} = L^T \mathbf{q}, \quad A^* = L^{-1} K L^{-T} = A^{*T} \qquad (b)$$

in which $\mathbf{q}$ is the displacement vector, K is the stiffness matrix and L is defined by Eq. (4.35). Inserting Eqs. (b) into Eq. (a) and considering Eq. (4.35), we can rewrite Rayleigh's quotient as

$$R(\mathbf{q}) = \lambda(\mathbf{q}) = \omega^2(\mathbf{q}) = \frac{\mathbf{q}^T K \mathbf{q}}{\mathbf{q}^T M \mathbf{q}} \qquad (c)$$

which is more convenient for the problem at hand. Indeed, using Eqs. (b) of Example 4.2, we obtain

$$\omega^2(\mathbf{q}) = \frac{k\mathbf{q}^T \begin{bmatrix} 5 & -2 & 0 & 0 \\ -2 & 4 & -2 & 0 \\ 0 & -2 & 3 & -1 \\ 0 & 0 & -1 & 1 \end{bmatrix} \mathbf{q}}{m\mathbf{q}^T \begin{bmatrix} 2 & 0 & 0 & 0 \\ 0 & 2 & 0 & 0 \\ 0 & 0 & 1 & 0 \\ 0 & 0 & 0 & 1 \end{bmatrix} \mathbf{q}} \qquad (d)$$

To estimate the lowest natural frequency, we subject the system to forces proportional to the associated masses and calculate a trial vector $\mathbf{q}$ equal to the static displacement of the masses. The static displacement vector is

$$\mathbf{q} = (m/k)[2 \quad 4 \quad 5 \quad 6]^T \qquad (e)$$

Inserting Eq. (c) into Eq. (d), we obtain

$$\omega_{1e}^2 = \frac{\begin{bmatrix}2\\4\\5\\6\end{bmatrix}^T \begin{bmatrix}5 & -2 & 0 & 0\\-2 & 4 & -2 & 0\\0 & -2 & 3 & -1\\0 & 0 & -1 & 1\end{bmatrix}\begin{bmatrix}2\\4\\5\\6\end{bmatrix}}{\begin{bmatrix}2\\4\\5\\6\end{bmatrix}^T\begin{bmatrix}2 & 0 & 0 & 0\\0 & 2 & 0 & 0\\0 & 0 & 1 & 0\\0 & 0 & 0 & 1\end{bmatrix}\begin{bmatrix}2\\4\\5\\6\end{bmatrix}} \frac{k}{m} = \frac{23}{101}\frac{k}{m} = 0.2277\frac{k}{m} \qquad (f)$$

so that the estimate of the lowest natural frequency is

$$\omega_{1e} = 0.4772\sqrt{k/m} \qquad (g)$$

The actual natural frequency is $\omega_{1a} = 0.4709\sqrt{k/m}$. Hence, the relative error is

$$\frac{\omega_{1e} - \omega_{1a}}{\omega_{1a}} = \frac{0.4772 - 0.4709}{0.4709} = 1.34\% \qquad (h)$$

which is quite small. The reason for such a good estimate of the lowest natural frequency is that the trial vector given by Eq. (e) is reasonably close to the first eigenvector.

Example 4.5 Verify that the inclusion principle is satisfied by the eigenvalues of the matrices

$$A = \begin{bmatrix} 5 & -2 & 0 & 0 \\ -2 & 4 & -2 & 0 \\ 0 & -2 & 3 & -1 \\ 0 & 0 & -1 & 1 \end{bmatrix} \qquad (a)$$

and

$$B = \begin{bmatrix} 5 & -2 & 0 \\ -2 & 4 & -1 \\ 0 & -2 & 3 \end{bmatrix} \qquad (b)$$

Matrix B is embedded in matrix A, in the sense that B can be obtained from A by removing the last row and column. The eigenvalues of A are

$$\lambda_1 = 0.4709, \quad \lambda_2 = 1.0157, \quad \lambda_3 = 1.6713, \quad \lambda_4 = 2.0926 \qquad (c)$$

and the eigenvalues of B are

$$\gamma_1 = 0.7904, \quad \gamma_2 = 1.6294, \quad \gamma_3 = 2.0543 \qquad (d)$$

It is clear that

$$\lambda_1 < \gamma_1 < \lambda_2 < \gamma_2 < \lambda_3 < \gamma_3 < \lambda_4 \qquad (e)$$

so that the inclusion principle holds.

It should be pointed out that this example is quite contrived, as it has no physical meaning. In the case of discrete modeling of distributed structures, however, application of the inclusion principle is more natural, as can be concluded from Section 7.4.

4.5 COMPUTATIONAL METHODS FOR THE EIGENSOLUTION

As indicated in Section 4.3, control of structures can be carried out by controlling its modes. This requires knowledge of the structure eigenvalues and eigenvectors. There is a large variety of computational algorithms for the eigensolution. For the most part, the algorithms are tailored so as to take advantage of the properties of the structure, where the latter are reflected in the matrix A. By far the most efficient algorithms exist for real symmetric matrices. Fortunately, most structures can be represented by such matrices. The algorithms for real symmetric matrices are faster, more stable and can handle matrices of much larger order than algorithms for arbitrary matrices. Although they may differ in appearance and details, all eigensolution algorithms are iterative in nature. Following is a brief review of the better known techniques.

4.5.1 Matrix Iteration by the Power Method

This is one of the oldest techniques. The eigenvalue problem is described by

$$A\mathbf{u}_r = \lambda_r \mathbf{u}_r, \quad r = 1, 2, \ldots, n \qquad (4.67)$$

We consider first the case in which A is real, symmetric and positive definite, so that all the eigenvalues are real and positive and the eigenvectors are real and mutually orthogonal. The eigenvalues are assumed to satisfy $\lambda_1 \geq \lambda_2 \geq \cdots \geq \lambda_n$, where $\lambda_r = 1/\omega_r^2$. The iteration sequence is defined by

$$\mathbf{w}_p = A\mathbf{w}_{p-1}, \quad p = 1, 2, \ldots \tag{4.68}$$

where $\mathbf{w}_0$ is an arbitrary initial trial vector. Of course, $\mathbf{w}_0$ is real. The iteration sequence (4.68) converges to the first mode. Upon convergence, we have

$$\lim_{p\to\infty} \mathbf{w}_p = \lambda_1^p \alpha_1 \mathbf{u}_1, \quad \lim_{p\to\infty} \mathbf{w}_{p+1} = \lambda_1^{p+1} \alpha_1 \mathbf{u}_1 \tag{4.69a, b}$$

where α_1 is some constant. Equations (4.69) imply that, upon convergence, two consecutive iterated vectors are proportional to the first eigenvector $\mathbf{u}_1$. Moreover, from Eqs. (4.69), we can write

$$\lim_{p\to\infty} \frac{w_{p+1,i}}{w_{p,i}} = \lambda_1 \tag{4.70}$$

or, the highest eigenvalue is obtained as the ratio of homologous (having the same relative position) elements of two consecutive iterated vectors. Proof of convergence is based on the expansion theorem, and is given by Meirovitch [M26]. The rate of convergence is faster for smaller ratios λ_2/λ_1. Note that, because λ_r is inversely proportional to the square of the natural frequency ω_r, the highest eigenvalue corresponds to the lowest natural frequency.

To obtain the lower eigenvalues and associated modes, it is necessary to modify the matrix A in a way that the higher eigenvalues are excluded automatically in a process known as matrix deflation. It can be easily verified that the deflated matrix

$$A_k = A_{k-1} - \lambda_{k-1}\mathbf{u}_{k-1}\mathbf{u}_{k-1}^T, \quad k = 2, 3, \ldots, n, \quad A_1 = A \tag{4.71}$$

has the eigenvalues $0, 0, \ldots, 0, \lambda_k, \lambda_{k+1}, \ldots, \lambda_n$ and the eigenvectors $\mathbf{u}_1, \mathbf{u}_2, \ldots, \mathbf{u}_{k-1}, \mathbf{u}_k, \mathbf{u}_{k+1}, \ldots, \mathbf{u}_n$, respectively, so that it can be used to iterate to the eigenvalue λ_k and the eigenvector $\mathbf{u}_k$. In the above it is assumed that the computed eigenvectors have been normalized so as to satisfy $\mathbf{u}_r^T \mathbf{u}_s = \delta_{rs}$ $(r, s = 1, 2, \ldots, k-1)$. Indeed, extreme care must be exercised to ensure that the above normalization is performed. Otherwise, the results will be in error.

A similar matrix iteration exists for an arbitrary real matrix A, not necessarily symmetric, but known to possess real eigenvalues and eigenvectors. Of course, in this case we must compute right and left eigenvectors $\mathbf{u}_r$ and $\mathbf{v}_r$ $(r = 1, 2, \ldots, n)$. To obtain lower eigensolutions, we iterate with the deflated matrices

$$A_k = A_{k-1} - \lambda_{k-1}\mathbf{u}_{k-1}\mathbf{v}_{k-1}^T, \quad k = 2, 3, \ldots, n \tag{4.72}$$

for λ_k, $\mathbf{u}_k$ and the deflated matrix A_k^T for λ_k, $\mathbf{v}_k$, where $\mathbf{u}_r^T \mathbf{v}_s = \delta_{rs}$ $(r, s = 1, 2, \ldots, k-1)$.

When an arbitrary real matrix A admits complex eigensolutions, then the eigensolutions occur in pairs of complex conjugates. For example, if λ_1 and $\mathbf{u}_1$ are an eigenvalue and eigenvector of A, then the complex conjugates $\lambda_2 = \bar{\lambda}_1$ and $\mathbf{u}_2 = \bar{\mathbf{u}}_1$ are also an eigenvalue and eigenvector, respectively. Because λ_1 and λ_2 have the same magnitude, the convergence criteria given by Eqs. (4.69) and (4.70) are no longer applicable and must be replaced by criteria taking into account the complex conjugate nature of the eigensolutions. It turns out that the iteration to complex eigensolutions can be carried out with real iterated vectors. Indeed, it can be shown [M26] that, using the same iteration process as that given by Eq. (4.68), convergence is achieved when

$$\lim_{p \to \infty} \xi = \lim_{p \to \infty} \frac{a_p a_{p+3} - a_{p+1} a_{p+2}}{a_{p+1}^2 - a_p a_{p+2}} = \text{constant}$$

$$\lim_{p \to \infty} \eta = \lim_{p \to \infty} \frac{a_{p+2}^2 - a_{p+1} a_{p+3}}{a_{p+1}^2 - a_p a_{p+2}} = \text{constant}$$

(4.73)

where a_p, a_{p+1}, a_{p+2} and a_{p+3} are entries of the matrices A^p, A^{p+1}, A^{p+2} and A^{p+3} corresponding to the same pair of subscripts, in which A^p represents the matrix A raised to the power p, etc. Beginning with the trial vector $\mathbf{w}_0 = \mathbf{e}_1$, where $\mathbf{e}_1$ is a standard unit vector with the top component equal to unity and the remaining components equal to zero, a_p, a_{p+1}, a_{p+2} and a_{p+3} can be identified as the top components of the iterated vectors $\mathbf{w}_p$, $\mathbf{w}_{p+1}$, $\mathbf{w}_{p+2}$ and $\mathbf{w}_{p+3}$, respectively. After Eqs. (4.73) establish convergence, the eigenvalues can be obtained from

$$\begin{matrix}\lambda_1 \\ \lambda_2\end{matrix} = \frac{1}{2}\left(-\xi \pm i(4\eta - \xi^2)^{1/2}\right)$$

(4.74)

and the eigenvectors from

$$\text{Re}\,\mathbf{u}_1 = \frac{1}{2}\mathbf{w}_{p+2}, \quad \text{Im}\,\mathbf{u}_1 = -\frac{\xi}{2(4\eta - \xi^2)^{1/2}}\mathbf{w}_{p+2} - \frac{1}{(4\eta - \xi^2)^{1/2}}\mathbf{w}_{p+3}$$

(4.75)

A similar iteration process using the matrix A^T yields the eigenvalues λ_1 and λ_2 once again, as well as the left eigenvector $\mathbf{v}_1$.

To iterate to subdominant eigensolutions, we use the deflated matrices

$$\begin{aligned}A_{2k-1} &= A_{2k-3} - \lambda_{2k-3}\mathbf{u}_{2k-3}\mathbf{v}_{2k-3}^T - \bar{\lambda}_{2k-3}\bar{\mathbf{u}}_{2k-3}\bar{\mathbf{v}}_{2k-3}^T \\ &= A_{2k-3} - 2\,\text{Re}\left(\lambda_{2k-3}\mathbf{u}_{2k-3}\mathbf{v}_{2k-3}^T\right), \quad k = 2, 3, \ldots, n/2\end{aligned}$$

(4.76)

for the eigenvalue λ_{2k-1} and right eigenvector $\mathbf{u}_{2k-1}$, and A_{2k-1}^T for the eigenvalue λ_{2k-1} and left eigenvector $\mathbf{v}_{2k-1}$ ($k = 2, 3, \ldots, n/2$). Note that this type of problems implies a state space formulation, so that n is an even integer.

The iteration process based on matrix deflation yields one pair of complex conjugate eigenvalues and eigenvectors at a time. Other algorithms iterate to all the eigenvalues and eigenvectors simultaneously.

4.5.2 The Jacobi Method

The method is applicable to real symmetric matrices A and it produces all the eigenvalues and eigenvectors simultaneously. It uses similarity transformations to diagonalize the matrix A, where the diagonal matrix is simply the matrix of the eigenvalues of A. The iteration sequence is given by the equation

$$A_k = R_k^T A_{k-1} R_k, \quad k = 1, 2, \ldots \quad (4.77)$$

where $A_0 = A$ and R_k is a rotation matrix [M26]. The process is convergent, and indeed

$$\lim_{k \to \infty} A_k = \Lambda, \quad \lim_{k \to \infty} R_1 R_2 \cdots R_{k-1} R_k = U \quad (4.78a, b)$$

where Λ and U are the matrices of eigenvalues and orthonormal eigenvectors of A, respectively. Convergence is based on the fact that the matrices A_{k-1} and A_k have the same Euclidean norm. Moreover, the sum of the diagonal elements squared of A_k is larger than the sum of the diagonal elements squared of A_{k-1} by $2(a_{pq}^{(k)})^2$. Hence, with every iteration step, the diagonal elements increase at the expense of the off-diagonal elements.

4.5.3 Methods for the Reduction of a Matrix to Hessenberg Form

Certain iteration procedures are very efficient if the matrix A has an upper Hessenberg form, namely, one in which all the elements below the first subdiagonal are zero. If the matrix is symmetric, then the Hessenberg matrix is actually tridiagonal. There are two techniques capable of reducing a matrix to Hessenberg form and they both involve transformations. The first is Givens' method and it involves rotations and the second is Householder's method and it involves reflections [M26].

4.5.4 Eigenvalues of a Tridiagonal Matrix. Sturm's Theorem

Let us assume that the original symmetric matrix A has been reduced to the tridiagonal form A_k by either Givens' or Householder's method. The characteristic determinant associated with A_k can be written in the form

$$\det[A_k - \lambda I] = \begin{vmatrix} \alpha_1 - \lambda & \beta_2 & 0 & \cdots & 0 & 0 \\ \beta_2 & \alpha_2 - \lambda & \beta_3 & \cdots & 0 & 0 \\ 0 & \beta_3 & \alpha_3 - \lambda & \cdots & 0 & 0 \\ \vdots & & & & & \vdots \\ 0 & 0 & 0 & \cdots & \alpha_{n-1} - \lambda & \beta_n \\ 0 & 0 & 0 & \cdots & \beta_n & \alpha_n - \lambda \end{vmatrix}$$

(4.79)

Then, it can be shown by induction that the principal minor determinants of the matrix $A_k - \lambda I$ have the expressions

$$p_1(\lambda) = \alpha_1 - \lambda$$
$$p_i(\lambda) = (\alpha_i - \lambda) p_{i-1}(\lambda) - \beta_i^2 p_{i-2}(\lambda), \quad i = 2, 3, \ldots, n$$

(4.80)

Moreover, we let $p_0(\lambda) \equiv 1$. We note that the characteristic equation is given by

$$p_n(\lambda) = 0 \qquad (4.81)$$

The polynomials $p_0(\lambda), p_1(\lambda), \ldots, p_n(\lambda)$ possess certain properties that are associated with the so-called Sturm sequence.

Our interest lies in the roots of $p_n(\lambda)$. These roots can be located by using *Sturm's theorem*, which can be stated as follows: *Let $p_0(\lambda), p_1(\lambda), \ldots, p_n(\lambda)$ be a Sturm sequence on the interval $a \le \lambda \le b$ and denote by $s(\mu)$ the number of sign changes in the consecutive sequence of numbers $p_0(\mu), p_1(\mu), \ldots, p_n(\mu)$. Then, the number of zeros of the polynomial $p_n(\lambda)$ in the interval $a \le \lambda \le b$ is equal to $s(b) - s(a)$.* To apply the theorem, we first select an interval (a, b) and calculate $s(b) - s(a)$. If $s(b) \ne s(a)$, then we know that there is at least one root in the interval (a, b). To isolate a root, we bisect the interval and consider the subintervals $(a, (a+b)/2)$ and $((a+b)/2, b)$. The method can be used to compute the roots to any degree of accuracy by narrowing the intervals containing roots. It should be pointed out that *the polynomials $p_i(\lambda)$ $(i = 1, 2, \ldots, n)$ are never derived explicitly*, as only the numbers $p_i(\mu)$ must be computed, which is done recursively by means of Eqs. (4.80).

The method based on Sturm's theorem yields only the eigenvalues. The eigenvectors must be obtained by another method, such as inverse iteration.

4.5.5 The QR Method

Every matrix A can be transformed into a triangular matrix T through a similarity transformation, where the diagonal elements of T are the eigenvalues of A. In general, the transformation can be taken as unitary [M26]. If the matrix A is real and its eigenvalues are real, then the transformation is orthogonal. In the special case in which A is symmetric, the matrix T is actually diagonal.

One of the most attractive methods for reducing a general matrix to triangular form is the QR method. If the matrix A to be reduced to triangular form is fully populated, the iteration process is quite laborious. For this reason, the QR method is used mainly in conjunction with a simpler form for A, such as an upper Hessenberg form or a tridiagonal form. As pointed out above, reduction to Hessenberg or tridiagonal form can be performed by Givens' method or Householder's method.

The iteration process is defined by the relations

$$A_s = Q_s R_s, \quad A_{s+1} = R_s Q_s, \quad s = 1, 2, \ldots \quad (4.82\text{a, b})$$

where $A_1 = A$ is the given matrix, Q_s is a unitary matrix and R_s is an upper triangular matrix. Premultiplying Eq. (4.82a) by $Q_s^H = \bar{Q}_s^T$ and recalling that for unitary matrices $Q_s^H Q_s = I$, we obtain

$$R_s = Q_s^H A_s, \quad s = 1, 2, \ldots \quad (4.83)$$

so that introducing Eq. (4.83) into Eq. (4.82b), we can write

$$A_{s+1} = Q_s^H A_s Q_s, \quad s = 1, 2, \ldots \quad (4.84)$$

from which we conclude that Eqs. (4.82b) represent a unitary similarity transfromation. As $s \to \infty$, A_s converges to an upper triangular matrix, with the eigenvalues on the main diagonal. If A_s is real and symmetric, then both Q_s and R_s are real, so that the transformation is actually orthonormal. In fact, the matrices Q_s can be taken as rotation matrices. In this case, A_s converges to a diagonal matrix, with the eigenvalues on the main diagonal.

Convergence can be accelerated appreciably by using shifts, in which case the algorithm is defined by the relations

$$A_s - \mu_s I = Q_s R_s, \quad A_{s+1} = R_s Q_s + \mu_s I, \quad s = 1, 2, \ldots \quad (4.85\text{a, b})$$

where the shifts μ_s are generally different for each iteration step.

A variant of the QR method uses lower triangular matrices instead of upper triangular matrices and is known as the QL method. When A is real and symmetric, so that the Hessenberg form is actually tridiagonal, the QR is particularly simple. Details of the method and shifting strategy are given by Meirovitch [M26].

The QR method produces only the eigenvalues of a matrix. The eigenvectors can be obtained by inverse iteration, as discussed below.

4.5.6 Inverse Iteration

The inverse iteration method is a procedure for computing the eigenvectors of a matrix whose eigenvalues are known. The iteration scheme is described by

$$[A - \lambda I]\mathbf{w}_p = \mathbf{w}_{p-1}, \quad p = 1, 2, \ldots \tag{4.86}$$

where $\mathbf{w}_0$ is an arbitrary initial trial vector. By the expansion theorem, $\mathbf{w}_0$ can be taken as a superposition of the eigenvectors of A. Then, it can be shown that if λ is close to λ_r, in the limit, as $p \to \infty$, we obtain

$$\lim_{p \to \infty} \mathbf{w}_p = \frac{1}{(\lambda_r - \lambda)^p} \alpha_r \mathbf{u}_r \tag{4.87}$$

so that the process iterates to $\mathbf{u}_r$.

In obtaining the iterated vectors $\mathbf{w}_1, \mathbf{w}_2, \ldots$ by means of Eq. (4.86), it is not necessary to invert the matrix $A - \lambda I$, which is close to being singular. In fact, Eqs. (4.86) can be regarded as a set of nonhomogeneous algebraic equations and can be solved by Gaussian elimination. To this end, one can reduce $A - \lambda I$ to upper triangular from by elementary transformations and solve the resulting equations by back substitution [M26].

The method is valid for arbitrary matrices but is particularly efficient for tridiagonal matrices. Note that if the original matrix A was reduced to tridiagonal form, then the eigenvectors of the tridiagonal form must undergo some transformation to obtain the eigenvectors of A [M26].

4.5.7 Subspace Iteration

In deriving discrete models of distributed structures, it is possible at times to end up with a model possessing an excessive number of degrees of freedom. The associated eigenvalue problem may be of such large dimensions that a complete solution may not be feasible. Moreover, because higher eigenvalues and eigenvectors corresponding to discrete models of distributed structures tend to be very inaccurate, and they are seldom excited, a complete solution may not be even necessary. Hence, the interest lies in a partial solution of an eigenvalue problem of high order, and in particular in a limited number of lower eigenvalues and eigenvectors. In this regard, we recall that the matrix iteration by the power method described earlier in this section is capable of producing such a solution. The method iterates to one eigenvalue and eigenvector at a time and, because the matrix used in a given iteration cycle depends on the previously computed eigenvalues and eigenvectors, accuracy is lost at each iteration cycle. Another method, known as subspace iteration, permits iteration to a given number of eigenvalues and eigenvectors simultaneously [M26].

Let us consider the eigenvalue problem

$$K\mathbf{x} = \lambda M\mathbf{x} \tag{4.88}$$

where K and M are $n \times n$ real symmetric mass and stiffness matrices. The iterest lies in m lower eigenvalues and eigenvectors, $m \ll n$. We assume that the eigenvectors have been normalized with respect to M, so that the m solutions satisfy

$$\hat{X}^T M \hat{X} = I^m, \quad \hat{X}^T K \hat{X} = \Lambda^m \tag{4.89}$$

where $\hat{X}$ is an $n \times m$ matrix of eigenvectors orthonormal with respect to M, I^m is an $m \times m$ unit matrix and Λ^m is an $m \times m$ diagonal matrix of eigenvalues.

The subspace iteration sequence is described by

$$K U_{k+1} = M \hat{U}_k, \quad k = 1, 2, \ldots \tag{4.90}$$

where $\hat{U}_k$ is a given $n \times m$ matrix orthonormal with respect to M. For every column of $\hat{U}_k$, Eq. (4.90) represents a set of n simultaneous algebraic equations with the components of the corresponding column of U_{k+1} acting as unknowns. Equations (4.90) can be solved for U_{k+1} by Gaussian elimination in conjunction with back substitution [M26]. The next task is to orthonormalize the $n \times m$ matrix U_{k+1}. This is accomplished by solving the reduced-order eigenvalue problem

$$K_{k+1} P_{k+1} = M_{k+1} P_{k+1} \Lambda^m_{k+1} \tag{4.91}$$

where

$$K_{k+1} = U^T_{k+1} K U_{k+1}, \quad M_{k+1} = U^T_{k+1} M U_{k+1} \tag{4.92a, b}$$

are real symmetric $m \times m$ matrices. The solution consists of the $m \times m$ matrix P_{k+1} of eigenvectors and the $m \times m$ diagonal matrix Λ^m_{k+1} of eigenvalues. The matrix of eigenvectors is normalized so as to satisfy

$$P^T_{k+1} M_{k+1} P_{k+1} = I^m \tag{4.93}$$

Then, for the next iteration step, we use the matrix

$$\hat{U}_{k+1} = U_{k+1} P_{k+1} \tag{4.94}$$

where $\hat{U}_{k+1}$ is orthonormal with respect to M. This can be verified by inserting Eq. (4.92b) into Eq. (4.93) and considering Eq. (4.94).

To begin the iteration process, we select m arbitrary n-dimensional vectors and orthonormalize them with respect to the mass matrix M, where the latter can be done by the Gram–Schmidt process [M26]. The resulting $n \times m$ orthonormal matrix represents $\hat{U}_1$. Then, Eqs. (4.90) with $k = 1$ are solved for U_2, which is inserted in Eqs. (4.92) to obtain K_2 and M_2. Solving the eigenvalue problem (4.91), we obtain P_2 and Λ^m_2. Inserting U_2 and P_2 into Eq. (4.94), we obtain $\hat{U}_2$. The process is repeated by solving Eqs. (4.90) with $k = 2$. The iteration process coverges, so that

$$\lim_{k\to\infty} \hat{U}_{k+1} = \hat{X}, \quad \lim_{k\to\infty} \Lambda^m_{k+1} = \Lambda^m \tag{4.95}$$

It should be pointed out that, although subspace iteration requires the solution of an eigenvalue problem for each iteration cycle, this eigenvalue problem is of considerably smaller order than that of the original one, as $m \ll n$. Moreover, the solution of the reduced eigenvalue problem need not be very accurate, as the results are fed back into an iteration process.

Example 4.6 Solve the eigenvalue problem for the undamped four-degree-of-freedom system of Example 4.2.

For undamped structures, the eigenvalue problem can be solved in the configuration space in terms of real quantities alone. The problem is described by two real symmetric matrices, Eq. (4.34a), and can be reduced to one in terms of a single real symmetric matrix, Eq. (4.37). From Example 4.2, we can write the eigenvalue problem

$$A^* \mathbf{p} = \lambda \mathbf{p}, \quad \lambda = \omega^2 m/k \tag{a}$$

where

$$\mathbf{p} = [p_1 \quad p_2 \quad p_3 \quad p_4]^T = m^{1/2}[\sqrt{2}q_1 \quad \sqrt{2}q_2 \quad q_3 \quad q_4]^T \tag{b}$$

and

$$A^* = \begin{bmatrix} 2.5 & -1 & 0 & 0 \\ -1 & 2 & -\sqrt{2} & 0 \\ 0 & -\sqrt{2} & 3 & -1 \\ 0 & 0 & -1 & 1 \end{bmatrix} \tag{c}$$

The solution of the eigenvalue problem consists of the matrix of eigenvalues

$$\Lambda = \text{diag}\,[0.4709 \quad 1.0517 \quad 1.6713 \quad 2.0926] \tag{d}$$

and the orthonormal matrix of eigenvectors

$$P = \begin{bmatrix} 0.2323 & -0.4200 & -0.8242 & 0.3005 \\ 0.5292 & -0.5855 & 0.2417 & -0.5645 \\ 0.5012 & -0.0731 & 0.4472 & 0.7372 \\ 0.6440 & 0.6895 & -0.2494 & -0.2182 \end{bmatrix} \tag{e}$$

The results were obtained by means of the IMSL routine EVCSF using the QL method in conjunction with tridiagonalization by orthogonal transformations. The tridiagonalization is based on the EISPACK routine TRED2 and the QL algorithm is based on the EISPACK routine IMTQL2.

Example 4.7 Solve the eigenvalue problem for the damped four-degree-of-freedom system of Example 4.3. Let $m = k = 1$, $c = 0.1$.

There are actually two eigenvalue problems, the first is given by

$$A\mathbf{u} = \lambda \mathbf{u} \tag{a}$$

and yields the eigenvalues and right eigenvectors and the second is given by

$$A^T \mathbf{v} = \lambda \mathbf{v} \tag{b}$$

and yields the eigenvalues and left eigenvectors.

From Example 4.3, we obtain

$$A = \begin{bmatrix} 0 & & & & I & & & \\ \hline -2.5 & 1 & 0 & 0 & -0.2 & 0.1 & 0 & 0 \\ 1 & -2 & 1 & 0 & 0.1 & -0.15 & 0.05 & 0 \\ 0 & 2 & -3 & 1 & 0 & 0.1 & -0.2 & 0.1 \\ 0 & 0 & 1 & -1 & 0 & 0 & 0.1 & -0.1 \end{bmatrix} \tag{c}$$

which represents an 8×8 real nonsymmetric matrix.

The solution of the eigenvalue problem (b) consists of the matrix of eigenvalues

$$\Lambda = \text{diag}\,[-0.0089 + i0.4708 \quad -0.0089 - i0.4708 \quad -0.0451 + i1.0511$$
$$-0.0451 - i1.0511 \quad -0.1189 + i1.6669 \quad -0.1189 - i1.6669$$
$$-0.1521 + i2.0864 \quad -0.1521 - i2.0864] \tag{d}$$

and the matrix of right eigenvectors

$$U = \begin{bmatrix} 0.1481 - i0.1310 & 0.1481 + i0.1310 & 0.1441 + i0.1942 & 0.1441 - i0.1942 \\ 0.3344 - i0.3016 & 0.3344 + i0.3016 & 0.2065 + i0.2657 & 0.2065 - i0.2657 \\ 0.4497 - i0.4020 & 0.4497 + i0.4020 & 0.0463 + i0.0388 & 0.0463 - i0.0388 \\ 0.5764 - i0.5181 & 0.5764 + i0.5181 & -0.3489 - i0.4387 & -0.3489 + i0.4387 \\ 0.0604 + i0.0709 & 0.0604 - i0.0709 & -0.2106 + i0.1427 & -0.2106 - i0.1427 \\ 0.1390 + i0.1602 & 0.1390 - i0.1602 & -0.2886 + i0.2050 & -0.2886 - i0.2050 \\ 0.1853 + i0.2153 & 0.1853 - i0.2153 & -0.0428 + i0.0469 & -0.0428 - i0.0469 \\ 0.2388 + i0.2760 & 0.2388 - i0.2760 & 0.4768 - i0.3469 & 0.4768 + i0.3469 \end{bmatrix}$$

$$\begin{bmatrix} 0.3260 - i0.1528 & 0.3260 + i0.1528 & -0.1660 - i0.0107 & -0.1660 + i0.0107 \\ -0.0980 + i0.0403 & -0.0980 - i0.0403 & 0.3094 - i0.0088 & 0.3094 + i0.0088 \\ -0.2581 + i0.1015 & -0.2581 - i0.1015 & -0.5712 + i0.0260 & -0.5712 - i0.0260 \\ 0.1465 - i0.0512 & 0.1465 + i0.0512 & 0.1706 + i0.0063 & 0.1706 - i0.0063 \\ 0.2159 + i0.5615 & 0.2159 - i0.5615 & 0.0475 - i0.3448 & 0.0475 + i0.3448 \\ -0.0556 - i0.1681 & -0.0556 + i0.1681 & -0.0288 + i0.6469 & -0.0288 - i0.6469 \\ -0.1385 - i0.4423 & -0.1385 + i0.4423 & 0.0327 - i1.1957 & 0.0327 + i1.1957 \\ 0.0679 + i0.2502 & 0.0679 - i0.2502 & -0.0390 + i0.3549 & -0.0390 - i0.3549 \end{bmatrix} \tag{e}$$

The solution of the eigenvalue problem (b) consists of the same eigenvalues as in Eq. (d) and the matrix of left eigenvectors

$$V = \begin{bmatrix} 0.0942 + i0.1002 & 0.0942 - i0.1002 & 0.2290 - i0.2873 & 0.2290 + i0.2873 \\ 0.2448 + i0.1920 & 0.2448 - i0.1920 & 0.3056 - i0.4082 & 0.3056 + i0.4082 \\ 0.1513 + i0.1436 & 0.1513 - i0.1436 & 0.0010 - i0.0537 & 0.0010 + i0.0537 \\ 0.2042 + i0.1730 & 0.2042 - i0.1730 & -0.2369 + i0.3522 & -0.2369 - i0.3522 \\ 0.1952 - i0.2144 & 0.1952 + i0.2144 & -0.2692 - i0.2212 & -0.2692 + i0.2212 \\ 0.4400 - i0.4925 & 0.4400 + i0.4925 & -0.3814 - i0.2995 & -0.3814 + i0.2995 \\ 0.2961 - i0.3285 & 0.2961 + i0.3285 & -0.0391 - i0.0190 & -0.0391 + i0.0190 \\ 0.3794 - i0.4231 & 0.3794 + i0.4231 & 0.3204 + i0.2458 & 0.3204 - i0.2458 \\ 0.8951 + i0.3048 & 0.8951 - i0.3048 & -0.2758 - i0.0011 & -0.2758 + i0.0011 \\ -0.2652 - i0.0813 & -0.2652 + i0.0813 & 0.5157 - i0.0232 & 0.5157 + i0.0232 \\ -0.3358 - i0.1415 & -0.3358 + i0.1415 & -0.4758 + i0.0268 & -0.4758 - i0.0268 \\ 0.1872 + i0.0792 & 0.1872 - i0.0792 & 0.1410 - i0.0067 & 0.1410 + i0.0067 \\ 0.2173 - i0.5222 & 0.2173 + i0.5222 & -0.0154 + i0.1316 & -0.0154 - i0.1316 \\ -0.0712 + i0.1505 & -0.0712 - i0.1505 & 0.0059 - i0.2464 & 0.0059 + i0.2464 \\ -0.0963 + i0.1954 & -0.0963 - i0.1954 & -0.0015 + i0.2276 & -0.0015 - i0.2276 \\ 0.0581 - i0.1072 & 0.0581 + i0.1072 & 0.0060 - i0.0677 & 0.0060 + i0.0677 \end{bmatrix}$$

(f)

The eigensolution was produced by means of the IMSL routine EVCRG, which involves reduction of A to an upper Hessenberg form and use of the QR method with shifts. The reduction routine is based on the EISPACK routines ORTHES and ORTRAN and the QR algorithm is based on the EISPACK routine HQR2.

Comparing the results obtained here with those obtained in Example 4.6 for an undamped four-degree-of-freedom system, we conclude that the presence of damping complicates matters appreciably. Not only does the order of the eigenvalue problem increase twofold but there are two eigenvalue problems to be solved. Moreover, the solutions for damped systems are in terms of complex quantities, which require substantially larger computational effort.

4.6 MODAL ANALYSIS FOR THE RESPONSE OF OPEN-LOOP SYSTEMS

We showed earlier that the motion of discrete (in space) systems is governed by a set of simultaneous second-order ordinary differential equations, Eqs. (4.16). In the case of open-loop systems, the force vector depends on time alone, $\mathbf{Q} = \mathbf{Q}(t)$. When the eigensolution is relatively easy to produce, the response of the system can be derived by modal analysis, whereby the motion is represented by a superposition of modal vectors multiplied by generalized coordinates. When the eigensolution is difficult to produce, the above approach is still a possibility, but an approach based on the transition matrix may be competitive.

4.6.1 Undamped Nongyroscopic Systems

The differential equations for an n-degree-of-freedom system can be written in the matrix form

$$M\ddot{q}(t) + Kq(t) = Q(t) \tag{4.96}$$

where M is positive definite and K is assumed to be only positive semidefinite, both matrices being real and symmetric. The eigenvalue problem is given by Eq. (4.34a), in which $\lambda = \omega^2$. Assuming that there are r rigid-body modes, the first r natural frequencies are zero $\omega_i = 0$ ($i = 1, 2, \ldots, r$). Using the expansion theorem, the solution of Eq. (4.96) can be written in the form

$$q(t) = \eta_1(t)q_1 + \eta_2(t)q_2 + \cdots + \eta_n(t)q_n = Q\boldsymbol{\eta}(t) \tag{4.97}$$

where Q is the $n \times n$ modal matrix and $\boldsymbol{\eta}(t)$ is an n-dimensional generalized coordinate vector known as the modal coordinate vector. The eigenvectors q_i ($i = 1, 2, \ldots, n$) are assumed to be normalized so that the modal matrix Q satisfies the orthonormality relations (4.44). Introducing Eq. (4.97) into Eq. (4.96), premultiplying by Q^T and using Eqs. (4.44), we obtain the modal equations

$$\ddot{\boldsymbol{\eta}}(t) + \Lambda\boldsymbol{\eta}(t) = N(t) \tag{4.98}$$

where

$$N(t) = Q^T Q(t) \tag{4.99}$$

is an n-vector of generalized forces known as modal forces. Equation (4.98) represents a set of n independent second-order ordinary differential equations having the explicit form

$$\ddot{\eta}_i(t) = N_i(t), \quad i = 1, 2, \ldots, r \tag{4.100a}$$

$$\ddot{\eta}_i(t) + \omega_i^2 \eta_i(t) = N_i(t), \quad i = r+1, r+2, \ldots, n \tag{4.100b}$$

The solution of Eqs. (4.100) can be shown to be [M26]

$$\eta_i(t) = \int_0^t \left[\int_0^\tau N_i(\sigma)\, d\sigma \right] d\tau + \eta_{i0} + \dot{\eta}_{i0} t, \quad i = 1, 2, \ldots, r \tag{4.101a}$$

$$\eta_i(t) = \frac{1}{\omega_i} \int_0^t N_i(t-\tau) \sin \omega_i \tau\, d\tau + \eta_{i0} \cos \omega_i t + \frac{\dot{\eta}_{i0}}{\omega_i} \sin \omega_i t,$$

$$i = r+1, r+2, \ldots, n \tag{4.101b}$$

where η_{i0} and $\dot{\eta}_{i0}$ are initial modal displacements and velocities, related to the actual initial displacements and velocities by the vector relations

$$\boldsymbol{\eta}_0 = Q^T M \mathbf{q}(0), \quad \dot{\boldsymbol{\eta}}_0 = Q^T M \dot{\mathbf{q}}(0) \tag{4.102}$$

Insertion of Eqs. (4.101) into Eq. (4.97) completes the formal solution.

4.6.2 Undamped Gyroscopic Systems

The system state equations have the matrix form

$$M^* \dot{\mathbf{x}}(t) + G^* \mathbf{x}(t) = \mathbf{X}(t) \tag{4.103}$$

where M^* and G^* are given by Eqs. (4.50) and

$$\mathbf{X}(t) = [\mathbf{0}^T \mid Q^T(t)]^T \tag{4.104}$$

The eigenvalue problem has the form (4.53), and was shown to have solutions consisting of the eigenvalues $s_r = i\omega_r$, $\bar{s}_r = -i\omega_r$ and the eigenvectors $\mathbf{u}_r = \mathbf{y}_r + i\mathbf{z}_r$, $\bar{\mathbf{u}}_r = \mathbf{y}_r - i\mathbf{z}_r$ $(r = 1, 2, \ldots, n)$. The eigenvectors are orthogonal and can be normalized. Introducing the matrix of eigenvectors

$$U = [\mathbf{y}_1 \ \mathbf{z}_1 \ \mathbf{y}_2 \ \mathbf{z}_2 \ \cdots \ \mathbf{y}_n \ \mathbf{z}_n] \tag{4.105}$$

the orthonormality relation can be written in the form

$$U^T M^* U = I \tag{4.106}$$

Moreover, it can be verified that [M19]

$$U^T G^* U = \Lambda = \text{block-diag} \begin{bmatrix} 0 & -\omega_r \\ \omega_r & 0 \end{bmatrix} \tag{4.107}$$

Next, we consider the linear transformation

$$\mathbf{x}(t) = \sum_{r=1}^{n} [\xi_r(t)\mathbf{y}_r + \eta_r(t)\mathbf{z}_r] = U\mathbf{w}(t) \tag{4.108}$$

where

$$\mathbf{w}(t) = [\xi_1(t) \ \eta_1(t) \ \xi_2(t) \ \eta_2(t) \ \cdots \ \xi_n(t) \ \eta_n(t)]^T \tag{4.109}$$

Introducing Eq. (4.108) into Eq. (4.103), premultiplying the result by U^T and considering Eqs. (4.106) and (4.107), we obtain

$$\dot{\mathbf{w}}(t) + \Lambda \mathbf{w}(t) = \mathbf{W}(t) \tag{4.110}$$

where

$$W(t) = UX(t) = [Y_1(t) \quad Z_1(t) \quad Y_2(t) \quad Z_2(t) \quad \cdots \quad Y_n(t) \quad Z_n(t)]$$
(4.111)

is a $2n$-dimensional generalized force vector. Equation (4.110) represents a system of n independent second-order systems of the form

$$\dot{\xi}_r(t) - \omega_r \eta_r(t) = Y_r(t) ,$$
$$\dot{\eta}_r(t) + \omega_r \xi_r(t) = Z_r(t) , \qquad r = 1, 2, \ldots, n \quad (4.112)$$

The solution of Eqs. (4.112) can be shown to be [M19]

$$\xi_r(t) = \int_0^t [\mathbf{y}_r^T \mathbf{X}(\tau) \cos \omega_r(t - \tau) + \mathbf{z}_r^T \mathbf{X}(\tau) \sin \omega_r(t - \tau)] \, d\tau$$
$$+ \mathbf{y}_r^T M^* \mathbf{x}(0) \cos \omega_r t + \mathbf{z}_r^T M^* \mathbf{x}(0) \sin \omega_r t ,$$

$$\eta_r(t) = \int_0^T [\mathbf{z}_r^T \mathbf{X}(\tau) \cos \omega_r(t - \tau) - \mathbf{y}_r^T \mathbf{X}(\tau) \sin \omega_r(t - \tau)] \, d\tau$$
$$+ \mathbf{z}_r^T M^* \mathbf{x}(0) \cos \omega_r t - \mathbf{y}_r^T M^* \mathbf{x}(0) \sin \omega_r t ,$$
$$r = 1, 2, \ldots, n \quad (4.113)$$

Introducing Eqs. (4.113) into Eq. (4.108), we obtain the state vector

$$\mathbf{x}(t) = \sum_{r=1}^n \left\{ \int_0^t [(\mathbf{y}_r \mathbf{y}_r^T + \mathbf{z}_r \mathbf{z}_r^T) \mathbf{X}(\tau) \cos \omega_r(t - \tau) \right.$$
$$+ (\mathbf{y}_r \mathbf{z}_r^T - \mathbf{z}_r \mathbf{y}_r^T) \mathbf{X}(\tau) \sin \omega_r(t - \tau)] \, d\tau$$
$$+ (\mathbf{y}_r \mathbf{y}_r^T + \mathbf{z}_r \mathbf{z}_r^T) M^* \mathbf{x}(0) \cos \omega_r t$$
$$\left. + (\mathbf{y}_r \mathbf{z}_r^T - \mathbf{z}_r \mathbf{y}_r^T) M^* \mathbf{x}(0) \sin \omega_r t \right\} \quad (4.114)$$

Note that $\mathbf{x}(0)$ denotes the initial state vector. Moreover, $\mathbf{X}(t)$ is defined only for $t > 0$ and is zero for $t < 0$.

4.6.3 General Dynamical Systems

The equations of motion can be written in the state form

$$\dot{\mathbf{x}}(t) = A\mathbf{x}(t) + \mathbf{X}(t) \quad (4.115)$$

where A is given by eq. (4.26) and $\mathbf{X}(t)$ has the form

$$\mathbf{X}(t) = [\mathbf{0}^T \mid (M^{-1} \mathbf{Q}(t))^T]^T \quad (4.116)$$

The response can be obtained by means of the approach based on the transition matrix discussed in Section 3.9. To this end, we first compute the transition matrix associated with A, or

$$\Phi(t, \tau) = \Phi(t - \tau) = e^{A(t-\tau)} = I + (t - \tau)A + \frac{(t - \tau)^2}{2!} A^2$$

$$+ \frac{(t - \tau)^3}{3!} A^3 + \cdots \qquad (4.117)$$

and then we use Eq. (3.94) to write

$$\mathbf{x}(t) = e^{At}\mathbf{x}(0) + \int_0^t e^{A(t-\tau)}\mathbf{X}(\tau)\, d\tau \qquad (4.118)$$

where $\mathbf{x}(0)$ is the initial state vector. The response can also be obtained by first solving the eigenvalue problem associated with A for the matrix Λ of eigenvalues, matrix U of right eigenvectors and matrix V of left eigenvectors, where the matrices of eigenvectors satisfy the biorthonormality relations (3.111). Then, from Eq. (3.118), we obtain the response

$$\mathbf{x}(t) = Ue^{\Lambda t}V^T\mathbf{x}(0) + \int_0^t Ue^{\Lambda(t-\tau)}V^T\mathbf{X}(\tau)\, d\tau \qquad (4.119)$$

Quite frequently, the nature of the excitation vector $\mathbf{X}(t)$ requires that the response be evaluated numerically. In such cases, it is convenient to regard the system as a discrete-time system. Then, as demonstrated in Section 3.15, Eq. (4.118) can be replaced by the sequence

$$\mathbf{x}(k + 1) = \Phi\mathbf{x}(k) + \Gamma\mathbf{X}(k), \quad k = 0, 1, 2, \ldots \qquad (4.120)$$

where

$$\Phi = e^{TA}, \quad \Gamma = \int_0^T e^{\sigma A}\, d\sigma \qquad (4.121)$$

The argument k in Eq. (4.120) indicates the discrete time $t = t_k = kT$, where T is the sampling period. Using the approach of Section 3.15, it is not difficult to produce the discrete-time counterpart of Eq. (4.119).

CHAPTER 5

CONTROL OF LUMPED-PARAMETER SYSTEMS. CLASSICAL APPROACH

For the most part, structures can be modeled by linear differential equations with constant coefficients. This type of equations can be treated conveniently by the Laplace transformation. This transformation permits relating the system output to the input by means of transfer functions, which are algebraic expressions. When the system involves more than one input and more than one output, then the inputs and outputs can be related by a matrix of transfer functions known as the transfer matrix.

Early work on control relied heavily on transform techniques. One of the main concerns was system stability, which was investigated by examining the poles of the closed-loop transfer function. Various graphical techniques were developed permitting determination of the poles of the closed-loop transfer function by working with the open-loop transfer function. Widely used techniques are the root-locus method, Nyquist plots, Bode diagrams and Nichols plots. The implication is that the system is characterized by a single input and a single output, and hence by a single transfer function. In the case of multiple inputs and outputs, there is a transfer function relating every input to every output. The theory concerned with transform techniques for the design of feedback control systems has come to be known as *classical control* and is the subject of this chapter.

5.1 FEEDBACK CONTROL SYSTEMS

As indicated in Section 3.1, the idea of control is to exert certain forces on a system so as to cause the system to exhibit satisfactory performance. The simplest type of control is shown in the block diagram of Fig. 5.1. The

130 CONTROL OF LUMPED-PARAMETER SYSTEMS. CLASSICAL APPROACH

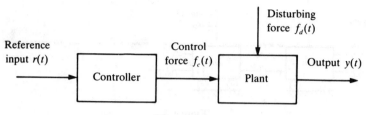

Figure 5.1

reference input, denoted by $r(t)$ and representing the *desired output*, is fed into the *controller*, which converts it into the *control force* $f_c(t)$. The control force $f_c(t)$ combines with the *disturbing force* $f_d(t)$ to give the total force $f(t)$ acting on the *plant*. The response of the plant to the force $f(t)$ is the *actual output* $y(t)$, where the latter plays the role of the *controlled variable*.

In the system described above, the control force does not depend on the actual output. Hence, if some unexpected factor causes the output to deviate from the desired output, there is no way of correcting the deviation. A system capable of redressing the problem just described is shown in Fig. 5.2, and it contains two additional elements, namely an *output sensor* and an *error detector*. The sensor output is the *feedback signal* $b(t)$ and the error detector produces the *actuating signal* $e(t)$ given by

$$e(t) = r(t) - b(t) \qquad (5.1)$$

where $e(t)$ is a measure of the error. A control system of this type is known as a *feedback control system*. Because the block diagram of Fig. 5.2 has the appearance of a closed loop, the system is called a *closed-loop system*. In contrast, the system of Fig. 5.1 is referred to as *open-loop control*.

As pointed out in Section 3.3, the Laplace transformation provides a useful tool for deriving the response of linear systems with constant coefficients. The feedback control system of Fig. 5.2 can be regarded as represent-

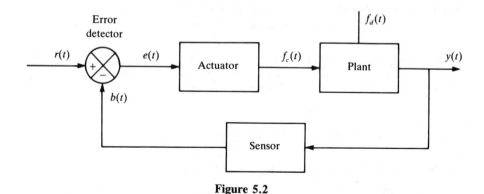

Figure 5.2

ing such a system, so that it is natural to use the Laplace transformation to analyze it. It is customary to treat the controller and plant as one entity, so that a typical block diagram of a feedback control system is as shown in Fig. 5.3, where $R(s) = \mathscr{L}r(t)$, $Y(s) = \mathscr{L}y(t)$, $B(s) = \mathscr{L}b(t)$ and $E(s) = \mathscr{L}e(t)$ are Laplace transformed variables. Moreover, $G(s)$ is known as the *forward-path transfer function* and $H(s)$ as the *feedback-path transfer function*, so that

$$Y(s) = G(s)E(s) \tag{5.2}$$

and

$$B(s) = H(s)Y(s) \tag{5.3}$$

The ratio of the transformed feedback signal $B(s)$ to the transformed actuating signal is called the *open-loop transfer function*. From Eqs. (5.2) and (5.3), the open-loop transfer function has the expression

$$\frac{B(s)}{E(s)} = G(s)H(s) \tag{5.4}$$

The term open-loop transfer function derives from the fact that $G(s)H(s)$ would be the transfer function if the transformed feedback signal $B(s)$ were not fed back, so that the loop would not be closed. When the feedback-path transfer function is equal to unity, $H(s) = 1$, the open-loop and the forward-path transfer functions coincide. In this case, the feedback signal reduces to the output signal and the actuating signal becomes the error signal.

The *closed-loop transfer function* is defined as the ratio of the transformed output to the transformed reference input. Hence, transforming Eq. (5.1) and using Eqs. (5.2) and (5.3), we obtain the closed-loop transfer function

$$\frac{Y(s)}{R(s)} = M(s) = \frac{G(s)}{1 + G(s)H(s)} \tag{5.5}$$

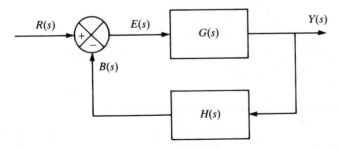

Figure 5.3

so that the transformed output can be written in terms of the transformed reference input as follows:

$$Y(s) = \frac{G(s)}{1 + G(s)H(s)} R(s) \qquad (5.6)$$

Equations (5.5) and (5.6) are representative of a single-input, single-output (SISO) system and are used extensively in classical control.

5.2 PERFORMANCE OF CONTROL SYSTEMS

To evaluate the control system performance, it is customary to use a given reference input, such as the unit step function, and examine how well the output follows the reference input. The performance is judged on the basis of several criteria designed to ascertain the extent to which the output matches the reference input.

The system of Fig. 5.4 represents a position control system. The plant consists of a mass m attached at the end of a massless rigid bar of length L. The other end is hinged in a way that the bar can undergo angular motion in the horizontal plane. The hinge possesses viscous damping so that it produces a resisting torque proportional to the angular velocity of the bar, where the constant of proportionality is the coefficient of viscous damping c.

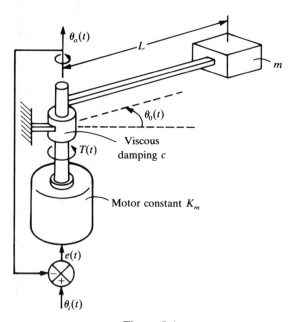

Figure 5.4

The control torque, denoted by $T(t)$, is proportional to the angular error $e(t)$, where the constant of proportionality is the motor constant K_m. The object is to maneuver the bar so as to follow a given reference input angle $\theta_i(t)$. Denoting the actual output angle by $\theta_o(t)$ and taking moments about the hinge, the equation of motion can be shown to be

$$J\ddot{\theta}_o(t) + c\dot{\theta}_o(t) = T(t) \tag{5.7}$$

where $J = mL^2$ is the mass moment of inertia of the system about the hinge. The error is given by

$$e(t) = \theta_i(t) - \theta_o(t) \tag{5.8}$$

and the torque by

$$T(t) = K_m e(t) \tag{5.9}$$

It is clear that the system is linear and that it possesses constant coefficients. Hence, we can use the approach of Section 5.1 to examine the system performance.

Laplace transforming Eq. (5.7) and letting the initial conditions be equal to zero, we obtain

$$(Js^2 + cs)\Theta_o(s) = T(s) \tag{5.10}$$

where $\Theta_o(s) = \mathscr{L}\theta_o(t)$ and $T(s) = \mathscr{L}T(t)$. Similarly, Eqs. (5.8) and (5.9) yield

$$E(s) = \Theta_i(s) - \Theta_o(s) \tag{5.11}$$

and

$$T(s) = K_m E(s) \tag{5.12}$$

respectively, where $\Theta_i(s) = \mathscr{L}\theta_i(t)$ and $E(s) = \mathscr{L}e(t)$. Equations (5.10)–(5.12) can be used to draw the block diagram shown in Fig. 5.5.

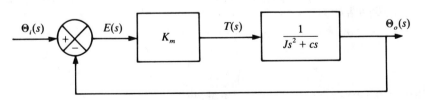

Figure 5.5

134 CONTROL OF LUMPED-PARAMETER SYSTEMS. CLASSICAL APPROACH

Considering Eq. (5.2) in conjunction with Eqs. (5.10) and (5.12), we obtain the forward-path transfer function

$$G(s) = \frac{\Theta_o(s)}{E(s)} = \frac{K_m}{Js^2 + cs} \qquad (5.13)$$

Moreover, from Fig. 5.5, we conclude that the feedback-path transfer function is equal to unity, $H(s) = 1$, so that considering Eq. (5.6) we obtain the relation between the transformed output and transformed reference input in the form

$$\Theta_o(s) = \frac{G(s)}{1 + G(s)} \Theta_i(s) = \frac{K_m}{Js^2 + cs + K_m} \Theta_i(s) \qquad (5.14)$$

Hence, the closed-loop transfer function is

$$\frac{\Theta_o(s)}{\Theta_i(s)} = \frac{K_m}{Js^2 + cs + K_m} \qquad (5.15)$$

which, except for the constant factor K_m in the numerator, is identical to the transfer function of a viscously damped single-degree-of-freedom system. Hence, the effect of closing the loop is to provide an effective torsional spring of stiffness K_m. Indeed, the open-loop transfer function, which for $H(s) = 1$ is equal to the forward-path transfer function $G(s)$, does not possess such restraining effect, as can be ascertained from Eq. (5.13).

We propose to use Eq. (5.14) to check the system performance. In particular, we wish to examine the ability of the system to follow a reference input in the form of a unit step function

$$\theta_i(t) = u(t) \qquad (5.16)$$

which implies an abrupt change in the angular position. It should be clear from the outset that it is physically impossible for the output to follow such a reference input exactly, because systems with inertia cannot undergo instantaneous displacements. Nevertheless the purpose here is to present certain criteria capable of assessing system performance. From Eq. (3.41), we conclude that the transformed reference input is

$$\Theta_i(s) = \frac{1}{s} \qquad (5.17)$$

so that, inserting Eq. (5.17) into Eq. (5.14) and carrying a partial fractions expansion, the transformed output can be written in the form

$$\Theta_o(s) = \frac{K_m}{J}\frac{1}{s(s^2+2\zeta\omega_n s+\omega_n^2)} = \frac{1}{s} - \frac{s+2\zeta\omega_n}{s^2+2\zeta\omega_n s+\omega_n^2} \tag{5.18}$$

where

$$2\zeta\omega_n = c/J, \quad \omega_n^2 = K_m/J \tag{5.19}$$

in which ζ is the *viscous damping factor* and ω_n is the *frequency of undamped oscillation*. Using Eq. (5.18) in conjunction with tables of Laplace transforms [M42], it can be verified that the output has the expression

$$\theta_o(t) = \mathscr{L}^{-1}\Theta_o(s) = \mathscr{L}^{-1}\left(\frac{1}{s} - \frac{s+2\zeta\omega_n}{s^2+2\zeta\omega_n s+\omega_n^2}\right)$$

$$= \left[1 - e^{-\zeta\omega_n t}\left(\cos\omega_d t + \frac{\zeta\omega_n}{\omega_d}\sin\omega_d t\right)\right]u(t) \tag{5.20}$$

where $\omega_d = (1-\zeta^2)^{1/2}\omega_n$ is the frequency of damped oscillation. It is implied in the above that $\zeta < 1$, which is ordinarily the case. Equation (5.20) indicates that the output is the difference between a unit step function and an exponentially decaying harmonic oscillation of unit initial amplitude. Hence, as expected, the output does not reproduce the reference input exactly. Indeed, inserting Eqs. (5.16) and (5.20) into Eq. (5.8), we obtain the error

$$e(t) = \theta_i(t) - \theta_o(t) = e^{-\zeta\omega_n t}\left(\cos\omega_d t + \frac{\zeta\omega_n}{\omega_d}\sin\omega_d t\right)u(t) \tag{5.21}$$

so that the error dies out eventually.

To introduce the performance criteria, it is convenient to plot the step response $\theta_o(t)$ versus t, as given by Eq. (5.20), with ζ and ω_n as parameters. A typical plot is shown in Fig. 5.6. The criteria provide a quantitative measure of the step response characteristics, and can be identified as follows:

1. **Peak Time T_p**: defined as the time necessary for the step response to reach its peak value.
2. **Maximum Overshoot M_p**: defined as the value of the peak step response minus the final value of the response. This can be expressed conveniently in terms of the *percent overshoot* (P.O.), given by

$$\text{P.O.} = \frac{\theta_o(T_p) - \theta_o(\infty)}{\theta_o(\infty)} \times 100\% \tag{5.22}$$

3. **Delay Time T_d**: defined as the time needed for the step response to reach 50% of its final value.

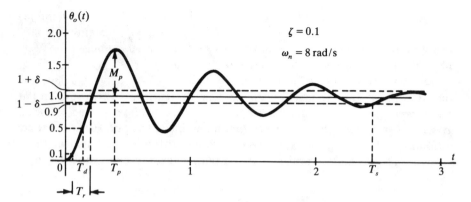

Figure 5.6

4. *Rise Time T_r*: defined as the time required for the step response to rise from 10% to 90%, or from 5% to 95%, or from 0% to 100% of its final value.
5. *Settling Time T_s*: defined as the time required for the step response to decrease to and remain within a certain percentage δ of the final value. A commonly used value for δ is 5%.
6. *Steady-State Error e_{ss}*: defined here as the difference between the steady-state values of the reference input and output, or

$$e_{ss} = \theta_i(\infty) - \theta_o(\infty) \tag{5.23}$$

This is a special definition, applicable in the case in which $H(s) = 1$. In the more general case, in which $H(s) \neq 1$, the steady-state error is defined as the difference between the steady-state values of the reference input and the feedback signal, or

$$e_{ss} = r(\infty) - b(\infty) \tag{5.24}$$

The steady-state error, regardless of which of the two definitions is used, can be evaluated conveniently by means of the final-value theorem, which states that [M44]

$$\lim_{t \to \infty} e(t) = \lim_{s \to 0} sE(s) \tag{5.25}$$

Ideally, the control system should be such that the output follows the reference input exactly. As pointed out earlier, this is not physically possible in the position control discussed above. Hence, to judge the merits of the system, one must select numerical values of the above quantities, thus establishing performance criteria. Some of the criteria tend to be more important than others. In particular, good performance of a feedback control system is characterized by small maximum overshoot, low settling time and small steady-state error. To obtain good performance, it is often necessary to change some of the system parameters. Care must be exercised, however, as the changes improving the performance as measured by some criteria can harm the performance as measured by others. A good control design is characterized by the satisfaction of the important performance criteria.

5.3 THE ROOT-LOCUS METHOD

A critical problem in control system design is stability. In Section 3.8, we studied ways of checking system stability, with special emphasis on linear systems with constant coefficients. In particular, it was demonstrated that stability is determined by the location of the roots of the characteristic equation in the complex plane. Before one can obtain these roots, it is necessary to assign numerical values to the system parameters. Quite often, the designer is interested not merely in a stability statement for a given set of parameters but in the manner in which the system behavior changes as the parameters change. To address this problem, several graphical techniques have been developed. One of these techniques, known as the *root-locus method*, consists of plotting the roots of the characteristic equation in the *s*-plane as a given parameter of the system changes. The roots lie on smooth curves known as *loci*, and the plots themselves are called *root-locus plots*.

In Section 5.1, we derived the closed-loop transfer function

$$M(s) = \frac{Y(s)}{R(s)} = \frac{G(s)}{1 + G(s)H(s)} \qquad (5.26)$$

where $Y(s)$ is the transformed output, $R(s)$ the transformed reference input, $G(s)$ the forward-path transfer function and $G(s)H(s)$ the open-loop transfer function. The characteristic equation is simply

$$1 + G(s)H(s) = 0 \qquad (5.27)$$

so that the root loci are plots of the roots of Eq. (5.27) as functions of a given parameter. To place the parameter in evidence, we express the open-loop transfer function in the form

$$G(s)H(s) = KF(s) \qquad (5.28)$$

where K is the parameter in question, so that Eq. (5.27) can be written as

$$1 + KF(s) = 0 \qquad (5.29)$$

Then, *the root loci are defined as the loci in the s-plane of the roots of Eq. (5.29) as K varies from 0 to ∞*. The loci obtained by letting K vary from 0 to $-\infty$ are called *complementary root loci*. We are concerned here only with positive values of K.

To establish the conditions under which Eq. (5.29) is satisfied, we recognize that $F(s)$ is a complex function and write

$$F(s) = -\frac{1}{K} = |F(s)|e^{i\psi_F} \qquad (5.30)$$

where $|F(s)|$ is the magnitude and ψ_F the phase angle of $F(s)$. Hence, the characteristic equation is satisfied if both equations

$$|F(s)| = \frac{1}{K} \qquad (5.31)$$

and

$$\psi_F = \underline{/F(s)} = (2k+1)\pi, \quad k = 0, \pm 1, \pm 2, \ldots \qquad (5.32)$$

are satisfied simultaneously, where $\underline{/F(s)}$ denotes the phase angle of the function $F(s)$. To construct the root loci, we must find all the points in the s-plane that satisfy Eq. (5.32). Then, the value of K corresponding to a given s on a locus can be determined by means of Eq. (5.31).

The root loci determine the poles of the closed-loop transfer function $M(s)$, and hence the stability of the system, by working with the open-loop transfer function $KF(s)$. It should be pointed out that the roots of the characteristic equation, Eq. (5.29), are not the only poles of $M(s)$. Indeed, from Eq. (5.26), we observe that $M(s)$ has a pole also where $G(s)$ has a pole and $G(s)H(s)$ is finite, or zero, or has a pole of lower order.

The construction of the root loci is basically a graphical problem. Quite often, particularly in preliminary design, accurate root-locus plots are not really necessary, and rough sketches suffice. To help with these sketches, there exist a number of rules of construction. These rules are based on the relation between the zeros and poles of the open-loop transfer function and the roots of the characteristic equation. To present the rules, it is convenient to rewrite the characteristic equation, Eq. (5.29), in the rational form

$$F(s) = \frac{\prod_{i=1}^{m}(s-z_i)}{\prod_{j=1}^{n}(s-p_j)} = -\frac{1}{K} \qquad (5.33)$$

where z_i ($i = 1, 2, \ldots, m$) are the zeros and p_j ($j = 1, 2, \ldots, n$) the poles of $F(s)$, i.e., they are the *open-loop zeros* and the *open-loop poles*, respectively. Without loss of generality, we assume that $n \geq m$.

The construction rules are as follows:
1. *The number of root loci.* There are as many root loci as there are open-loop poles, i.e., there are n loci.
2. *Origination of root loci.* The loci originate for $K = 0$ at the open-loop poles.
3. *Termination of the root loci.* As $K \to \infty$, m of the loci approach the open-loop zeros and the remaining $n - m$ loci approach infinity.
4. *Symmetry.* The root loci are symmetric with respect to the real axis.
5. *Root loci on the real axis.* A locus always exists on a given section of the real axis when the total number of real open-loop poles and open-loop zeros to the right of the section is odd.
6. *Asymptotes.* For large values of s, the $n - m$ loci that tend to infinity approach straight lines asymptotically, where the angles of the asymptotes are given by

$$\theta_k = \frac{(2k+1)\pi}{n-m}, \quad k = 0, 1, \ldots, n - m - 1 \tag{5.34}$$

7. *Intersection of the asymptotes.* The $n - m$ asymptotes intersect on the real axis at a distance

$$c = \frac{\sum_{j=1}^{n} p_j - \sum_{i=1}^{m} z_i}{n - m} \tag{5.35}$$

from the origin of the s-plane. The intersection point is sometimes called the *centroid*.

8. *Breakaway points.* The breakaway points $s = s_b$ are points on the root loci at which multiple roots occur, and they must satisfy the equation

$$\left.\frac{dF(s)}{ds}\right|_{s=s_b} = 0 \tag{5.36}$$

Equation (5.36) represents only a necessary condition. In addition, a breakaway point must satisfy the characteristic equation, Eq. (5.29), which determines the value of K corresponding to the point $s = s_b$.

A more detailed discussion of the construction rules, including various proofs, is provided by Meirovitch [M42].

Example 5.1 Consider a feedback control system with the open-loop transfer function

$$G(s)H(s) = \frac{K}{s(s+2)(s+3)} \tag{a}$$

Sketch the root loci, describe their main characteristics and determine the value of K for which the system becomes unstable.

The root loci are displayed in Fig. 5.7. The characteristics of the loci are as follows:

i. There are three loci, one root locus starting from each of the poles

$$p_1 = 0, \quad p_2 = -2, \quad p_3 = -3$$

of $F(s) = G(s)H(s)/K$.

ii. All three roots loci terminate at infinity, which is consistent with the fact that $F(s)$ has no zeros.

iii. There is one root locus on the real axis between p_1 and p_2 and another one on the real axis to the left of p_3.

iv. There are three asymptotes with angles $\theta_1 = \pi/3$, $\theta_2 = \pi$ and $\theta_3 = -\pi/3$. The asymptotes intersect on the real axis at $c = -5/3$.

v. There is a breakaway point at $s = -0.785$.

The system becomes unstable when the root loci cross into the right half of the s-plane. At such a point, $s = i\omega$ satisfies the characteristic equation. Hence, ω and K must be such that

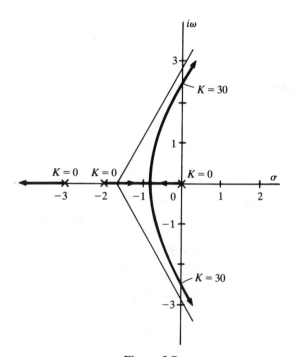

Figure 5.7

$$1 + \frac{K}{i\omega(i\omega + 2)(i\omega + 3)} = 0 \qquad \text{(b)}$$

yielding

$$-i\omega^3 - 5\omega^2 + 6i\omega + K = (K - 5\omega^2) + i\omega(6 - \omega^2) = 0 \qquad \text{(c)}$$

Setting the real part and imaginary part to zero, we obtain $\omega = \pm\sqrt{6}$ and

$$K = 5\omega^2 = 30 \qquad \text{(d)}$$

5.4 THE NYQUIST METHOD

The root-locus method gives the closed-loop poles of a system, provided the poles and zeros of the open-loop transfer function $G(s)H(s)$ are known. At times, the function $G(s)H(s)$ is not available and must be determined experimentally. The Nyquist method is a technique permitting the determination of the degree of stability of a closed-loop system on the basis of the open-loop frequency response, where the latter can be obtained experimentally by exciting the open-loop system harmonically.

The Nyquist method is based on the concept of mapping. It involves a single-valued rational function $F_1(s)$ that is analytic everywhere in a given region in the s-plane, except at a finite number of points. In addition to the s-plane, we consider an $F_1(s)$-plane, so that to each value s_i in the complex s-plane corresponds a value $F_1(s_i)$ in the complex $F_1(s)$-plane. If s moves clockwise around a closed contour C_s in the s-plane, then $F_1(s)$ moves around a closed contour C_F in the $F_1(s)$-plane (Fig. 5.8) in either the clockwise or counterclockwise sense, depending on the nature of $F_1(s)$. According to the *principle of the argument* [M42], if $F_1(s)$ has Z zeros and P poles inside a closed contour and s moves once clockwise around the

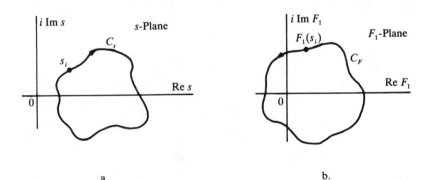

Figure 5.8

contour, then $F_1(s)$ experience a clockwise angle change of $2\pi(Z-P)$. Clearly, if $P > Z$, the angular change is in the counterclockwise sense.

Next, we consider the function

$$F_1(s) = 1 + G(s)H(s) \tag{5.37}$$

which is recognized as the denominator of the closed-loop transfer function. Letting s move clockwise around a closed contour in the s-plane, we can determine the difference between the number of zeros and poles of $F_1(s)$ inside the contour by measuring the angle change in $F_1(s)$. Because an angle change of 2π represents a clockwise encirclement of the origin in the $F_1(s)$-plane, we can determine the difference between the number of zeros and poles of $F_1(s)$ inside the closed contour by counting the number of encirclements of the origin in the $F_1(s)$-plane. Choosing a clockwise contour C_s enclosing the entire right half of the s-plane and assuming that the number P of poles of $F_1(s)$ in the right half of the s-plane is known, we can use the principle of the argument and obtain the number Z of zeros of $F_1(s)$ in the right half of the s-plane by writing

$$Z = P + N \tag{5.38}$$

where N is the number of clockwise encirclements of the origin of the $F_1(s)$-plane. But, the zeros of $F_1(s)$ are the poles of the closed-loop transfer function, so that Eq. (5.38) permits us to check the stability of the system.

Instead of working with the $F_1(s)$-plane, it is more convenient to work with the $G(s)H(s)$-plane. From Eq. (5.37), we conclude that the origin of the $F_1(s)$-plane corresponds to the point $-1 + i0$ in the $G(s)H(s)$-plane, as shown in Fig. 5.9. Hence, we can check stability of the closed-loop system by counting the number of encirclements of the point $-1 + i0$ in the $G(s)H(s)$-plane, which implies working with the open-loop transfer function.

The method calls for a particular contour C_s, namely one extending along the imaginary axis from $-iR$ to iR and then along the semicircle of radius R in the right half of the s-plane, as shown in Fig. 5.10. By letting $R \to \infty$, the

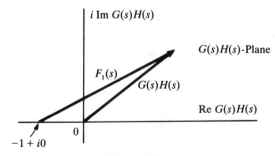

Figure 5.9

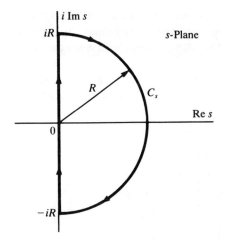

Figure 5.10

closed contour C_s will enclose the entire right half of the plane. If there are poles of $G(s)H(s)$ on the imaginary axis, then the contour C_s can be modified so as to exclude them. Such a contour C_s is known as a *Nyquist path*. If we substitute the values of s along the contour C_s in the open-loop transfer function $G(s)H(s)$, then the resulting plot in the $G(s)H(s)$-plane is referred to as a *Nyquist plot*.

The Nyquist plot can be expedited by considering certain features of the contour C_s in the s-plane. Because C_s is symmetric with respect to the real axis, the Nyquist plot is also symmetric with respect to the real axis. The large semicircle on C_s maps as a single point in the $G(s)H(s)$-plane, so that when the number of zeros of $G(s)H(s)$ does not exceed the number of poles, the large semicircle in the s-plane can be ignored. Moreover, if there is a pole of $G(s)H(s)$ on the imaginary axis, then the pole can be excluded by means of a small semicircle with the center at the pole and with a radius δ; the semicircle lies in the right side of the s-plane. Then, for s on the small semicircle, if the pole has multiplicity r, then we have the approximation

$$G(s)H(s) \cong \frac{K}{\delta^r} e^{-ir\phi} \qquad (5.39)$$

so that, as $\delta \to 0$, the magnitude of $G(s)H(s)$ becomes increasingly large. Moreover, as the angle in the s-plane changes counterclockwise from $\phi = -\pi/2$ to $\phi = \pi/2$, the angle of $G(s)H(s)$ changes clockwise from $r\pi/2$ to $-r\pi/2$. Although the curve in the $G(s)H(s)$-plane corresponding to the small semicircle is likely to be far removed from the point $-1 + i0$, it may help in establishing whether the Nyquist plot encircles $-1 + i0$.

Finally, over the imaginary axis, the open-loop transfer function $G(s)H(s)$ takes the value $G(i\omega)H(i\omega)$, which is recognized as the frequency response of the open-loop system. The advantage of the Nyquist method lies

144 CONTROL OF LUMPED-PARAMETER SYSTEMS. CLASSICAL APPROACH

in the fact that $G(i\omega)H(i\omega)$ can be obtained experimentally, as pointed out earlier. Moreover, the plots of $G(i\omega)H(i\omega)$ are the most significant parts of the Nyquist plot as far as the determination of the number of encirclements of the point $-1+i0$ is concerned.

For the closed-loop system to be stable, the number of zeros of $1+G(s)H(s)$ in the right half of the s-plane must be zero. Hence, for stability, Eq. (5.38) must reduce to

$$P = -N \tag{5.40}$$

where P is the number of poles of $G(s)H(s)$ in the right half of the s-plane and N is the number of clockwise encirclements of the point $-1+i0$ in the $G(s)H(s)$-plane. But, a clockwise encirclement is the negative of a counterclockwise encirclement. In view of this, Eq. (5.39) permits us to state that: *A closed-loop system is stable if the Nyquist plot of $G(s)H(s)$ encircles the point $-1+i0$ in a counterclockwise sense as many times as there are poles of $G(s)H(s)$ in the right half of the s-plane.* This statement is known as the *Nyquist criterion*. If $G(s)H(s)$ has no poles in the right side of the s-plane, then the closed-loop system is stable if the Nyquist plot does not encircle the point $-1+i0$. A system that has no open-loop poles or zeros in the right half of the s-plane is known as a *minimum-phase* system. Many practical systems fall in the minimum-phase category. Still, the Nyquist criterion is applicable to both minimum-phase and nonminimum-phase systems.

Example 5.2 Sketch the Nyquist plots for the feedback system with the open-loop transfer function

$$G(s)H(s) = \frac{K}{s(s^2+2s+2)} \tag{a}$$

for the two cases: (1) $K=2$ and (2) $K=6$. Determine the system stability in each case.

There are no poles of $G(s)H(s)$ in the right half of the s-plane. Hence, for stability, there must be no encirclements of the point $-1+i0$ in the $G(s)H(s)$-plane. Because the open-loop transfer function $G(s)H(s)$ has a pole at the origin, we choose the contour C_s shown in Fig. 5.11. For identification purposes, the segment $-R<\omega<-\delta$ on the imaginary axis is denoted by 1, the semicircle of radius δ with the center at the origin by 2, the segment $\delta<\omega<R$ by 3 and the semicircle of radius R by 4. The mappings in the $G(s)H(s)$-plane of these portions of C_s are denoted by corresponding numbers.

For any point on the imaginary axis, we replace s by $i\omega$ in Eq. (a) and obtain

$$G(i\omega)H(i\omega) = \frac{K}{i\omega(-\omega^2+i2\omega+2)} = \frac{K[-2\omega^2+i\omega(\omega^2-2)]}{\omega^2(\omega^4+4)} \tag{b}$$

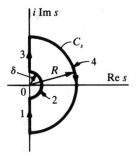

Figure 5.11

Corresponding to negative values of ω we obtain curve 1 of the Nyquist plot, and corresponding to positive values we obtain curve 3. Corresponding to the small semicircle around the origin, letting $s = \delta e^{i\phi}$ in Eq. (a), we obtain

$$G(s)H(s)\bigg|_{\text{on 2}} = \lim_{\delta \to 0} \frac{K}{\delta e^{i\phi}(\delta^2 e^{2i\phi} + 2\delta e^{i\phi} + 2)} = \lim_{\delta \to 0} \frac{K}{2\delta e^{i\phi}} = \infty e^{-i\phi} \quad \text{(c)}$$

As ϕ varies from $-\pi/2$ to $\pi/2$, a point on the small semicircle maps into a point on a semicircle of very large radius centered at the origin and with angle varying from $\pi/2$ or $-\pi/2$. This circle is denoted by 2. Of course, the semicircle of very large radius on C_s maps into a point at the origin of the $G(s)H(s)$-plane, where the point is denoted by 4. The Nyquist plots for $K = 2$ and $K = 6$ are displayed in Fig. 5.12.

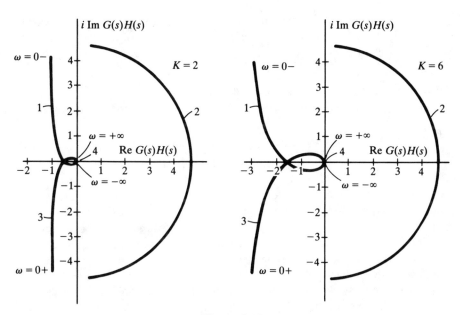

Figure 5.12

From Eq. (b), we conclude that the intersection of the Nyquist plots with the real axis occurs when $\omega^2 = 2$ at the points

$$G(\pm i\sqrt{2})H(\pm i\sqrt{2}) = \frac{-4K}{2(4+4)} = -\frac{K}{4} \qquad (d)$$

Hence, when $K = 2$ the point $-1 + i0$ is not encircled by the Nyquist plot and, because this is a minimum-phase system, the closed-loop system is stable. On the other hand, when $K = 6$ the point $-1 + i0$ is encircled by the Nyquist plot, so that the closed-loop system is unstable.

It is clear that the above conclusions were reached solely on the basis of the open-loop frequency response $G(i\omega)H(i\omega)$.

5.5 FREQUENCY RESPONSE PLOTS

The frequency response was defined in a general way in Section 3.2 and was shown in Section 5.4 to be the centerpiece in the construction of Nyquist plots. The Nyquist plots are essentially polar plots, with the frequency response being plotted as the curve generated by the tip of a complex vector in the $G(s)H(s)$-plane as ω changes from $-\infty$ to $+\infty$. The frequency response is more commonly plotted in rectangular coordinates and there are two plots, magnitude versus frequency and phase angle versus frequency. The frequency response plots are widely used in vibrations and control. In this section, we discuss the concept by means of a second-order system and then use the plots to introduce certain frequency-domain performance measures.

From Section 5.2, the closed-loop transfer function of a position control system can be written in the form

$$M(s) = \frac{\Theta_o(s)}{\Theta_i(s)} = \frac{K_m}{Js^2 + cs + K_m} = \frac{\omega_n^2}{s^2 + 2\zeta\omega_n s + \omega_n^2} \qquad (5.41)$$

where

$$2\zeta\omega_n = c/J, \quad \omega_n^2 = K_m/J \qquad (5.42)$$

in which ζ is known as the viscous damping factor and ω_n as the frequency of undamped oscillation. For undamped systems, ω_n represents the natural frequency. It was shown in Section 3.3 that the frequency response can be obtained from the transfer function by replacing s by $i\omega$. In the case of open-loop response to harmonic excitation of a damped single-degree-of-freedom system, such as that shown in Fig. 5.13, one obtains the same transfer function as that given by the right side of Eq. (5.41). In this case, ω represents the excitation frequency. In the case of closed-loop response,

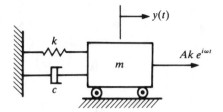

Figure 5.13

such as the position control system considered here, ω is the frequency of a harmonic reference input.

Inserting $s = i\omega$ into Eq. (5.41) and dividing the numerator and denominator by ω_n^2, we obtain the frequency response function

$$M(i\omega) = \frac{1}{1 - (\omega/\omega_n)^2 + i2\zeta\omega/\omega_n} \tag{5.43}$$

Equation (5.43) can be expressed as

$$M(i\omega) = |M(i\omega)|e^{i\phi(\omega)} \tag{5.44}$$

where

$$|M(i\omega)| = \frac{1}{\{[1 - (\omega/\omega_n)^2]^2 + (2\zeta\omega/\omega_n)^2\}^{1/2}} \tag{5.45}$$

is the magnitude and

$$\phi(\omega) = \underline{/M(i\omega)} = \tan^{-1}\left[-\frac{2\zeta\omega/\omega_n}{1 - (\omega/\omega_n)^2}\right] = -\tan^{-1}\frac{2\zeta\omega/\omega_n}{1 - (\omega/\omega_n)^2} \tag{5.46}$$

is the phase angle of the frequency response.

A great deal can be learned about the nature of the response by examining how the magnitude and phase angle of the frequency response change as the excitation frequency varies. Figure 5.14 shows a plot of the magnitude $|M(i\omega)|$ as a function of the ratio ω/ω_n and Fig. 5.15 depicts the phase angle $\phi(\omega)$ as a function of ω/ω_n, both for a given value of the parameter ζ.

The magnitude plot can be used to define certain performance criteria in the frequency domain. To this end, we consider the magnitude plot shown in Fig. 5.16 for a given damping factor ζ. The frequency-domain characteristics can be identified as follows:

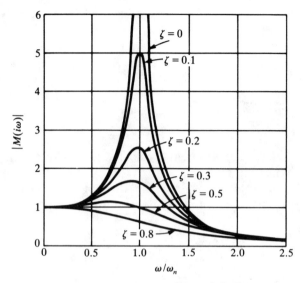

Figure 5.14 Adapted from [M42].

1. *Peak Amplitude* $|M(I\omega)|_p$: defined as the maximum value of $|M(i\omega)|$. The peak amplitude gives a measure of the relative stability of the feedback control system. A system exhibiting a larger maximum overshoot in the step response (see Section 5.2) is likely to exhibit a large peak amplitude in the frequency response. A value of $|M(i\omega)|$ lying between 1.1 and 1.5 is considered acceptable. The peak amplitude is sometimes referred to as *peak resonance*.
2. *Resonant Frequency* ω_p: defined as the frequency corresponding to the peak amplitude.

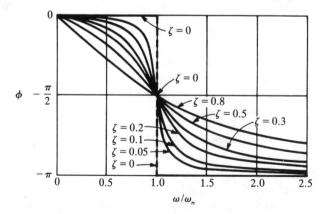

Figure 5.15 Adapted from [M42]. Reprinted with permission.

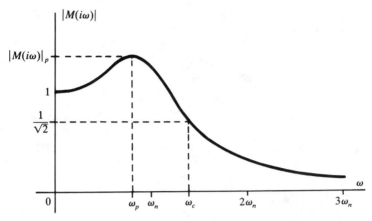

Figure 5.16

3. *Bandwidth BW.* There are several definitions of the bandwidth, and there is a certain degree of arbitrariness in every one of them. The most commonly used one defines the bandwidth as the range of frequencies between zero and the frequency corresponding to a value of the magnitude of the frequency response equal to $|M(0)|/\sqrt{2}$. In the case of the second-order system under consideration this value is $1/\sqrt{2}$. The bandwidth provides a measure of the filtering characteristics of the system. For example, a large bandwidth permits higher frequencies to be passed to the output, and vice versa. The bandwidth is also a measure of the speed of response of a system. Indeed, because higher frequencies can pass through, a large bandwidth implies a faster rise time.
4. *Cutoff Frequency ω_c.* The cutoff frequency corresponds to the frequency defining the bandwidth.
5. *Cutoff Rates*: defined as the frequency rate at which the magnitude of the frequency response decreases beyond the cutoff frequency. A steep cutoff rate tends to imply a large peak amplitude.

It is not difficult to verify that for the second-order system under consideration, and for which the magnitude of the frequency response is given by Eq. (5.45), the peak amplitude is given by

$$|M(i\omega)|_p = \frac{1}{2\zeta(1-\zeta^2)^{1/2}} \qquad (5.47)$$

and the resonant frequency by

$$\omega_p = (1-2\zeta^2)^{1/2}\omega_n \qquad (5.48)$$

which implies that $\zeta < 1/\sqrt{2}$. Indeed, if $\zeta > 1/\sqrt{2}$, there is no resonant frequency, as the maximum amplitude occurs at $\omega = 0$. Moreover, the cutoff frequency can be shown to have the expression

$$\omega_c = (1 - 2\zeta^2 + \sqrt{2 - 4\zeta^2 + 4\zeta^4})^{1/2}\omega_n \qquad (5.49)$$

We observe from Eq. (5.47) that the peak amplitude depends on the damping factor ζ only, and not on ω_n. This characteristic is shared with the maximum overshoot in the step response (Section 5.2). On the other hand, the resonant frequency and the cutoff frequency depend on both ζ and ω_n.

5.6 BODE DIAGRAMS

Frequency response can also be plotted in terms of logarithmic coordinates, in which case the plots are called *Bode plots*. As the plots in terms of ordinary rectangular coordinates discussed in Section 5.5, there are two such plots, namely magnitude versus frequency and phase angle versus frequency. There are several advantages to logarithmic plots over ordinary rectangular plots. In the first place, logarithmic plots permit a much larger range of the excitation frequency. More important, however, is the fact that logarithmic plots are easier to construct. This fact is of special significance in feedback control, where the frequency response tends to be more complicated than in open-loop vibration. Indeed, in controls the frequency response is ordinarily a rational function with the numerator and denominator consisting of products of various factors. The logarithm of such a function is simply the sum of logarithms of the factors in the numerator minus the logarithms of the factors in the denominator. Hence, in plotting the logarithm of the frequency response, each factor can be plotted separately and then added or subtracted, depending on whether the factor is in the numerator or denominator, respectively. As with the root-locus plots and Nyquist plots, Bode plots enable us to draw conclusions concerning the stability of closed-loop systems by working with open-loop frequency response functions.

Let us consider for simplicity the case in which the feedback-path transfer function $H(s)$ is equal to unity, in which case the open-loop transfer function $G(s)H(s)$ reduces to the forward-path transfer function $G(s)$. Hence, the open-loop frequency response is simply $G(i\omega)$. As in Section 5.5, we express the frequency response in the form

$$G(i\omega) = |G(i\omega)|e^{i\phi(\omega)} \qquad (5.50)$$

where

$$|G(i\omega)| = [\text{Re}^2\, G(i\omega) + \text{Im}^2\, G(i\omega)]^{1/2} \qquad (5.51)$$

is the magnitude of the frequency response, often referred to as the *gain*, and

$$\phi(\omega) = \tan^{-1} \frac{\text{Im } G(i\omega)}{\text{Re } G(i\omega)} \quad (5.52)$$

is the phase angle. In constructing the Bode diagrams, it is customary to work with logarithms to the base 10, denoted by log, instead of natural logarithms. Using Eq. (5.50), the logarithm to the base 10 of $G(i\omega)$ is

$$\log G(i\omega) = \log |G(i\omega)| + i0.434\phi(\omega) \quad (5.53)$$

where $\log |G(i\omega)|$ is called the *logarithmic gain*. The factor 0.434 is ordinarily omitted from the complex part in Eq. (5.53).

The logarithmic gain is commonly expressed in *decibels*, where the decibel (dB) represents a unit such that, for a given number N, $N_{dB} = 20 \log N$ (dB). Hence, the logarithmic gain in decibels is

$$|G(i\omega)|_{dB} = 20 \log |G(i\omega)| \quad (dB) \quad (5.54)$$

There are two units used to express the frequency band from ω_1 to ω_2 or the ratio ω_2/ω_1 of the two frequencies. If $\omega_2/\omega_1 = 2$, the frequency ω_1 and ω_2 are said to be separated by an *octave*. Then, the number of octaves between any two given frequencies ω_1 and ω_2 is

$$\frac{\log (\omega_2/\omega_1)}{\log 2} = \frac{1}{0.301} \log \frac{\omega_2}{\omega_1} = 3.32 \log \frac{\omega_2}{\omega_1} \quad \text{(octaves)} \quad (5.55)$$

Similarly, if $\omega_2/\omega_1 = 10$, the frequencies ω_1 and ω_2 are said to be separated by a *decade*. Then, the number of decades between any two given frequencies ω_1 and ω_2 is

$$\frac{\log (\omega_2/\omega_1)}{\log 10} = \log \frac{\omega_2}{\omega_1} \quad \text{(decades)} \quad (5.56)$$

The units of octave and decade are commonly used to express the slope of straight lines in Bode plots.

Next, we consider a frequency response function in the general form

$$G(i\omega) = \frac{K\Pi_k(1 + i\omega\tau_k)}{(i\omega)^l \Pi_m(1 + i\omega\tau_m)\Pi_n[1 - (\omega/\omega_n)^2 + i2\zeta_n\omega/\omega_n]} \quad (5.57)$$

where K is a positive real gain constant, τ_k and τ_m are real time constants, ζ_n are real damping factors and ω_n are natural frequencies. Then, the logarithmic gain of $G(i\omega)$ in decibels has the simple expression

$$|G(i\omega)|_{dB} = 20\bigg[\log K + \sum_k \log|1 + i\omega\tau_k| - l\log|i\omega|$$

$$-\sum_m \log|1 + i\omega\tau_m| - \sum_n \log|1 - (\omega/\omega_n)^2 + i2\zeta_n\omega/\omega_n|\bigg] \quad (5.58)$$

so that the logarithmic gain is a linear combination of the logarithms of the individual factors. Moreover, the phase angle of $G(i\omega)$ has the form

$$\phi(\omega) = \underline{/G(i\omega)} = \underline{/K} + \sum_k \underline{/1 + i\omega\tau_k} - l\underline{/i\omega} - \sum_m \underline{/1 + i\omega\tau_m}$$

$$-\sum_n \underline{/[1 - (\omega/\omega_n)^2 + i2\zeta_n\omega/\omega_n]} \quad (5.59)$$

In plotting the logarithmic gain, one can plot the logarithms of the individual factors separately and then add or subtract them according to their sign in the summation. There are four types of factors: (i) a constant factor, K, (ii) poles at the origin, $(i\omega)^{-1}$, (iii) poles or zeros at $\omega \neq 0$, $(1 + i\omega\tau)^{\pm 1}$ and (iv) complex poles, $[1 - (\omega/\omega_n)^2 + i2\zeta_n\omega/\omega_n]^{-1}$. Similarly, the phase angle can be constructed by plotting the phase angles of the individual factors separately and then combining them linearly In the following, we construct Bode plots for the four types of factors. As customary, we use $\log \omega$ as abscissa in these plots.

i. Constant Factor

Because K is a positive constant,

$$K_{dB} = 20 \log K = \text{constant} \quad (5.60)$$

so that the plot of the logarithmic gain is a horizontal straight line. The role of the constant logarithmic gain is to raise the logarithmic gain of the complete frequency response by that amount. Moreover, because K is real, the phase angle plot is the horizontal straight line

$$\underline{/K} = \tan^{-1}\frac{0}{K} = 0° \quad (5.61)$$

ii. Poles at the Origin

The logarithmic gain of the factor $(i\omega)^{-l}$ is simply

$$|(i\omega)^{-l}|_{dB} = -20l \log|i\omega| = -20l \log \omega \text{ (dB)} \quad (5.62)$$

which is a straight line through the point 0 dB at $\omega = 1$ with the slope

$$\frac{d(-20l \log \omega)}{d(\log \omega)} = -20l \text{ (dB/decade)} \qquad (5.63)$$

The phase angle is

$$\underline{/(i\omega)^{-l}} = -l\underline{/i\omega} = -l \tan^{-1}\frac{\omega}{0} = -l \times 90° \qquad (5.64)$$

iii. Factors $(1 + i\omega\tau)^{\pm 1}$

The logarithmic gain of the factor $1 + i\omega\tau$ is

$$|1 + i\omega\tau|_{dB} = 20 \log |1 + i\omega\tau| = 20 \log \sqrt{1 + (\omega\tau)^2} \qquad (5.65)$$

The curve has two asymptotes, given by

$$\lim_{\omega\tau \to 0} |1 + i\omega\tau|_{dB} = \lim_{\omega\tau \to 0} 20 \log \sqrt{1 + (\omega\tau)^2} = 20 \log 1 = 0 \qquad (5.66a)$$

and

$$\lim_{\omega\tau \to \infty} |1 + i\omega\tau|_{dB} = \lim_{\omega\tau \to \infty} 20 \log \sqrt{1 + (\omega\tau)^2} = 20 \log \sqrt{(\omega\tau)^2} = 20 \log \omega\tau \qquad (5.66b)$$

The intersection of the two asymptotes, obtained by equating Eqs. (5.66a) and (5.66b), has the value

$$\omega_{cf} = \frac{1}{\tau} \qquad (5.67)$$

and is known as the *corner frequency*. The Bode plot is shown in Fig. 5.17a for $\tau = 1$. Using Eq. (5.52), the phase angle of $1 + i\omega\tau$ is simply

$$\underline{/1 + i\omega\tau} = \tan^{-1} \omega\tau \qquad (5.68)$$

It is plotted in Fig. 5.17b.

The logarithmic gain of the factor $(1 + i\omega\tau)^{-1}$ is simply the negative of that given by Eq. (5.65), so that the Bode plot of the logarithmic gain of $(1 + i\omega\tau)^{-1}$ is the mirror image of the Bode plot for the factor $1 + i\omega\tau$. Similarly, it is easy to verify that the Bode plot of the phase angle of $(1 + i\omega\tau)^{-1}$ is the mirror image of the plot for $1 + i\omega\tau$. The Bode plots for $(1 + i\omega\tau)^{-1}$ are also shown in Fig. 5.17.

iv. Factor $[1 - (\omega/\omega_n)^2 + i2\zeta\omega/\omega_n]^{-1}$

Note that the factor considered here is the same as the open-loop frequency response $M(i\omega)$ considered in Section 5.5. The logarithmic gain of $[1 - (\omega/\omega_n)^2 + i2\zeta\omega/\omega_n]^{-1}$ is

154 CONTROL OF LUMPED-PARAMETER SYSTEMS. CLASSICAL APPROACH

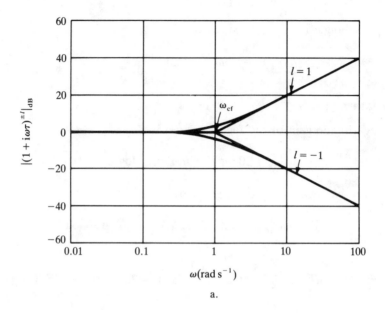

a.

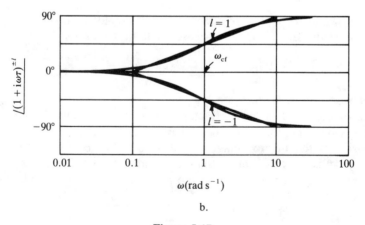

b.

Figure 5.17

$$|[1 - (\omega/\omega_n)^2 + i2\zeta\omega/\omega_n]^{-1}|_{\text{dB}} = -20 \log |1 - (\omega/\omega_n)^2 + i2\zeta\omega_n/\omega_n|$$
$$= -20 \log \{[1 - (\omega/\omega_n)^2]^2 + (2\zeta\omega/\omega_n)^2\}^{1/2} \quad (5.69)$$

The plot of the logarithmic gain versus ω/ω_n has two asymptotes, as follows:

$$\lim_{\omega/\omega_n \to 0} |[1 - (\omega/\omega_n)^2 + i2\zeta\omega/\omega_n]^{-1}|_{\text{dB}}$$
$$= \lim_{\omega/\omega_n \to 0} [-20 \log \{[1 - (\omega/\omega_n)^2]^2 + (2\zeta\omega/\omega_n)^2\}^{1/2}]$$
$$= -20 \log 1 = 0 \text{ (dB)} \quad (5.70a)$$

and

$$\lim_{\omega/\omega_n \to \infty} |[1 - (\omega/\omega_n)^2 + i2\zeta\omega/\omega_n]^{-1}|_{\text{dB}}$$
$$= \lim_{\omega/\omega_n \to \infty} [-20 \log \{[1 - (\omega/\omega_n)^2]^2 + (2\zeta\omega/\omega_n)^2\}^{1/2}]$$
$$= -20 \log [(\omega/\omega_n)^4]^{1/2} = -40 \log (\omega/\omega_n) \quad (5.70b)$$

Hence, the low-frequency asymptote is a horizontal straight line through the origin and the high-frequency asymptote is a straight line with the slope of -40 dB/decade. The corner frequency lies at the intersection of the two asymptotes and has the value

$$\omega_{cf} = \omega_n \quad (5.71)$$

The logarithmic gain is plotted in Fig. 5.18a for several values of ζ. The plot in Fig. 5.18a is essentially the same as that in Fig. 5.14, the only difference being in that Fig. 5.18a is on a logarithmic scale.

The phase angle of the factor $[1 - (\omega/\omega_n)^2 + i2\zeta\omega/\omega_n]^{-1}$ was already evaluated in Section 5.5. Hence, from Eq. (5.47), we can write

$$\underline{/[1 - (\omega/\omega_n)^2 + i2\zeta\omega/\omega_n]^{-1}} = -\tan^{-1} \frac{2\zeta\omega/\omega_n}{1 - (\omega/\omega_n)^2} \quad (5.72)$$

The phase angle is plotted in Fig. 5.18b. It differs from the plot in Fig. 5.15 in that it is on a logarithmic scale.

Example 5.3 Construct the Bode plots for the system of Example 5.2 for the same two cases.

Letting $s = i\omega$ in Eq. (a) of Example 5.2, we obtain the frequency response

156 CONTROL OF LUMPED-PARAMETER SYSTEMS. CLASSICAL APPROACH

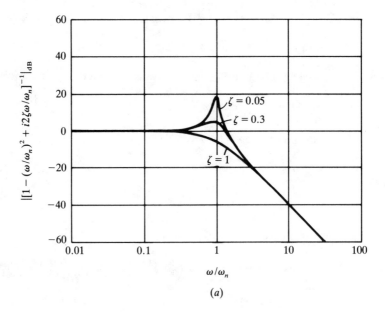

(a)

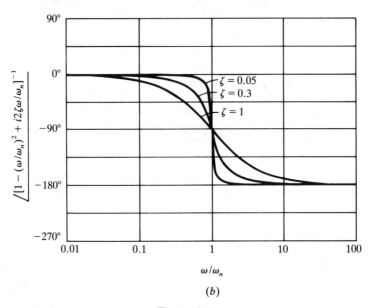

(b)

Figure 5.18

$$G(i\omega)H(i\omega) = \frac{K}{i\omega(2-\omega^2+i2\omega)} = \frac{K/2}{i\omega[1-(\omega/\sqrt{2})^2+i\omega]} \quad \text{(a)}$$

Hence, from Eqs. (5.58), (5.62) and (5.69), the logarithmic gain is

$$|G(i\omega)H(i\omega)|_{dB} = 20\left[\log\frac{K}{2} - \log\omega - \frac{1}{2}\log\left\{\left[1-\left(\frac{\omega}{\sqrt{2}}\right)^2\right]^2 + \omega^2\right\}\right] \quad \text{(b)}$$

and from Eqs. (5.59), (5.64) and (5.72), the phase angle is

$$\phi(\omega) = \underline{/G(i\omega)H(i\omega)} = \underline{/K/2} - \underline{/i\omega} - \underline{/1-(\omega/\sqrt{2})^2+i\omega}$$
$$= 0 - 90° - \tan^{-1}\frac{\omega}{1-(\omega/\sqrt{2})^2} \quad \text{(c)}$$

The Bode plots for $K = 2$ and $K = 6$ are shown in Fig. 5.19.

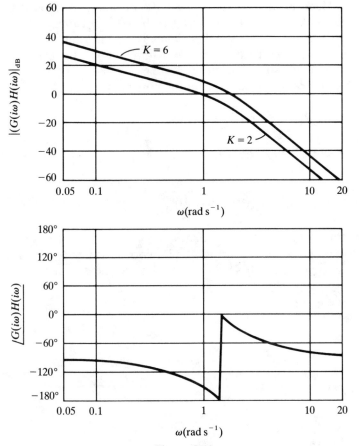

Figure 5.19

5.7 RELATIVE STABILITY. GAIN MARGIN AND PHASE MARGIN

In Section 3.8, we examined various procedures for checking system stability analytically. The object of this section is to investigate ways of obtaining additional information concerning the system stability, such as the *degree of stability* or the *relative stability*. Indeed, assuming that the system is asymptotically stable, a question of interest is how close the closed-loop poles are to the imaginary axis in the s-plane, because this has a direct effect on the settling time. A measure of the degree of stability of a system can be obtained graphically by means of the Nyquist plot or by means of Bode plots.

The Nyquist criterion defines the stability of a closed-loop system in terms of the number of encirclements of the point $-1 + i0$ in the $G(s)H(s)$-plane by the Nyquist plot, where $G(s)H(s)$ is the open-loop transfer function. As it turns out, in the region of interest the Nyquist plot involves the open-loop frequency response $G(i\omega)H(i\omega)$ and not the transfer function $G(s)H(s)$. Hence, the degree of stability of the closed-loop system is determined by the proximity of the $G(i\omega)H(i\omega)$ plot to the point $-1 + i0$. Quite often, the degree of stability depends on the gain constant K. The relative stability is measured in terms of two quantities, the *gain margin* and *phase margin*, as illustrated in the following.

Let us consider the open-loop transfer function

$$G(s)H(s) = \frac{K}{s(s^2 + 3s + 2)} \tag{5.73}$$

so that the open-loop frequency response is

$$G(i\omega)H(i\omega) = \frac{K}{i\omega[(i\omega)^2 + i3\omega + 2]} = K\frac{-3\omega^2 - i\omega(2 - \omega^2)}{\omega^2[(2 - \omega^2)^2 + 9\omega^2]} \tag{5.74}$$

The essential portion of the frequency response is plotted in Fig. 5.20 for $K = 4$, $K = 6$ and $K = 8$. Clearly, the plot corresponding to $K = 4$ does not encircle the point $-1 + i0$ so that, according to the Nyquist criterion, the

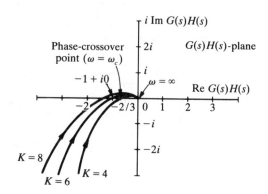

Figure 5.20

RELATIVITY STABILITY. GAIN MARGIN AND PHASE MARGIN

system is stable for $K = 4$. The plot intersects the real axis at a point called the *phase-crossover point*, and the intersection occurs for a value of ω known as the *phase-crossover frequency* and denoted by ω_c. The distance between the phase-crossover point and the point $-1 + i0$ is a quantitative measure of the relative stability. This quantity represents the *gain margin* and is defined mathematically as

$$\text{gain margin} = 20 \log \frac{1}{|G(i\omega_c)H(i\omega_c)|} \quad \text{(dB)} \tag{5.75}$$

By increasing the gain constant K, the phase-crossover point approaches the point $-1 + i0$. Hence, *the gain margin is the amount in decibels by which the gain can be increased before the closed-loop system becomes unstable.* In the example under consideration, we see from Eq. (5.74) that the phase-crossover frequency is $\omega_c = \sqrt{2}$ rad/s, so that the gain margin is

$$20 \log \frac{1}{|G(i\sqrt{2})H(i\sqrt{2})|} = 20 \log \frac{1}{2/3} = 3.5214 \text{ (dB)} \tag{5.76}$$

This implies that the gain can be increased by 3.5214 dB, i.e., by 50%, before instability is reached. Indeed, it is not difficult to verify that the Nyquist plot corresponding to $K = 6$ passes through the point $-1 + i0$, so that for $K = 6$ the gain margin reduces to zero. Of course, for $K > 6$ the Nyquist plot encircles the point $-1 + i0$ twice in the clockwise sense, and there is no pole of $G(s)H(s)$ in the right half of the s-plane. Hence, according to Eq. (5.38), there must be two closed-loop poles in the right half of the s-plane, so the system is unstable for $K > 6$.

The gain margin defined above it not always a reliable measure of the relative stability of the closed-loop system. The reason for this lies in the fact that the gain margin does not provide a complete picture of how close the Nyquist plot comes to the point $-1 + i0$. Indeed, there can be two different systems with the same gain margin, but with Nyquist plots making different angle with the real axis at the phase-crossover point. It is easy to see that the Nyquist plot making a smaller angles with the real axis is likely to get closer to the point $-1 + i0$ than the one making a larger angle, as can be seen from Fig. 5.21. Hence, a supplementary way of ascertaining relative

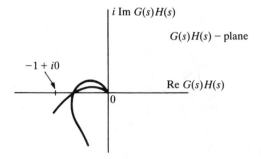

Figure 5.21

stability appears necessary. Such a measure can be obtained by defining the *gain-crossover point* as a point on the Nyquist plot of unit magnitude. The point lies at the intersection of the Nyquist plot with a circle of unit radius and centered at the origin of the $G(s)H(s)$-plane, as shown in Fig. 5.22. Then, another quantitative measure of relative stability is the *phase margin* defined as

$$\text{phase margin} = \underline{/G(i\omega_g)H(i\omega_g)} - 180° \tag{5.77}$$

where ω_g is the frequency corresponding to the gain-crossover point and is known as the *gain-crossover frequency*. From Fig. 5.22, we conclude that the *phase margin is the angle in degrees through which the Nyquist plot must be rotated about the origin to bring the gain-crossover point into coincidence with the point* $-1 + i0$. The phase margin can be interpreted as the additional phase lag necessary to render the system unstable. The phase margin is shown in Fig. 5.22.

The gain and phase margins can be evaluated with greater ease by means of Bode plots than Nyquist plots. To the critical point $-1 + i0$ in the Nyquist plot corresponds a logarithmic gain of 0 dB and a phase angle of $-180°$ in the Bode plots. Hence, *the gain margin represents the portion between the logarithmic gain plot and the 0-dB line corresponding to a phase angle of* $-180°$. On the other hand, *the phase margin represents the portion between the phase angle plot and the* $-180°$ *line corresponding to a logarithmic gain of* 0 dB. The gain margin and phase margin for the system with the open-loop frequency response given by Eq. (5.74) in which $K = 4$ are shown in Fig. 5.23; they are approximately 3.5 dB and 11.5°, respectively. We recall from Section 5.6 that the effect of increasing the gain constant K is to raise the entire logarithmic gain plot. Hence, from Fig. 5.23, we conclude that as K increases, the gain margin decreases. At the same time, the gain-crossover point moves to the right, which causes the phase margin to decrease as well. When K becomes equal to 6, both the gain margin and phase margin reduce to zero.

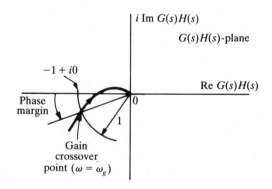

Figure 5.22

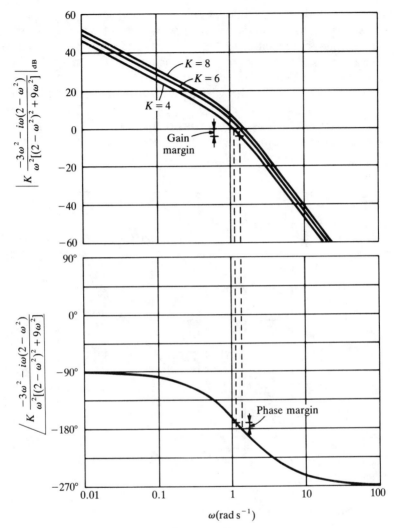

Figure 5.23

5.8 LOG MAGNITUDE-PHASE DIAGRAMS

In Section 5.6, we were concerned with Bode plots, which are open-loop frequency response plots in terms of logarithmic coordinates. There are two Bode plots, namely, the logarithm of the magnitude (measured in decibels) versus frequency and phase angle versus frequency, where logarithmic scales are used for the frequency.

The same information as in the Bode plots can be presented in a single plot known as the log magnitude-phase diagram. The diagram is a plot of

162   CONTROL OF LUMPED-PARAMETER SYSTEMS. CLASSICAL APPROACH

the logarithm of the magnitude of the open-loop frequency response in decibels versus the phase angle in degrees, with the frequency playing the role of a parameter. By changing the gain K, the plot moves up or down, as in the case of the Bode plot of the logarithm of the magnitude versus frequency. But, the unique property of the Bode plots which permits first drawing individual plots for the various factors of the frequency response and then combining them does not carry over to the log magnitude-phase diagram. However, it is possible to construct the curve by first computing the log magnitudes and phase angles of the individual factors corresponding

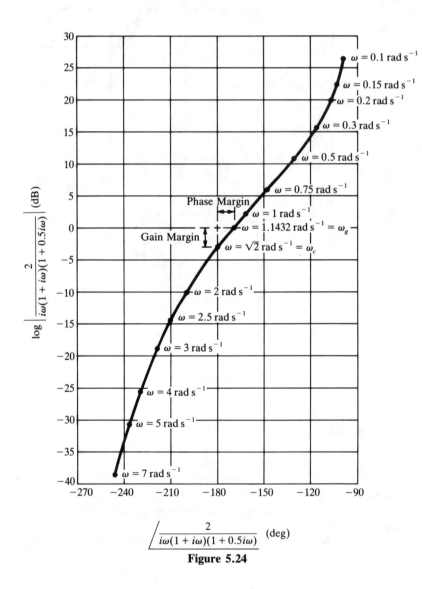

Figure 5.24

to given frequencies and then summing them. As an alternative, one can produce the two Bode plots first and then use the values of the log magnitudes and phase angles corresponding to given frequencies to produce the log magnitude-phase diagram.

As an illustration, let us consider the open-loop frequency response

$$G(i\omega)H(i\omega) = \frac{2}{i\omega(1 + i\omega)(1 + i0.5\omega)} \qquad (5.78)$$

which is the same function as that given by Eq. (5.74) with $K = 4$. The log magnitude-phase diagram is shown in Fig. 5.24 and the corresponding Bode plots are given in Fig. 5.23. The log magnitude-phase diagram has several noteworthy features. In particular, we observe that the curve crosses the $-180°$ axis at the phase-crossover point defined by the phase-crossover frequency $\omega_c = \sqrt{2}$ rad s^{-1}. At this point the log magnitude is equal to -3.52 dB, which implies that the system has a gain margin of 3.52 dB. Similarly, the curve crosses the 0 dB axis at the gain-crossover point defined by the gain-crossover frequency $\omega_g = 1.1432$ rad s^{-1}. At this point the phase is equal to $-168.6°$, so that there is a phase margin of 11.4°. By increasing the constant K, the plot moves up, thus reducing the gain margin and the phase margin, and vice versa. In fact, for $K = 6$ the plot passes through the point defined by the intersection of 0 dB axis and $-180°$ axis, so that when K reaches the value 6 the gain margin and phase margin reduce to zero.

Because the log magnitude-phase diagrams contain the same information as the Bode plots, and Bode plots are easier to generate, the question arises as to the purpose of log magnitude-phase diagrams. It turns out that log magnitude-phase diagrams can be used in conjunction with Nichols charts to obtain the closed-loop frequency response, as demonstrated in Section 5.9.

5.9 THE CLOSED-LOOP FREQUENCY RESPONSE. NICHOLS CHARTS

In Section 5.1, we defined the closed-loop transfer function as the ratio of the (Laplace) transformed output to the transformed reference input. As shown in Section 5.1, the closed-loop transfer function has the form

$$M(s) = \frac{G(s)}{1 + G(s)H(s)} \qquad (5.79)$$

where $G(s)$ is the forward-path transfer function and $G(s)H(s)$ is the open-loop transfer function. The frequency response can be obtained from the transfer function by simply replacing s by $i\omega$. Hence, the closed-loop frequency response has the expression

164 CONTROL OF LUMPED-PARAMETER SYSTEMS. CLASSICAL APPROACH

$$M(i\omega) = \frac{G(i\omega)}{1 + G(i\omega)H(i\omega)} \tag{5.80}$$

where $G(i\omega)H(i\omega)$ is the open-loop frequency response.

In Sections 5.3–5.8, we studied the stability characteristics of closed-loop systems by means of the open-loop transfer function, or by means of the open-loop frequency response. In many design problems, the interest lies in certain performance characteristics, such as the maximum magnitude of the closed-loop frequency response. Hence, at times it is desirable to be able to determine the closed-loop frequency response from the open-loop frequency response. To this end, we consider the case in which $H(i\omega) = 1$, in which case the closed-loop frequency response reduces to

$$M(i\omega) = \frac{G(i\omega)}{1 + G(i\omega)} \tag{5.81}$$

The closed-loop frequency response can be expressed in terms of magnitude and phase angle as follows:

$$M(i\omega) = |M(i\omega)|e^{i\phi(\omega)} \tag{5.82}$$

One of our objects is to determine the maximum value of the magnitude $|M(i\omega)|$ from the plot of $G(i\omega)$, as well as to design a feedback control system with a given $|M(i\omega)|_{\max}$.

The relationship between $G(i\omega)$ and $M(i\omega)$ can be obtained by considering the complex $G(i\omega)$-plane. Recalling that $H(i\omega) = 1$, this is essentially the same plane as that used for Nyquist plots. Letting $u = \text{Re } G(i\omega)$ and $v = \text{Im } G(i\omega)$, we have

$$G(i\omega) = u + iv \tag{5.83}$$

To derive the desired relationship, we wish to generate curves of constant magnitude $|M(i\omega)|$ in the $G(i\omega)$-plane. For simplicity of notation, we let $|M(i\omega)| = M$ and consider Eqs. (5.81)–(5.83) to write

$$M = \left|\frac{G(i\omega)}{1 + G(i\omega)}\right| = \left|\frac{u + iv}{1 + u + iv}\right| = \left|\frac{u^2 + v^2}{(1 + u)^2 + v^2}\right|^{1/2} \tag{5.84}$$

Squaring Eq. (5.84) and rearranging, we obtain

$$\left(u - \frac{M^2}{1 - M^2}\right)^2 + v^2 = \left(\frac{M}{1 - M^2}\right)^2 \tag{5.85}$$

Equation (5.85) represents the equation of a circle with the center at $u = M^2/(1 - M^2)$, $v = 0$ and with the radius $r = |M/(1 - M^2)|$. Figure 5.25

THE CLOSED-LOOP FREQUENCY RESPONSE. NICHOLS CHARTS

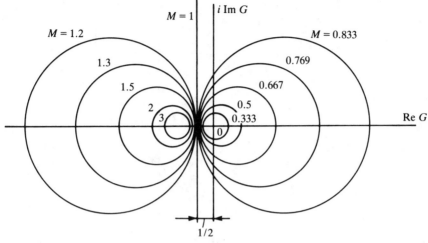

Figure 5.25

shows a family of constant M circles in the $G(i\omega)$-plane. In the special case in which $M = 1$, Eq. (5.85) is not valid. However, from Eq. (5.85), it is not difficult to show that $M = 1$ defines a straight line parallel to the v-axis and passing through the point $u = -1/2$, $v = 0$.

To obtain the magnitude of the closed-loop frequency response for a given system, we plot appropriate constant M circles on the $G(i\omega)$-plane, as well as the corresponding Nyquist plot of the open-loop frequency response. Then, the intersection points of the Nyquist plot and the constant M circles represent points on the closed-loop frequency response. As an illustration, we consider the open-loop frequency response

$$G(i\omega) = G(i\omega)H(i\omega) = \frac{K}{2i\omega(1 + i\omega)(1 + i0.5\omega)} \quad (5.86)$$

As shown in Fig. 5.26 the Nyquist plot for $K = K_2$ is tangent to the circle $M = M_1$ at a point corresponding to the frequency $\omega = \omega_{p2}$, it intersects the circle $M = M_2$ at ω_3 and ω_6 and the circle M_3 at ω_1 and ω_7. This implies that the closed-loop frequency response has the maximum magnitude M_1 at ω_{p2}, the magnitude M_2 at ω_3 and ω_6 and the magnitude M_3 at ω_1 and ω_7. Moreover, as $M \to 1$ the intersections of the circles with $K = K_2$ tend to occur at frequencies approaching zero on the one hand and at points corresponding to Re $G(i\omega) = -1/2$ on the other. The frequency corresponding to the latter is denoted by ω_8. The seven values of M and the associated seven frequencies can be used to sketch the plot of magnitude of the closed-loop frequency response versus the frequency corresponding to the gain $K = K_2$. The plot is shown in Fig. 5.27. Also shown in Fig. 5.27 is the plot corresponding to the $K = K_1$. One drawback in this approach lies in the

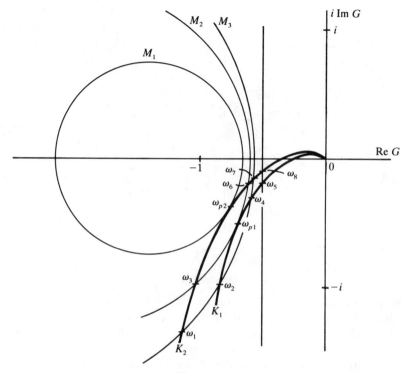

Figure 5.26

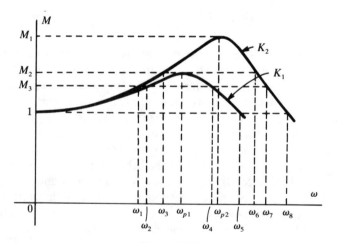

Figure 5.27

THE CLOSED-LOOP FREQUENCY RESPONSE. NICHOLS CHARTS

need to estimate the value of the frequency at a given intersection point, as the frequency is only implicit in the Nyquist plot.

The same approach can be used to produce the phase angle of the frequency response. To this end, we use Eqs. (5.81) and (5.83) and write the expression for the phase angle

$$\phi = \underline{/M(i\omega)} = \underline{/G(i\omega)/[1 + G(i\omega)]} = \underline{/(u + iv)/(1 + u + iv)}$$
$$= \underline{/[u(1 + u) + v^2 + iv]/[(1 + u)^2 + v^2]} \qquad (5.87)$$

Then, introducing the notation

$$N = \tan \phi = \frac{v}{u(1 + u) + v^2} \qquad (5.88)$$

and rearranging, we obtain

$$\left(u + \frac{1}{2}\right)^2 + \left(v - \frac{1}{2N}\right)^2 = \frac{1}{4}\left(1 + \frac{1}{N^2}\right) \qquad (5.89)$$

Equation (5.89) represents the equation of a circle with the center at $u = -1/2$, $v = 1/2N$ and with the radius $r = (1 + 1/N^2)^{1/2}/2$. Hence, we can plot constant N circles in the $G(i\omega)$-plane, in the same manner as we plotted constant M circles, and follow the same procedure as above to generate plots of the phase angle of the closed-loop frequency response versus the frequency. We shall not pursue the subject here.

Nyquist plots have the disadvantage that a simple change in the gain causes a change in the shape of the plot. By contrast, in Bode plots or in log magnitude-phase diagrams, a change in gain does not alter the shape but merely shifts the plot up or down. Moreover, Bode plots can accommodate modifications in the open-loop frequency response more readily than Nyquist plots. Hence, the question arises whether some benefits can be derived by working with logarithmic plots instead of polar plots. It turns out that the closed-loop frequency response can be generated more conveniently by working with the log magnitude-phase plane rather than the *G-plane*. This requires plotting the constant M loci and constant N loci in the log magnitude-phase plane. The loci are no longer circles in the log magnitude-phase plane, but this presents no particular problem as such plots already exist and are known as *Nichols charts* after their originator. The constant M contours in Nichols charts are given in decibels and the constant N contours in degrees.

Figure 5.28 shows two log magnitude-phase diagrams for the open-loop frequency response given by Eq. (5.86) superimposed on a Nichols chart. They correspond to the gains $K = 4$ and $K = 2$, and note that the plot for $K = 4$ is the same as that in Fig. 5.24. The plot of the magnitude of the closed-loop frequency response in decibels versus the frequency can be

168 CONTROL OF LUMPED-PARAMETER SYSTEMS. CLASSICAL APPROACH

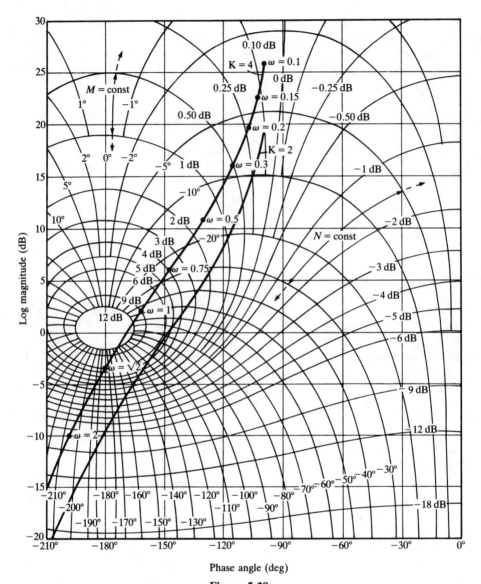

Figure 5.28

constructed by picking the intersections of the constant M contours and the log magnitude-phase diagram of the open-loop frequency response. Here again, we encounter the difficulty of estimating the frequency at these intersection points. Plots of the phase angle of the closed-loop frequency response versus the frequency can be generated by picking the intersections of the constant N contours and the same log magnitude-phase diagram.

One advantage of the Nichols plots is that they can accommodate a much larger range of frequencies. Another advantage is that the effect of changing the gain K on the gain margin and phase margin can be readily estimated. For example, in decreasing the gain from $K = 4$ to $K = 2$ the gain margin is increased from 3.52 dB to 9.54 dB. Perhaps a better measure of the degree of stability is the closest approach to the critical point 0 dB, $-180°$, a measure easy to observe on a Nichols plot. Of course, similar information is available on a Nyquist plot, but the effect of design changes on the frequency response can be more readily assessed on Nichols plots. Nyquist plots and Nichols plots can be regarded as being complimentary, with Nichols plots being more useful for shaping the closed-loop frequency response at low frequencies, once the high-frequency behavior (roll off) is known.

5.10 SENSITIVITY OF CONTROL SYSTEMS TO VARIATIONS IN PARAMETERS

On occasions, the system parameters differ in value from the values used in the original design. These variations can be brought about by aging or result from the fact that the original values of the parameters were not known very well. Then, the question arises whether the control system is capable of performing satisfactorily in spite of the variations or it is sensitive to these variations. Clearly, good control system design requires that the system be relatively insensitive to variations in the parameters, i.e., that it perform well and remain stable in spite of the parameter variations. The subject of system sensitivity was discussed in Section 3.14 in connection with the eigenvalue problem. The discussion here is more consistent with the classical approach to control.

A quantitative measure of the *system sensitivity* is the ratio of the percentage change in the system output to the percentage change in the system parameters. For comparison purposes, we consider first an open-loop system. Letting $H(s) = 0$ in Eq. (5.5), we obtain the transformed output

$$Y(s) = G(s)R(s) \qquad (5.90)$$

Then, assuming that the parameter variations bring about an incremental change $\Delta G(s)$ in the transfer function $G(s)$, Eq. (5.90) yields

$$Y(s) + \Delta Y(s) = [G(s) + \Delta G(s)]R(s) \qquad (5.91)$$

Using Eq. (5.90), Eq. (5.91) yields the sensitivity of an open-loop system

$$S = \frac{\Delta Y(s)/Y(s)}{\Delta G(s)/G(s)} = 1 \qquad (5.92)$$

Hence, a change in the plant transfer function causes a proportional change in the transformed output.

Next, we consider a closed-loop system. To examine the sensitivity, we distinguish parameters entering into the plant transfer function $G(s)$ and into the sensor transfer function $H(s)$ separately, and define sensitivity functions accordingly. Hence, considering first a change in $G(s)$, we use Eq. (5.5) and write

$$Y(s) + \Delta Y(s) = \frac{G(s) + \Delta G(s)}{1 + [G(s) + \Delta G(s)]H(s)} R(s)$$

$$= \frac{G(s) + \Delta G(s)}{1 + G(s)H(s) + \Delta G(s)H(s)} R(s) \quad (5.93)$$

Assuming that $\Delta G(s)$ is small relative to $G(s)$ and using the approximation $(1 + \varepsilon)^{-1} \cong 1 - \varepsilon$, Eq. (5.93) can be rewritten as

$$Y(s) + \Delta Y(s) \cong \frac{1}{[1 + G(s)H(s)]^2} \{G(s)[1 + G(s)H(s)] + \Delta G(s)\} R(s) \quad (5.94)$$

Then, using Eq. (5.5), we can define the sensitivity of the closed-loop system to changes in $G(s)$ as

$$S_G = \frac{\Delta Y(s)/Y(s)}{\Delta G(s)/G(s)} = \frac{1}{1 + G(s)H(s)} \quad (5.95)$$

so that a closed-loop system is less sensitive to variations in parameters affecting $G(s)$ than an open-loop system. In fact, the sensitivity can be improved by increasing $G(s)H(s)$.

Following the same procedure, we consider a change in $H(s)$ in Eq. (5.5) and obtain

$$Y(s) + \Delta Y(s) = \frac{G(s)}{1 + G(s)[H(s) + \Delta H(s)]} R(s)$$

$$= \frac{G(s)}{1 + G(s)H(s) + \Delta H(s)G(s)} R(s) \quad (5.96)$$

Once again, using the earlier approximation to the binomial expansion, we can rewrite Eq. (5.96) as

$$Y(s) + \Delta Y(s) \cong \frac{G(s)}{1 + G(s)H(s)} \left[1 - \frac{\Delta H(s)G(s)}{1 + G(s)H(s)} \right] R(s) \quad (5.97)$$

Hence, using Eq. (5.5), we can define the sensitivity of the closed-loop system to changes in $H(s)$ as

$$S_H = \frac{\Delta Y(s)/Y(s)}{\Delta H(s)/H(s)} = -\frac{G(s)H(s)}{1 + G(s)H(s)} \tag{5.98}$$

Equation (5.98) illustrates a typical difficulty encountered in control design. In particular, we observe that to reduce S_H we must reduce $G(s)H(s)$, which is in conflict with the requirement that we increase $G(s)H(s)$ to reduce S_G.

The above developments bring into sharp focus the fact that there are limits to altering the parameters to improve the control system performance. Indeed, to improve the performance beyond anything that can be achieved by modifying the system parameters, it is necessary to add some components not included in the block diagram of Fig. 5.3.

5.11 COMPENSATORS

The principal object of a feedback control system is performance. The subject of system performance was discussed in Section 5.2, in which several criteria for measuring the performance in the time domain were introduced. In particular, the performance can be regarded as good if the system exhibits small maximum overshoot M_p, low settling time T_s and small steady-state error e_{ss}. The performance was also discussed in Section 5.5 in terms of criteria in the frequency domain. Good performance is characterized by small peak amplitudes $|M(i\omega)|_p$ and large cutoff rate. Finally, the performance was discussed in a different context in Section 5.10, in which the sensitivity of the control system to variations in the parameters was investigated. To improve the performance of a control system it is often necessary to modify the system parameters. However, it is impossible to change parameters so as to produce improvement in all performance measures. In fact, as demonstrated in Section 5.10, changes in parameters improving the performance as measured by one criterion can degrade the performance as measured by another criterion. Hence, there are limits to the performance improvement that can be achieved by modifying parameters for a given configuration. In such cases, it is advisable to redesign the control system by inserting in the control loop a device capable of eliminating deficiencies from the system performance. The process of redesigning the control system to render the performance satisfactory is known as *compensation*, and the newly added device is called a *compensator*. More often than not the compensator is an electrical circuit, although it can be a mechanical device. The compensator can be placed in the forward path, in which case it is called a *cascade* or *series compensator*, or it can be placed in the feedback path, in which case it is called a *feedback* or *parallel compensator*. We will be concerned with the first type only.

A commonly used compensator is the electrical circuit shown in Fig. 5.29, in which $v_i(t)$ and $v_o(t)$ are input and output voltages, respectively, $i_1(t)$, $i_2(t)$ and $i(t)$ are currents, C is the capacitance and R_1 and R_2 are

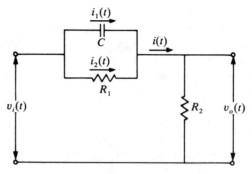

Figure 5.29

resistances. Using Kirchhoff's current and voltage laws, it can be shown that the relation between the input and output voltages is given by the differential equation

$$C\frac{dv_i(t)}{dt} + \frac{1}{R_1}v_i(t) = C\frac{dv_o(t)}{dt} + \frac{R_1 + R_2}{R_1 R_2}v_o(t) \qquad (5.99)$$

It is not difficult to verify that the transfer function of the compensator is

$$G_c(s) = \frac{V_o(s)}{V_i(s)} = \frac{Cs + 1/R_1}{Cs + (R_1 + R_2)/R_1 R_2} \qquad (5.100)$$

Introducing the notation

$$\frac{1}{CR_1} = a, \quad \frac{R_1 + R_2}{CR_1 R_2} = b \qquad (5.101)$$

the transfer function can be written in the form

$$G_c(s) = \frac{s + a}{s + b} \qquad (5.102)$$

The frequency response is obtained by substituting $s = i\omega$ in Eq. (5.102), or

$$G_c(i\omega) = \frac{a}{b}\frac{1 + i\omega/a}{1 + i\omega/b} = \frac{a}{b}\frac{1 + (\omega^2/ab) + i\omega(b - a)/ab}{1 + \omega^2/b^2} \qquad (5.103)$$

From Eqs. (5.101), we observe that $b > a$. Hence, recalling Eq. (5.52), we conclude that the phase angle of the compensator is positive. For this reason the compensator shown in Fig. 5.29 is referred to as a *phase-lead compensator*, or simply as a *lead compensator*.

COMPENSATORS   173

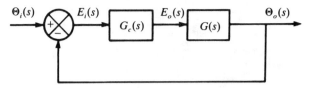

Figure 5.30

Next, let us use the above lead compensator to improve the performance of a closed-loop system. The block diagram for the compensated system is shown in Fig. 5.30. The forward-path transfer function of the uncompensated system is assumed to have the form

$$G(s) = \frac{4}{s(s+1)(s+2)} \qquad (5.104)$$

From Fig. 5.30, the open-loop transfer function of the compensated system is

$$\frac{\Theta_o(s)}{E_i(s)} = G_c(s)G(s) = \frac{s+a}{s+b} \frac{4}{s(s+1)(s+2)} \qquad (5.105)$$

so that the effect of adding the compensator is to add one zero and one pole to the open-loop transfer function. The problem of designing a compensator to produce satisfactory performance reduces to selecting the parameters a and b.

The compensator design can be carried out by means of Nyquist plots or by means of Bode diagrams. Because Bode diagrams of the compensated system can be obtained by drawing Bode diagrams of the uncompensated system and of the compensator separately and adding them graphically, we choose the later approach. Of course, compensator design can also be carried out by means of Nichols plots.

As can be seen from Fig. 5.31, the uncompensated system has a gain margin of approximately 3.5 dB and a phase margin of approximately 11.5°. The object is to design a lead compensator so that the compensated system will have a gain margin exceeding 15 dB and a phase margin exceeding 70°. The design amounts to trying a variety of pairs of values for a and b. To illustrate the procedure, we choose $a = 0.5$ and $b = 1$, $a = 0.5$ and $b = 2$ and $a = 0.5$ and $b = 5$. From the figure, we see that the gain and phase margin for $a = 0.5$ and $b = 2$ are approximately 16 dB and 75°, respectively, so that a lead compensator with these values for a and b yields satisfactory performance.

Another type of compensating circuit is displayed in Fig. 5.32. It can be shown to have the transfer function

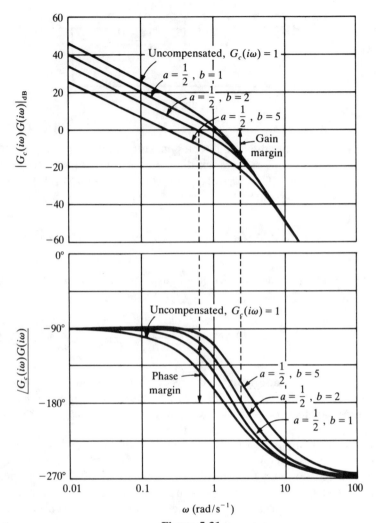

Figure 5.31

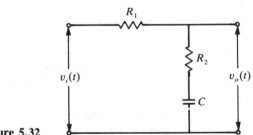

Figure 5.32

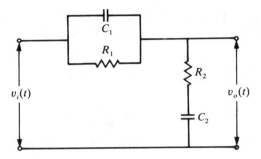

Figure 5.33

$$G_c(s) = \frac{a}{b}\frac{s+b}{s+a} \tag{5.106}$$

where a and b are as defined by Eqs. (5.101). This compensator is known as a *lag compensator*. Yet another type is the *lead-lag compensator* shown in Fig. 5.33. It has the transfer function

$$G_c(s) = \frac{(s+a_1)(s+b_2)}{(s+a_2)(s+b_1)} \tag{5.107}$$

where $a_1 = 1/R_1C_1$, $b_2 = 1/R_2C_2$, $b_1 + a_2 = a_1 + b_2 + 1/R_2C_1$, $a_2b_1 = a_1b_2$, and we note that $b_1 > a_1$ and $b_2 > a_2$.

A more extensive discussion of compensators is presented in the book by D'Azzo and Houpis [D5].

5.12 SOLUTION OF THE STATE EQUATIONS BY THE LAPLACE TRANSFORMATION

Implicit throughout this chapter has been the assumption of a single control variable. At this point, we turn out attention to multivariable systems. To this end, we consider the state description of the equations of motion.

In Section 3.7, we have shown that the state equations for a linear system can be written in the matrix form

$$\dot{\mathbf{x}}(t) = A\mathbf{x}(t) + B\mathbf{u}(t) \tag{5.108}$$

where $\mathbf{x}(t)$ is the n-dimensional state vector, $\mathbf{u}(t)$ is the r-dimensional input vector and A and B are $n \times n$ and $n \times r$ matrices of coefficients, respectively. Moreover, in Section 3.13, we have shown that the output equations can be written as

$$\mathbf{y}(t) = C\mathbf{x}(t) \tag{5.109}$$

where $y(t)$ is the q-dimensional output vector and C is a $q \times n$ matrix of coefficients. We consider the case in which the matrices A, B and C are constant.

We propose to solve Eqs. (5.108) and (5.109) by means of the Laplace transformation. Hence, transforming Eq. (5.108), we obtain

$$s\mathbf{X}(s) - \mathbf{x}(0) = A\mathbf{X}(s) + B\mathbf{U}(s) \qquad (5.110)$$

where $\mathbf{X}(s) = \mathscr{L}\mathbf{x}(t)$ and $\mathbf{U}(s) = \mathscr{L}\mathbf{u}(t)$ are the transformed state and input vectors, respectively, and $\mathbf{x}(0)$ is the initial state vector. Solving for $\mathbf{X}(s)$, we have

$$\mathbf{X}(s) = [sI - A]^{-1}\mathbf{x}(0) + [sI - A]^{-1}B\mathbf{U}(s) \qquad (5.111)$$

Similarly, letting $\mathbf{Y}(s) = \mathscr{L}\mathbf{y}(t)$ be the transformed output vector, transforming Eq. (5.109) and using Eq. (5.111), we can write

$$\mathbf{Y}(s) = C\Phi(s)\mathbf{x}(0) + C\Phi(s)B\mathbf{U}(s) \qquad (5.112)$$

where

$$\Phi(s) = [sI - A]^{-1} \qquad (5.113)$$

is the resolvent of A. Equation (5.112) is the Laplace domain counterpart of the equation

$$\mathbf{y}(t) = C\Phi(t)\mathbf{x}(0) + C\int_0^t \Phi(t - \tau)B\mathbf{u}(\tau)\,d\tau$$

$$= Ce^{At}\mathbf{x}(0) + C\int_0^t e^{A(t-\tau)}B\mathbf{u}(\tau)\,d\tau \qquad (5.114)$$

which can be obtained by inserting Eq. (3.94) into Eq. (3.133). Comparing Eqs. (5.112) and (5.114), we conclude that the transition matrix can be obtained from

$$e^{At} = \mathscr{L}^{-1}\{[sI - A]^{-1}\} \qquad (5.115)$$

For $\mathbf{x}(0) = \mathbf{0}$, Eq. (5.111) reduces to

$$\mathbf{Y}(s) = G(s)\mathbf{U}(s) \qquad (5.116)$$

where

$$G(s) = C\Phi(s)B = C[sI - A]^{-1}B \qquad (5.117)$$

SOLUTION OF THE STATE EQUATIONS BY THE LAPLACE TRANSFORMATION

is the matrix of transfer functions of the system and is sometimes referred to as the *transfer matrix*. Note that the entry $G_{ij}(s)$ of the transfer matrix is the transfer function relating the ith component of the output to the jth component of the input.

The transfer matrix can be written in the form

$$G(s) = \frac{1}{\det[sI - A]} P(s) \qquad (5.118)$$

where $P(s)$ is a matrix whose entries are polynomials in s. Hence, the elements of $G(s)$ are rational functions of s. The equation

$$\det[sI - A] = 0 \qquad (5.119)$$

is recognized as the characteristic equation of A and its roots are the poles of the transfer matrix. Unless cancellation of factors of the forms $s - \lambda_i$ occurs in *all* the entries of $G(s)$, where λ_i is an eigenvalue of A, the poles of the transfer matrix $G(s)$ are the poles of the system.

Computation of the transition matrix by means of Eq. (5.115) is feasible for relatively small values of n. For larger values of n, one may wish to consider a procedure based on the Leverrier algorithm. To this end, we write the characteristic polynomial associated with the matrix A in the form

$$\det[sI - A] = s^n + a_1 s^{n-1} + \cdots + a_{n-1} s + a_n \qquad (5.120)$$

Then, the resolvent of A can be shown to have the form [V6]

$$[sI - A]^{-1} = \frac{1}{\det(sI - A)} \sum_{i=1}^{n} s^{n-i} H_{i-1}, \quad H_0 = I \qquad (5.121)$$

where the matrices H_i and the coefficients a_i ($i = 1, 2, \ldots, n$) can be obtained recursively from

$$a_i = -\frac{1}{i} \operatorname{tr}[AH_{i-1}], \quad i = 1, 2, \ldots, n \qquad (5.122a)$$

$$H_i = AH_{i-1} + a_i I, \quad i = 1, 2, \ldots, n - 1 \qquad (5.122b)$$

Designing compensators for systems with multiple inputs and outputs in the frequency domain is considerably more involved than for systems with a single input and output. For a discussion of the use of frequency-domain methods for multivariable systems the reader is referred to the book by Rosenbrock [R14].

CHAPTER 6

CONTROL OF LUMPED-PARAMETER SYSTEMS. MODERN APPROACH

In Chapter 5, we studied classical control of lumped-parameter systems. The principal tool in classical control is the Laplace transformation, with the analysis and design being carried out for the most part in the s-domain or in the frequency domain. The approach makes extensive use of the concept of transfer functions and is suitable for single-input, single-output systems. For multi-input, multi-output systems, one must consider a transfer function for every pair of inputs and outputs and the complexity increases significantly beyond a few inputs and outputs. Moreover, the approach is limited to linear systems with constant coefficients.

Classical control uses time-domain and frequency-domain performance criteria such as rise time, maximum overshoot, settling time, steady-state error, gain margin, phase margin, maximum magnitude and bandwidth. With the advent of the space age in the 1950s, the interest turned to a new class of performance criteria, such as minimum time and minimum fuel. These performance criteria lead to nonlinear control laws. Other prerformance criteria attracting a great deal of attention are quadratic performance criteria, leading to linear control laws. The approach consisting of the minimization of a performance index is known as optimal control and falls in the domain of *modern control*. Major advances in optimal control are Bellman's dynamic programming and Pontryagin's minimum principle, where the latter can be regarded as a generalization of the fundamental theorem of the calculus of variations. A control approach developed in the late 1960s is known as pole allocation, and can also be considered as part of modern control. Modern control should not be regarded as replacing classical control but complementing it, as a blend of the two can enhance understanding of the problem.

CONTROL OF LUMPED-PARAMETER SYSTEMS. MODERN APPROACH

Modern control is essentially a time-domain approach, based on the state-space description of the behavior of dynamical systems, where the concept of state space can be regarded as an extension of the concept of phase space of the Hamiltonian mechanics introduced in Chapter 2. The mathematical tools of modern control are linear algebra, including matrix theory, and calculus of variations. This chapter is devoted to modern control.

6.1 FEEDBACK CONTROL SYSTEMS

Recalling Eqs. (3.60) and (3.132), we can write the state equations and output equations in the matrix form

$$\dot{\mathbf{x}}(t) = A\mathbf{x}(t) + B\mathbf{u}(t) \tag{6.1}$$

and

$$\mathbf{y}(t) = C\mathbf{x}(t) + D\mathbf{u}(t) \tag{6.2}$$

respectively, where the various quantities are as defined earlier. Equations (6.1) and (6.2) can be represented by the block diagram of Fig. 6.1. The diagram is for the case of *open-loop control*, i.e., the case in which the system input $\mathbf{u}(t)$ is independent of the state $\mathbf{x}(t)$ or the output $\mathbf{y}(t)$. In the case of open-loop control, the input $\mathbf{u}(t)$ depends solely on the expected behavior of the system and not on the actual behavior.

When the external disturbances are relatively insignificant, open-loop control can perform very well. Quite often, however, better results can be obtained with *feedback control*, in which the control $\mathbf{u}(t)$ takes into consideration the actual response of the system instead of the expected response. We distinguish two types of feedback control. In the first case, the input depends on the state, or

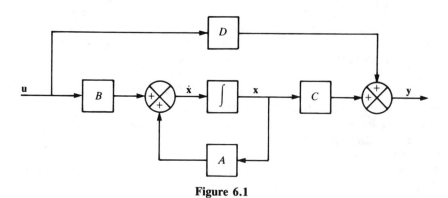

Figure 6.1

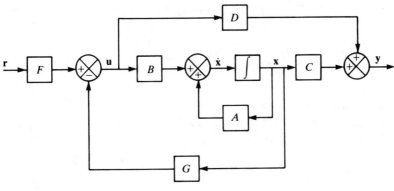

Figure 6.2

$$\mathbf{u}(t) = -G\mathbf{x}(t) + F\mathbf{r}(t) \tag{6.3}$$

where G is known as the *feedback gain matrix*, or the *control gain matrix*, F is a *feedforward matrix* and $\mathbf{r}(t)$ is a *reference input*, or *command input*. This case is called *state feedback control* and is displayed in the block diagram of Fig. 6.2. In the second case, the input depends on the output, or

$$\mathbf{u}(t) = -G\mathbf{y}(t) + F\mathbf{r}(t) \tag{6.4}$$

This case is known as *output feedback control*, and the corresponding block diagram is shown in Fig. 6.3. In contrast to Fig. 6.1, Figs. 6.2 and 6.3 represent *closed-loop control*.

Introducing Eq. (6.3) into Eqs. (6.1) and (6.2), we obtain for state feedback control

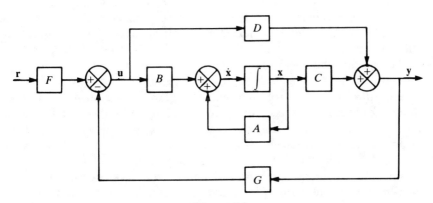

Figure 6.3

$$\dot{\mathbf{x}}(t) = (A - BG)\mathbf{x}(t) + B F \mathbf{r}(t) \tag{6.5}$$

and

$$\mathbf{y}(t) = (C - DG)\mathbf{x}(t) + DF\mathbf{r}(t) \tag{6.6}$$

On the other hand, inserting Eq. (6.4) into Eqs. (6.1) and (6.2) and considering the identity $I - G(I + DG)^{-1}D \equiv (I + GD)^{-1}$, the equations for output feedback control reduce to

$$\dot{\mathbf{x}}(t) = [A - BG(I + DG)^{-1}C]\mathbf{x}(t) + B(I + GD)^{-1}F\mathbf{r}(t) \tag{6.7}$$

and

$$\mathbf{y}(t) = (I + DG)^{-1}[C\mathbf{x}(t) + DF\mathbf{r}(t)] \tag{6.8}$$

One of the most important characteristics in control is stability. As pointed out in Section 3.8, the stability of a linear system is determined by the eigenvalues of the matrix of coefficients. More specifically, the system is merely stable if all the eigenvalues are pure imaginary, asymptotically stable if the eigenvalues are real and negative, or complex with negative real part, and unstable if at least one eigenvalue is real and positive, or complex with positive real part. If the eigenvalues are plotted on a complex plane, then the left half of the plane implies asymptotic stability, the imaginary axis mere stability and the right half of the plane instability. In the case of open-loop control, stability is governed by the eigenvalues of A. On the other hand, in the case of state feedback control stability is governed by the eigenvalues of $A - BG$ and in the case of output feedback control by the eigenvalues of $A - BG(I + DG)^{-1}C$. Consistent with this, the eigenvalues of A are known as *open-loop eigenvalues* and the eigenvalues of $A - BG$, or the eigenvalues of $A - BG(I + DG)^{-1}C$, are known as *closed-loop eigenvalues*. Hence, the object of linear feedback control is to ensure that the closed-loop eigenvalues lie in the left half of the complex plane, thus guaranteeing asymptotic stability. Note that open-loop and closed-loop eigenvalues are also known as *open-loop* and *closed loop poles*, respectively.

Assuming that the coefficient matrices are given, we conclude from Eqs. (6.5) and (6.7) that the closed-loop poles depend on the control gains, i.e., on the entries of the control gain matrix G. Two of the most widely used methods for computing control gains are pole allocation and optimal control. We discuss these methods in the following sections. To this end, we concentrate on state feedback. Note that, in the case in which the output is of lower dimension than the state, it is possible to estimate the state by means of a state estimator, or observer and to feed back the estimated state instead of the actual state. We discuss deterministic observers in Section 6.12 and stochastic observers in Section 6.13.

6.2 POLE ALLOCATION METHOD

As pointed out in Section 6.1, the object of linear feedback control is to place the closed-loop poles on the left half of the complex plane of the eigenvalues so as to ensure asymptotic stability of the closed-loop system. One approach consists of prescribing first the closed-loop poles associated with the modes to be controlled and then computing the control gains required to produce these poles. Because this amounts to controlling a system by controlling its modes, this approach is known as *modal control*. The algorithm for producing the control gains is known as *pole allocation*, *pole assignment*, or *pole placement*. The pole allocation method has been presented in a variety of forms [see, e.g., B37, D2, P5, S31, W12]. In this text, we follow the approach of Porter and Crossley [P5].

The complexity of the procedure depends on the number of inputs. In fact, the number of inputs affects the details of the procedure to such an extent that there are in essence different algorithms corresponding to different cases. We identify the cases: single-input control, dyadic control, which represents a special case of multi-input control, and general multi-input control.

6.2.1 Single-Input Control

The state equations can be written in the vector form

$$\dot{\mathbf{x}}(t) = A\mathbf{x}(t) + \mathbf{b}u(t) \tag{6.9}$$

where $\mathbf{b}$ is a constant n-vector and $u(t)$ is the control input. From Section 3.11, the open-loop eigensolution consists of the eigenvalues $\lambda_1, \lambda_2, \ldots, \lambda_n$ and the right and left eigenvectors $\mathbf{u}_1, \mathbf{u}_2, \ldots, \mathbf{u}_n$ and $\mathbf{v}_1, \mathbf{v}_2, \ldots, \mathbf{v}_n$, respectively. The two sets of eigenvectors are assumed to be normalized, so that they satisfy the biorthonormality relations $\mathbf{v}_j^T \mathbf{u}_i = \delta_{ij}$ $(i, j = 1, 2, \ldots, n)$.

We consider control of m modes and assume that the control force has the form

$$u(t) = -\sum_{j=1}^{m} g_j \mathbf{v}_j^T \mathbf{x}(t) \tag{6.10}$$

where g_j $(j = 1, 2, \ldots, m)$ are modal control gains. Inserting Eq. (6.10) into Eq. (6.9), we obtain the closed-loop equation

$$\dot{\mathbf{x}}(t) = C\mathbf{x}(t) \tag{6.11}$$

where

$$C = A - \mathbf{b} \sum_{j=1}^{m} g_j \mathbf{v}_j^T \tag{6.12}$$

and we observe that

$$C\mathbf{u}_j = A\mathbf{u}_j = \lambda_j \mathbf{u}_j, \quad j = m+1, m+2, \ldots, n \tag{6.13}$$

so that the control given by Eq. (6.10) is such that the closed-loop eigenvalues and eigenvectors corresponding to the uncontrolled modes are equal to the open-loop eigenvalues and eigenvectors, respectively. It can be verified that the same is not true for the eigenvalues and eigenvectors associated with the controlled modes, $1 \le j \le m$.

Next, we denote the closed-loop eigenvalues and right eigenvectors of C associated with the controlled modes by ρ_j and $\mathbf{w}_j$ ($j = 1, 2, \ldots, m$), respectively. But, because the open-loop right eigenvectors are linearly independent, they can be used as a basis for an n-vector space, so that the closed-loop eigenvectors can be expanded in terms of the open-loop eigenvectors as follows:

$$\mathbf{w}_j = \sum_{k=1}^{n} d_{jk} \mathbf{u}_k, \quad j = 1, 2, \ldots, m \tag{6.14}$$

Recalling that $\mathbf{v}_j^T \mathbf{u}_k = \delta_{jk}$, the closed-loop eigenvalue problem can be written in the form

$$C\mathbf{w}_j = \left(A - \mathbf{b} \sum_{l=1}^{m} g_l \mathbf{v}_l^T \right) \sum_{k=1}^{n} d_{jk} \mathbf{u}_k = \sum_{k=1}^{n} d_{jk} \lambda_k \mathbf{u}_k - \mathbf{b} \sum_{l=1}^{m} g_l d_{jl}$$

$$= \rho_j \mathbf{w}_j = \rho_j \sum_{k=1}^{n} d_{jk} \mathbf{u}_k, \quad j = 1, 2, \ldots, m \tag{6.15}$$

Moreover, letting

$$\mathbf{b} = \sum_{k=1}^{n} p_k \mathbf{u}_k \tag{6.16}$$

Eqs. (6.15) become

$$\sum_{k=1}^{n} (\rho_j - \lambda_k) d_{jk} \mathbf{u}_k + \sum_{k=1}^{n} p_k \sum_{l=1}^{m} d_{jl} g_l \mathbf{u}_k = \mathbf{0}, \quad j = 1, 2, \ldots, m \tag{6.17}$$

which are equivalent to the $n \times m$ scalar equations

$$(\rho_j - \lambda_k)d_{jk} + p_k \sum_{l=1}^{m} d_{jl}g_l = 0, \quad j=1,2,\ldots,m; \quad k=1,2,\ldots,n \tag{6.18}$$

It can be shown [P5] that Eqs. (6.18) have a solution if

$$\begin{vmatrix} \rho_j - \lambda_1 + p_1 g_1 & p_1 g_2 & \cdots & p_1 g_m \\ p_2 g_1 & \rho_j - \lambda_2 + p_2 g_2 & \cdots & p_2 g_m \\ \vdots & \vdots & & \vdots \\ p_m g_1 & p_m g_2 & \cdots & \rho_j - \lambda_m + p_m g_m \end{vmatrix} = 0,$$

$$j = 1, 2, \ldots, m \tag{6.19}$$

Expanding the determinant, we obtain

$$\prod_{k=1}^{m}(\rho_j - \lambda_k) = -\sum_{l=1}^{m} p_l g_l \prod_{\substack{k=1 \\ k \neq l}}^{m}(\rho_j - \lambda_k), \quad j=1,2,\ldots,m \tag{6.20}$$

which reduce to

$$-\sum_{k=1}^{m} \frac{p_k g_k}{\rho_j - \lambda_k} = 1, \quad j=1,2,\ldots,m \tag{6.21}$$

where it is implied that the closed-loop eigenvalues have been selected so that $\rho_j \neq \lambda_k$ ($j, k = 1, 2, \ldots, m$). Solving Eqs. (6.21), we obtain the gains

$$g_j = -\frac{\prod_{k=1}^{m}(\rho_k - \lambda_j)}{p_j \prod_{\substack{k=1 \\ k \neq j}}^{m}(\lambda_k - \lambda_j)}, \quad j=1,2,\ldots,m \tag{6.22}$$

Clearly, for g_j to exist, we must have $p_j \neq 0$. If any one of the p_j is zero, then the associated mode is not controllable. Finally, the control law is obtained by inserting Eqs. (6.22) into Eq. (6.10).

Equation (6.22) can be used to compute the gains for control of any arbitrary number of modes, $m = 1, 2, \ldots, n$. If the interest lies in controlling a single mode, say mode j, then the control gain reduces to

$$g_j = -\frac{\rho_j - \lambda_j}{p_j} \tag{6.23}$$

This assumes, of course, that the jth mode is controllable, which requires that the right eigenvector $\mathbf{u}_j$ be represented in the vector $\mathbf{b}$, Eq. (6.16).

6.2.2 Dyadic Control

In this case, multi-input control is treated as if it were single-input control. We consider the case in which the state equations have the form (6.1) and assume that the control vector has the expression

$$\mathbf{u}(t) = -G\mathbf{x}(t) \tag{6.24}$$

In general, the computation of the $r \times n$ control gain matrix G requires the solution of nonlinear algebraic equations. This case will be discussed later in this section. At this point, we consider the special case in which G is assumed to have the dyadic form

$$G = \mathbf{f}\mathbf{d}^T \tag{6.25}$$

where $\mathbf{f}$ is an r-vector and $\mathbf{d}$ is an n-vector. Inserting Eqs. (6.24) and (6.25) into Eq. (6.1), we obtain

$$\dot{\mathbf{x}} = A\mathbf{x} - B\mathbf{f}\mathbf{d}^T\mathbf{x} \tag{6.26}$$

Comparing Eqs. (6.9) and (6.26), we conclude that the case in which the gain matrix has the form (6.25) can be treated as single-input control by writing

$$B\mathbf{f} = \mathbf{b} \tag{6.27}$$

and

$$-\mathbf{d}^T\mathbf{x}(t) = u(t) \tag{6.28}$$

Then, comparing Eqs. (6.10) and (6.28), we obtain

$$\mathbf{d}^T = \sum_{j=1}^{m} g_j \mathbf{v}_j^T \tag{6.29}$$

Next, recalling Eq. (6.16) and using Eq. (6.27), we can write

$$B\mathbf{f} = \sum_{k=1}^{n} p_k \mathbf{u}_k \tag{6.30}$$

Multiplying Eq. (6.30) on the left by $\mathbf{v}_j^T$ and recalling the biorthonormality conditions on the right and left eigenvectors, we obtain

$$p_j = \mathbf{v}_j^T B\mathbf{f}, \quad j = 1, 2, \ldots, n \tag{6.31}$$

Finally, using Eq. (6.22), (6.25), (6.29) and (6.31), we obtain the gain matrix

$$G = -\mathbf{f} \sum_{j=1}^{m} \frac{\prod_{k=1}^{m}(\rho_k - \lambda_j)\mathbf{v}_j^T}{\mathbf{v}_j^T B \mathbf{f} \prod_{\substack{k=1 \\ k \neq j}}^{m}(\lambda_k - \lambda_j)} \qquad (6.32)$$

It should be pointed out that dyadic control is not a genuine multi-input control, because the gains are constrained according to Eq. (6.25). Indeed, Eq. (6.25) represents an outer product of two vectors. As a result, every row of G is proportional to $\mathbf{d}^T$. But, $\mathbf{d}^T\mathbf{x}(t)$ represents a given linear combination of the states. It follows that every component of the control vector $\mathbf{u}(t)$ is proportional to the same linear combination of the states, where the proportionality constant is the corresponding component of the vector $\mathbf{f}$. Hence, the vector $\mathbf{f}$ is selected according to the relative emphasis one wishes to place on the components of the control vector $\mathbf{u}(t)$.

6.2.3 General Multi-Input Control

Let us consider once again Eq. (6.1) and write it as follows:

$$\dot{\mathbf{x}}(t) = A\mathbf{x}(t) + B\mathbf{u}(t) = A\mathbf{x}(t) + \sum_{i=1}^{r} \mathbf{b}_i u_i(t) \qquad (6.33)$$

where $\mathbf{b}_i$ are n-vectors and $u_i(t)$ are associated control inputs ($i = 1, 2, \ldots, r$). Then, by analogy with Eq. (6.10) for single-input control, if the interest lies in controlling m modes, we assume that the inputs have the form

$$u_i(t) = -\sum_{j=1}^{m} g_{ij} \mathbf{v}_j^T \mathbf{x}(t), \quad i = 1, 2, \ldots, r \qquad (6.34)$$

Inserting Eqs. (6.34) into Eq. (6.33), we obtain the closed-loop equation

$$\dot{\mathbf{x}}(t) = A\mathbf{x}(t) - \sum_{i=1}^{r} \mathbf{b}_i \sum_{j=1}^{m} g_{ij} \mathbf{v}_j^T \mathbf{x}(t) = C\mathbf{x}(t) \qquad (6.35)$$

where

$$C = A - \sum_{i=1}^{r} \mathbf{b}_i \sum_{j=1}^{m} g_{ij} \mathbf{v}_j^T \qquad (6.36)$$

is the closed-loop coefficient matrix.

As in the preceding cases, we denote the closed-loop eigenvalues associated with the controlled modes by ρ_j and the corresponding closed-loop eigenvectors by $\mathbf{w}_j$ ($j = 1, 2, \ldots, m$), where $\mathbf{w}_j$ are given by Eqs. (6.14). Introducing Eqs. (6.14) into the closed-loop eigenvalue problem and using Eq. (6.36), we obtain

$$C\mathbf{w}_j = \left(A - \sum_{r=1}^{r} \mathbf{b}_i \sum_{l=1}^{m} g_{il}\mathbf{v}_l^T\right) \sum_{k=1}^{n} d_{jk}\mathbf{u}_k = \sum_{k=1}^{n} \lambda_k d_{jk}\mathbf{u}_k - \sum_{i=1}^{r} \mathbf{b}_i \sum_{l=1}^{m} g_{il} d_{jl} \quad (6.37)$$

Using the analogy with Eq. (6.16), we can write the expansions

$$\mathbf{b}_i = \sum_{k=1}^{n} p_{ki}\mathbf{u}_k, \quad i = 1, 2, \ldots, r \quad (6.38)$$

so that, inserting Eqs. (6.38) into Eq. (6.37), we have

$$\sum_{k=1}^{n} (\rho_j - \lambda_k) d_{jk}\mathbf{u}_k + \sum_{k=1}^{n} \sum_{i=1}^{r} \sum_{l=1}^{m} p_{ki} g_{il} d_{jl}\mathbf{u}_k = \mathbf{0}, \quad j = 1, 2, \ldots, m \quad (6.39)$$

which are equivalent to the $n \times m$ scalar equations

$$(\rho_j - \lambda_k) d_{jk} - \sum_{i=1}^{r} \sum_{l=1}^{m} p_{ki} g_{il} d_{jl} = 0, \quad j = 1, 2, \ldots, m; \quad k = 1, 2, \ldots, n \quad (6.40)$$

It can be shown [P5] that Eqs. (6.40) have a solution if

$$\begin{vmatrix} \rho_j - \lambda_1 + \sum_{i=1}^{r} p_{1i}g_{i1} & \sum_{i=1}^{r} p_{1i}g_{i2} & \cdots & \sum_{i=1}^{r} p_{1i}g_{im} \\ \sum_{i=1}^{r} p_{2i}g_{i1} & \rho_j - \lambda_2 + \sum_{i=1}^{r} p_{2i}g_{i2} & \cdots & \sum_{i=1}^{r} p_{2i}g_{im} \\ \vdots & \vdots & & \vdots \\ \sum_{i=1}^{r} p_{mi}g_{i1} & \sum_{i=1}^{r} p_{mi}g_{i2} & \cdots & \rho_j - \lambda_m + \sum_{i=1}^{r} p_{mi}g_{im} \end{vmatrix} = 0,$$

$$j = 1, 2, \ldots, m \quad (6.41)$$

Equations (6.41) represent m equations in $m \times r$ unknowns, namely the modal control gains g_{ij} ($i = 1, 2, \ldots, r$; $j = 1, 2, \ldots, m$). Because Eqs. (6.41) are in general nonlinear and, moreover, they are underdetermined, no unique solution exists.

It may be possible to obtain a solution of Eqs. (6.41) by augmenting the equations with a set of $m \times (r-1)$ algebraic equations involving the gains g_{ij}, thus yielding $m \times r$ equations. Two schemes representing minimum-gain modal controllers can be found in the book by Porter and Crossley [P5].

The problem may be also approached by placing one pole at a time. This approach is likely to be tedious, as it requires repeated solution of an

POLE ALLOCATION METHOD

eigenvalue problem. This can be explained by the fact that, as soon as a pole has been placed, the system is regarded as open-loop as far as the placing of the next pole is concerned. Hence, before a new pole is placed, it is necessary to solve eigenvalue problems for the new open-loop poles and the associated right and left eigenvectors.

Example 6.1 Consider the fourth-order system shown in Fig. 6.4 and shift the first two poles by an amount -0.1000 in the complex plane by means of a single input acting on m_2. Use the values $m_1 = 1$, $m_2 = 2$, $k_1 = k_2 = k$. Compute the control gains and the closed-loop coefficient matrix.

The mass and stiffness matrices are

$$M = \begin{bmatrix} m_1 & 0 \\ 0 & m_2 \end{bmatrix} = \begin{bmatrix} 1 & 0 \\ 0 & 2 \end{bmatrix} \quad (a)$$

$$K = \begin{bmatrix} k_1 + k_2 & -k_2 \\ -k_2 & k_2 \end{bmatrix} = \begin{bmatrix} 2 & -1 \\ -1 & 1 \end{bmatrix} \quad (b)$$

so that the coefficient matrix is

$$A = \begin{bmatrix} 0 & | & I \\ \hline -M^{-1}K & | & 0 \end{bmatrix} = \begin{bmatrix} 0 & 0 & 1 & 0 \\ 0 & 0 & 0 & 1 \\ -2 & 1 & 0 & 0 \\ 0.5 & -0.5 & 0 & 0 \end{bmatrix} \quad (c)$$

The state vector is

$$\mathbf{x}(t) = [q_1(t) \quad q_2(t) \quad \dot{q}_1(t) \quad \dot{q}_2(t)]^T \quad (d)$$

and the vector **b** in Eq. (6.9) is

$$\mathbf{b} = \begin{bmatrix} 0 \\ \hline M^{-1} \end{bmatrix} \begin{bmatrix} 0 \\ 1 \end{bmatrix} = \begin{bmatrix} 0 & 0 \\ 0 & 0 \\ 1 & 0 \\ 0 & 0.5 \end{bmatrix} \begin{bmatrix} 0 \\ 1 \end{bmatrix} = [0 \quad 0 \quad 0 \quad 0.5]^T \quad (e)$$

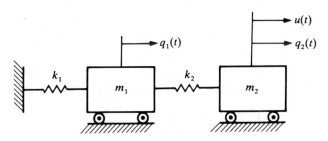

Figure 6.4

The eigenvalues of the matrix A can be shown to be

$$\lambda_1 = i0.4682, \quad \lambda_2 = -i0.4682, \quad \lambda_3 = i1.5102, \quad \lambda_4 = -i1.5102 \quad \text{(f)}$$

and the matrices of right and left eigenvectors are

$$U = \begin{bmatrix} -i0.7501 & i0.7501 & -i0.4573 & i0.4573 \\ -i1.3358 & i1.3358 & i0.1284 & -i0.1284 \\ 0.3512 & 0.3512 & 0.6906 & 0.6906 \\ 0.6254 & 0.6254 & -0.1939 & -0.1939 \end{bmatrix} \quad \text{(g)}$$

$$V = \begin{bmatrix} i0.0908 & -i0.0908 & i0.9445 & -i0.9445 \\ i0.3233 & -i0.3233 & -i0.5304 & i0.5304 \\ 0.1939 & 0.1939 & 0.6254 & 0.6254 \\ 0.6906 & 0.6906 & -0.3512 & -0.3512 \end{bmatrix} \quad \text{(h)}$$

respectively, where the eigenvectors have been normalized so that U and V satisfy Eqs. (3.111).

Equation (6.16) can be rewritten in the form

$$\mathbf{b} = U\mathbf{p} \quad \text{(i)}$$

so that

$$\mathbf{p} = V^T\mathbf{b} = [0.3453 \quad 0.3453 \quad -1.1756 \quad -0.1756]^T \quad \text{(j)}$$

Moreover, the closed-loop poles are

$$p_1 = -0.1000 + i0.4682, \quad p_2 = \bar{p}_1, \quad p_3 = \lambda_3, \quad p_4 = \lambda_4 \quad \text{(k)}$$

Using Eqs. (6.22), the gains corresponding to the placed poles have the values

$$\begin{aligned} g_1 &= -\frac{(p_1 - \lambda_1)(p_2 - \lambda_1)}{p_1(\lambda_2 - \lambda_1)} = -\frac{(-0.1000)(-0.1000 - i0.9364)}{0.3453(-i0.9364)} \\ &= -0.2896 + i0.0309 \\ g_2 &= \bar{g}_1 = -0.2896 - i0.0309 \end{aligned} \quad \text{(l)}$$

Finally, using Eq. (6.10), the control input is

$$\begin{aligned} u(t) &= -(g_1\mathbf{v}_1^T + g_2\mathbf{v}_2^T)\mathbf{x}(t) = -2\,\text{Re}\,g_1\mathbf{v}_1^T\mathbf{x}(t) \\ &= -2\,\text{Re}\,(-0.2896 + i0.0309)\begin{bmatrix} i0.0908 \\ i0.3233 \\ 0.1939 \\ 0.6906 \end{bmatrix}^T \mathbf{x}(t) \\ &= [0.0056 \quad 0.0200 \quad 0.1123 \quad 0.4000]\mathbf{x}(t) \end{aligned} \quad \text{(m)}$$

and, using Eq. (6.12), the closed-loop coefficient matrix is

$$C = A - \mathbf{b}(g_1 \mathbf{v}_1^T + g_2 \mathbf{v}_2^T) = \begin{bmatrix} 0 & 0 & 1 & 0 \\ 0 & 0 & 0 & 1 \\ -2 & 1 & 0 & 0 \\ 0.4972 & -0.5100 & -0.0562 & -0.2000 \end{bmatrix} \quad (n)$$

It is not difficult to verify that the above matrix C has the eigenvalues given by Eqs. (k).

Example 6.2 Consider the system of Example 6.1 and use dyadic control to compute the gains corresponding to the closed-loop poles

$$p_1 = -0.1000 + i0.4682, \quad p_2 = \bar{p}_1$$
$$p_3 = -0.1500 + i1.5102, \quad p_4 = \bar{p}_3 \quad (a)$$

Use the vector

$$\mathbf{f} = [1 \quad 1]^T \quad (b)$$

which implies that the components $u_1(t)$ and $u_2(t)$ of the control vector $\mathbf{u}(t)$ are to be equal to one another.

Because in this case there is an input on each mass,

$$B = \left[\frac{0}{M^{-1}} \right] = \begin{bmatrix} 0 & 0 \\ 0 & 0 \\ 1 & 0 \\ 0 & 0.5 \end{bmatrix} \quad (c)$$

so that, using Eqs. (a), (b) and (c) above in conjunction with the open-loop eigenvalues and left eigenvectors from Example 6.1, Eq. (6.32) yields the feedback control gain matrix

$$G = \begin{bmatrix} 0.1247 & -0.0645 & 0.4853 & 0.0294 \\ 0.1247 & -0.0645 & 0.4853 & 0.0294 \end{bmatrix} \quad (d)$$

6.3 OPTIMAL CONTROL

In Section 3.7, we showed that the state equations can be written in the general form

$$\dot{\mathbf{x}}(t) = \mathbf{a}(\mathbf{x}(t), \mathbf{u}(t), t) \quad (6.42)$$

where the n-vector $\mathbf{a}$ is a given function of the state vector $\mathbf{x}(t)$, the control vector $\mathbf{u}(t)$ and the time t. Then, the optimal control problem can be defined as follows: *Determine an admissible control* $\mathbf{u}^*(t)$ *causing the system* (6.42) *to follow an admissible trajectory* $\mathbf{x}^*(t)$ *in the state space that minimizes the performance measure*

$$J = h(\mathbf{x}(t_f), t_f) + \int_{t_0}^{t_f} g(\mathbf{x}(t), \mathbf{u}(t), t)\, dt \qquad (6.43)$$

in which h and g are given functions, t_0 is the *initial time* and t_f is the *final time*. The control $\mathbf{u}(t)$ and state time history $\mathbf{x}(t)$ are known as *optimal control* and *optimal trajectory*, respectively.

There is no guarantee that an optimal control exists, and if it does exist there is no guarantee that it is unique. Moreover, it is possible to define a variety of performance measures for a given system by choosing different functions h and g.

As soon as a performance measure has been selected, the task of minimizing it begins. There are two approaches to the minimization, namely, the method of *dynamic programming* due to Bellman [B23] and the *minimum principle* of Pontryagin [P4]. Dynamic programming is based on the *principle of optimality* [A18], which in effect states that a control policy that is optimal over a given time interval (t_0, t_f) is optimal over all subintervals (t, t_f). On the other hand, Pontryagin's minimum principle is a variational principle and it essentially states that an optimal control must minimize a given function known as the Hamiltonian. In this regard, it must be pointed out that the Hamiltonian arising in optimal control has a different definition than Eq. (2.39), as we will see shortly. In this text, we are following the variational approach.

The derivation of the necessary conditions for optimal control is relatively lengthy and is not presented here; for details of the derivations, the reader may wish to consult the work of Kirk [K5]. Here we summarize the main results only. To this end, we introduce the *Hamiltonian*

$$\mathcal{H}(\mathbf{x}(t), \mathbf{u}(t), \mathbf{p}(t), t) = g(\mathbf{x}(t), \mathbf{u}(t), t) + \mathbf{p}^T(t)[\mathbf{a}(\mathbf{x}(t), \mathbf{u}(t), t)] \qquad (6.44)$$

where the various quantities are as defined in Eqs. (6.42) and (6.43), with the exception of $\mathbf{p}(t)$, which is an n-vector of Lagrange's multipliers known as the *costate vector* and whose purpose is to ensure that Eq. (6.42) is taken into account in the minimization process. Then, the *necessary conditions* for optimal control are as follows:

$$\dot{\mathbf{x}}^*(t) = \frac{\partial \mathcal{H}(\mathbf{x}^*(t), \mathbf{u}^*(t), \mathbf{p}^*(t), t)}{\partial \mathbf{p}}, \quad t_0 < t < t_f \qquad (6.45a)$$

$$\dot{\mathbf{p}}^*(t) = -\frac{\partial \mathcal{H}(\mathbf{x}^*(t), \mathbf{u}^*(t), \mathbf{p}^*(t), t)}{\partial \mathbf{x}}, \quad t_0 < t < t_f \qquad (6.45b)$$

$$\frac{\partial \mathcal{H}(\mathbf{x}^*(t), \mathbf{u}^*(t), \mathbf{p}^*(t), t)}{\partial \mathbf{u}} = \mathbf{0}, \quad t_0 < t < t_f \qquad (6.45c)$$

and

$$\left[\frac{\partial h(\mathbf{x}^*(t_f), t_f)}{\partial \mathbf{x}} - \mathbf{p}^*(t_f)\right]^T \delta \mathbf{x}_f$$

$$+ \left[\mathcal{H}(\mathbf{x}^*(t_f), \mathbf{u}^*(t_f), \mathbf{p}^*(t_f), t_f) + \frac{\partial h(\mathbf{x}^*(t_f), t_f)}{\partial t}\right] \delta t_f = 0 \quad (6.45d)$$

Equation (6.45a) ensures that the optimal control and the optimal trajectory satisfy the state equations, Eq. (6.42). Equation (6.45b) represents the *costate equations* and results from the arbitrariness of the costate vector. Equation (6.45c) ensures the stationarity of the Hamiltonian. The equation is not valid when the admissible controls are bounded. Finally, Eq. (6.45d) implies that the final state $\mathbf{x}(t_f)$ and final time t_f are free. The equation must be modified if either $\mathbf{x}(t_f)$ or t_f is not free. Equation (6.45d) can be used to derive the boundary condition at the final time. We note in passing that Eqs. (6.45a) and (6.45b) resemble Hamilton's canonical equations, Eqs. (2.44).

6.4 THE LINEAR REGULATOR PROBLEM

The regulator problem is defined as the problem of designing a control input so as to drive the plant from some initial state to a constant final state. Another way of defining the regulator problem is as one in which the reference input is constant. Because a simple coordinate transformation can translate the origin of the state space to any constant point in the space, the regulator problem can be redefined as the problem of designing a control input so as to drive the plant to the zero state. The linear regulator problem is the one in which the control is a linear function of the state.

Let us consider a system described by the linear state equations

$$\dot{\mathbf{x}}(t) = A(t)\mathbf{x}(t) + B(t)\mathbf{u}(t) \quad (6.46)$$

The object is to determine an optimal control minimizing the quadratic performance measure

$$J = \frac{1}{2}\mathbf{x}^T(t_f)H\mathbf{x}(t_f) + \frac{1}{2}\int_{t_0}^{t_f}[\mathbf{x}^T(t)Q(t)\mathbf{x}(t) + \mathbf{u}^T(t)R(t)\mathbf{u}(t)]\,dt \quad (6.47)$$

where H and Q are real symmetric positive semidefinite matrices and R is a real symmetric positive definite matrix. Moreover, we assume that $\mathbf{x}(t_f)$ is free and t_f is fixed. The optimal control problem using the performance measure (6.47) can be interpreted as the problem of driving the initial state as close as possible to zero while placing a penalty on the control effort.

Comparing Eqs. (6.42) and (6.46) on the one hand and Eqs. (6.43) and (6.47) on the other hand and using Eq. (6.44), we obtain the Hamiltonian

$$\mathcal{H}(\mathbf{x}(t), \mathbf{u}(t), \mathbf{p}(t), t) = \frac{1}{2}[\mathbf{x}^T(t)Q(t)\mathbf{x}(t) + \mathbf{u}^T(t)R(t)\mathbf{u}(t)]$$
$$+ \mathbf{p}^T(t)[A(t)\mathbf{x}(t) + B(t)\mathbf{u}(t)] \qquad (6.48)$$

so that, from Eqs. (6.45a–c), the necessary conditions for optimality are

$$\dot{\mathbf{x}}^*(t) = \frac{\partial \mathcal{H}}{\partial \mathbf{p}} = A(t)\mathbf{x}^*(t) + B(t)\mathbf{u}^*(t) \qquad (6.49)$$

$$\dot{\mathbf{p}}^*(t) = -\frac{\partial \mathcal{H}}{\partial \mathbf{x}} = -Q(t)\mathbf{x}^*(t) - A^T(t)\mathbf{p}^*(t) \qquad (6.50)$$

$$0 = \frac{\partial \mathcal{H}}{\partial \mathbf{u}} = R(t)\mathbf{u}^*(t) + B^T(t)\mathbf{p}^*(t) \qquad (6.51)$$

Moreover, recognizing that $\delta t_f = 0$, Eq. (6.45d) in conjunction with the first term on the right side of Eq. (6.47) yields

$$\mathbf{p}^*(t_f) = H(t_f)\mathbf{x}^*(t_f) \qquad (6.52)$$

From Eq. (6.51), we can solve for the optimal control in terms of the costate to obtain

$$\mathbf{u}^*(t) = -R^{-1}(t)B^T(t)\mathbf{p}^*(t) \qquad (6.53)$$

It remains to determine the relation between the costate and the state. We assume that this relation is linear and of the form

$$\mathbf{p}^*(t) = K(t)\mathbf{x}^*(t) \qquad (6.54)$$

Differentiating Eq. (6.54) with respect to time, we have

$$\dot{\mathbf{p}}^*(t) = \dot{K}(t)\mathbf{x}^*(t) + K(t)\dot{\mathbf{x}}^*(t) \qquad (6.55)$$

so that, inserting Eqs. (6.49), (6.50), (6.53) and (6.54) into Eq. (6.55), we obtain

$$-Q(t)\mathbf{x}^*(t) - A^T(t)K(t)\mathbf{x}^*(t)$$
$$= \dot{K}(t)\mathbf{x}^*(t) + K(t)A(t)\mathbf{x}^*(t) - K(t)B(t)R^{-1}(t)B^T(t)K(t)\mathbf{x}^*(t) \qquad (6.56)$$

Equation (6.56) can be satisfied for all times provided

$$\dot{K}(t) = -Q(t) - A^T(t)K(t) - K(t)A(t) + K(t)B(t)R^{-1}(t)B^T(t)K(t) \qquad (6.57)$$

Equation (6.57) represents a matrix differential equation known as the *Riccati equation*. The solution $K(t)$, called the *Riccati matrix*, is subject to the boundary condition

$$K(t_f) = H(t_f) = H \qquad (6.58)$$

as can be verified from Eqs. (6.52) and (6.54). Finally, using Eqs. (6.53) and (6.54), we conclude that the optimal feedback control gain matrix has the form

$$G(t) = R^{-1}(t) B^T(t) K(t) \qquad (6.59)$$

Because $K(t)$ is an $n \times n$ matrix, Eq. (6.57) represents a set of n^2 nonlinear differential equations. Observing, however, that $K(t)$ and $K^T(t)$ satisfy the same equation, we conclude that the Riccati matrix is symmetric, $K^T(t) = K(t)$, so that we must solve only $n(n+1)/2$ equations. These equations can be integrated backward in time from $t = t_f$ to $t = t_0$ using the boundary condition (6.58).

The preceding developments are applicable to any linear dynamical system. In the case of structures, in which the upper half of the state vector is equal to the displacement vector and the lower half is equal to the velocity vector, as can be seen from Eq. (3.51), the upper half of the state equations represent mere kinematical identitites. Consistent with this, the upper half of the matrix B is a null matrix, which is a mere reflection of the fact that kinematical identities cannot be subjected to control inputs. Then, regarding the matrix $K(t)$ as consisting of four $(n/2) \times (n/2)$ submatrices $K_{ij}(t)$ $(i, j = 1, 2)$, we conclude from Eq. (6.59) that the submatrix $K_{11}(t)$ is not really needed, although it is routinely computed in solving the matrix Riccati equation.

6.5 ALGORITHMS FOR SOLVING THE RICCATI EQUATION

The Riccati equation, Eq. (6.57), is a nonlinear matrix differential equation and is likely to cause computational difficulties. It is possible, however, to use a matrix transformation obviating the need for solving nonlinear equations. To this end, we introduce the transformation

$$K(t) = E(t) F^{-1}(t) \qquad (6.60)$$

so that

$$\dot{K}(t) = \dot{E}(t) F^{-1}(t) + E(t) \dot{F}^{-1}(t) \qquad (6.61)$$

To compute $\dot{F}^{-1}(t)$, we consider

196 CONTROL OF LUMPED-PARAMETER SYSTEMS. MODERN APPROACH

$$F^{-1}(t)F(t) = I \qquad (6.62)$$

so that

$$\dot{F}^{-1}(t)F(t) + F^{-1}(t)\dot{F}(t) = 0 \qquad (6.63)$$

from which it follows that

$$\dot{F}^{-1}(t) = -F^{-1}(t)\dot{F}(t)F^{-1}(t) \qquad (6.64)$$

Hence,

$$\dot{K}(t) = \dot{E}(t)F^{-1}(t) - E(t)F^{-1}(t)\dot{F}(t)F^{-1}(t) \qquad (6.65)$$

Introducing Eqs. (6.60) and (6.65) into Eq. (6.57), we obtain

$$\dot{E}(t)F^{-1}(t) - E(t)F^{-1}(t)\dot{F}(t)F^{-1}(t) = -Q(t) - A^T(t)E(t)F^{-1}(t)$$
$$- E(t)F^{-1}(t)A(t) + E(t)F^{-1}(t)B(t)R^{-1}(t)B^T(t)E(t)F^{-1}(t) \qquad (6.66)$$

so that, multiplying on the right by $F(t)$, we have

$$\dot{E}(t) - E(t)F^{-1}(t)\dot{F}(t) = -Q(t)F(t) - A^T(t)E(t) - E(t)F^{-1}(t)A(t)F(t)$$
$$+ E(t)F^{-1}(t)B(t)R^{-1}(t)B^T(t)E(t) \qquad (6.67)$$

Next, we assume that $E(t)$ and $F(t)$ are such that

$$\dot{E}(t) = -A^T(t)E(t) - Q(t)F(t) \qquad (6.68)$$

$$E(t)F^{-1}(t)\dot{F}(t) = -E(t)F^{-1}(t)[B(t)R^{-1}(t)B^T(t)E(t) - A(t)F(t)] \qquad (6.69)$$

Then, multiplying Eq. (6.69) on the left by $F(t)E^{-1}(t)$, we obtain

$$\dot{F}(t) = -B(t)R^{-1}(t)B^T(t)E(t) + A(t)F(t) \qquad (6.70)$$

Recalling the boundary condition (6.58), we can write

$$K(t_f) = E(t_f)F^{-1}(t_f) = H(t_f) = H \qquad (6.71)$$

yielding the boundary conditions on $E(t)$ and $F(t)$

$$E(t_f) = H, \quad F(t_f) = I \qquad (6.72a, b)$$

Equations (6.68) and (6.70) represent a set of $2n^2$ *linear* differential equations. They can be arranged in the form

$$\left[\begin{array}{c} \dot{E}(t) \\ \hline \dot{F}(t) \end{array}\right] = \left[\begin{array}{c|c} -A^T(t) & -Q(t) \\ \hline -B(t)R^{-1}(t)B^T(t) & A(t) \end{array}\right]\left[\begin{array}{c} E(t) \\ \hline F(t) \end{array}\right] \quad (6.73)$$

which can be integrated backward in time using conditions (6.72). Hence, the price for replacing the solution of a set of nonlinear equations by the solution of a set of linear equations is to double the order of the set.

In most cases encountered, the matrices A, B, Q and R are constant. In such cases, the solution of Eq. (6.73) can be carried out by the approach based on the transition matrix discussed in Section 3.9. In subsequent discussions, we assume that A, B, Q and R are constant.

If the system is controllable, $H = 0$ and A, B, Q and R are constant, the Riccati matrix approaches a constant value as the final time increases without bounds, $K(t) \to K = \text{constant}$ as $t_f \to \infty$. In this case, the matrix Riccati equation, Eq. (6.57), reduces to

$$-Q - A^T K - KA + KBR^{-1}B^T K = 0 \quad (6.74)$$

which constitutes a set of algebraic equations representing the so-called *steady-state matrix Riccati equation*. In contrast, when the Riccati matrix does depend on t, Eq. (6.57) is called the *transient matrix Riccati equation*, even when A, B, Q and R are constant.

As in the transient case, the solution of nonlinear equations can be avoided also in the steady-state case. In fact, Eq. (6.74) can be reduced to an algebraic eigenvalue problem of order $2n$, a procedure known as *Potter's algorithm* [P6]. To this end, we consider the matrix

$$C = BR^{-1}B^T K - A \quad (6.75)$$

and write the eigenvalue problem associated with C in the form

$$F^{-1}CF = J \quad (6.76)$$

where J is the matrix of eigenvalues of C, assumed to be diagonal, and F is the matrix of eigenvectors. Multiplying Eq. (6.76) on the left by KF and considering Eqs. (6.74) and (6.75), we obtain

$$KFF^{-1}CF = KCF = KBR^{-1}B^T KF - KAF = QF + A^T KF = KFJ \quad (6.77)$$

Moreover, multiplying Eq. (6.76) on the left by F and considering Eq. (6.75), we have

$$CF = BR^{-1}B^T KF - AF = FJ \quad (6.78)$$

Next, introduce the transformation

$$KF = E \qquad (6.79)$$

so that Eqs. (6.77) and (6.78) can be combined into

$$M \begin{bmatrix} E \\ \hline F \end{bmatrix} = \begin{bmatrix} E \\ \hline F \end{bmatrix} J \qquad (6.80)$$

where

$$M = \begin{bmatrix} A^T & | & Q \\ \hline BR^{-1}B^T & | & A \end{bmatrix} \qquad (6.81)$$

Equation (6.80) represents an eigenvalue problem of order $2n$. The matrix M has in general $2n$ eigenvalues and eigenvectors, but we are interested only in n of them, as can be concluded from Eq. (6.80). To determine which are the solutions to be retained, let us consider Eqs. (6.74) and (6.75) and form

$$C^T K + KC = -A^T K + KBR^{-1}B^T K + KBR^{-1}B^T K - KA$$
$$= Q + KBR^{-1}B^T K \qquad (6.82)$$

If the right side of Eq. (6.82) is positive definite, then the eigenvalues of C have positive real parts [P6]. Moreover, if Q and $BR^{-1}B^T$ are real symmetric positive semidefinite matrices and λ is an eigenvalue of M, then $-\lambda$ is also an eigenvalue of M [P6]. Hence, M has n eigenvalues with positive real parts. Then, comparing Eqs. (6.78) and (6.80), we conclude that these are the same eigenvalues as the eigenvalues of C. It follows that from the eigensolutions of M we must retain those with positive real parts. It also follows that the matrices E and F entering into Eq. (6.79) consist of the upper and lower halves of the eigenvectors associated with the eigenvalues with positive real parts. Finally, from Eq. (6.79), we obtain the steady-state solution of the matrix Riccati equation, Eq. (6.74), by writing simply

$$K = EF^{-1} \qquad (6.83)$$

Example 6.3 Consider the two-degree-of-freedom, single-input system of Example 6.1 and design an optimal control using the performance measure given by Eq. (6.47) in conjunction with

$$H = 0, \quad Q = \begin{bmatrix} K & | & 0 \\ \hline 0 & | & M \end{bmatrix} = \begin{bmatrix} 2 & -1 & 0 & 0 \\ -1 & 1 & 0 & 0 \\ 0 & 0 & 1 & 0 \\ 0 & 0 & 0 & 2 \end{bmatrix}, \quad R = 1 \qquad (a)$$

Plot the optimal control gains and the optimal components of the state vector as functions of time.

To obtain the gains, we solve Eq. (6.73) backward in time for the coefficient matrix

$$\left[\begin{array}{c|c} -A^T & -Q \\ \hline -BR^{-1}B^T & A \end{array}\right] = \begin{bmatrix} 0 & 0 & 2 & -0.5 & -2 & 1 & 0 & 0 \\ 0 & 0 & -1 & 0.5 & 1 & -1 & 0 & 0 \\ -1 & 0 & 0 & 0 & 0 & 0 & -1 & 0 \\ 0 & -1 & 0 & 0 & 0 & 0 & 0 & -2 \\ 0 & 0 & 0 & 0 & 0 & 0 & 1 & 0 \\ 0 & 0 & 0 & 0 & 0 & 0 & 0 & 1 \\ 0 & 0 & 0 & 0 & -2 & 1 & 0 & 0 \\ 0 & 0 & 0 & -0.25 & 0.5 & -0.5 & 0 & 0 \end{bmatrix} \quad \text{(b)}$$

where the solution is subject to the condition

$$E(t_f) = 0, \quad F(t_f) = I \tag{c}$$

Because the matrix in Eq. (b) is constant, we can solve Eq. (6.73) by the transition matrix approach. Then, the Riccati matrix is obtained by using Eq. (6.60) and the gain matrix by using Eq. (6.59), where in this case the gain matrix is a row matrix. The gains are plotted in Fig. 6.5. Having the gain matrix, we obtain the state vector from

$$\dot{x}(t) = [A(t) - BG(t)]x(t) \tag{d}$$

which can be solved by a discrete-time approach (Section 3.15). The components of the state vector are plotted in Fig. 6.6.

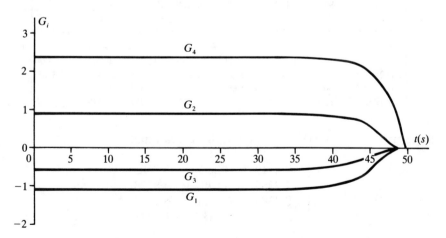

Figure 6.5

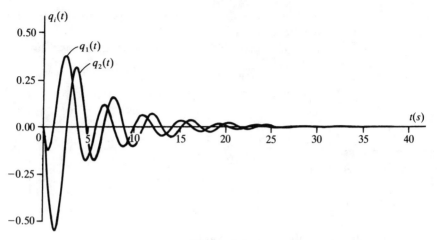

Figure 6.6

Example 6.4 Consider the system of Example 6.3 and obtain the steady-state control gains by Potter's algorithm.

From Eqs. (6.73) and (6.81), we observe that M is the negative of the matrix in Eq. (b) of Example 6.3. Solving the eigenvalue problem associated with M, we obtain the eigenvalues with positive real parts

$$\lambda_1 = 0.1652 + i1.5056, \quad \lambda_2 = \bar{\lambda}_1, \quad \lambda_3 = 0.4283 + i0.4405, \quad \lambda_4 = \bar{\lambda}_3 \tag{a}$$

and the associated eigenvectors

$$\begin{bmatrix} E \\ \hline F \end{bmatrix} = \begin{bmatrix} -1.9721 - i6.5073 & -1.9721 + i6.5073 & 0.1839 - i0.3059 & 0.1839 + i0.3059 \\ 1.1107 + i3.4998 & 1.1107 - i3.4998 & 0.8729 - i0.6717 & 0.8729 + i0.6717 \\ -4.3991 + i1.3097 & -4.3991 - i1.3097 & -0.5158 - i0.3042 & -0.5158 + i0.3042 \\ 1.8886 - i0.6951 & 1.8886 + i0.6951 & -1.4497 - i1.0340 & -1.4497 + i1.0340 \\ \hline -0.0136 - i0.4842 & -0.0136 + i0.4842 & 0.3675 - i0.2574 & 0.3675 + i0.2574 \\ 0.2442 + i0.1091 & 0.2442 - i0.1091 & 0.8282 - i0.3733 & 0.8282 + i0.3733 \\ -0.7267 + i0.1005 & -0.7267 - i0.1005 & -0.2708 - i0.0517 & -0.2708 + i0.0517 \\ 0.1239 - i0.3856 & 0.1239 + i0.3856 & -0.5192 - i0.2049 & -0.5192 + i0.2049 \end{bmatrix} \tag{b}$$

Hence, from Eq. (6.83), we compute the Riccati matrix

$$K = EF^{-1} = \begin{bmatrix} 13.4719 & -7.2285 & -0.3383 & -2.1728 \\ -7.2285 & 5.6704 & 0.8226 & 1.8185 \\ -0.3386 & 0.8226 & 6.1427 & -1.1374 \\ -2.1728 & 1.8185 & -1.1374 & 4.7484 \end{bmatrix} \tag{c}$$

Finally, using Eq. (6.59), we obtain the gain vector

$$G = R^{-1}B^T K = [-1.0864 \quad 0.9092 \quad -0.5687 \quad 2.3742]^T \qquad (d)$$

Note that the time-dependent gains obtained in Example 6.3 and plotted in Fig. 6.5 approach the values given in Eq. (d).

6.6 THE LINEAR TRACKING PROBLEM

The tracking problem is defined as the problem of designing a control input so as to cause the plant state to follow a given reference state $r(t)$. The linear tracking problem is the one in which the control is a linear function of the state.

Let us consider once again the linear state equations

$$\dot{x}(t) = A(t)x(t) + B(t)u(t) \qquad (6.84)$$

Because in this case we wish the plant to follow the reference input, we use the performance measure

$$J = \frac{1}{2}[x(t_f) - r(t_f)]^T H[x(t_f) - r(t_f)]$$
$$+ \frac{1}{2}\int_{t_0}^{t_f} \{[x(t) - r(t)]^T Q(t)[x(t) - r(t)] + u^T(t)R(t)u(t)\} \, dt \qquad (6.85)$$

As in the case of the regulator problem, we assume that the matrices H and Q are real symmetric and positive semidefinite, the matrix R is real symmetric and positive definite, the final time t_f is fixed and the final state $x(t_f)$ is free.

Consistent with Eq. (6.48), the Hamiltonian has the form

$$\mathcal{H}(x(t), u(t), p(t), t) = \frac{1}{2}[x(t) - r(t)]^T Q(t)[x(t) - r(t)]$$
$$+ \frac{1}{2} u^T(t)R(t)u(t) + p^T(t)[A(t)x(t) + B(t)u(t)] \qquad (6.86)$$

so that, using Eqs. (6.45a–c), the necessary condition for optimality are

$$\dot{x}^*(t) = \frac{\partial \mathcal{H}}{\partial p} = A(t)x^*(t) + B(t)u^*(t) \qquad (6.87)$$

$$\dot{p}^*(t) = -\frac{\partial \mathcal{H}}{\partial x} = -Q(t)x^*(t) - A^T(t)p^*(t) + Q(t)r(t) \qquad (6.88)$$

$$0 = \frac{\partial \mathcal{H}}{\partial u} = R(t)u^*(t) + B^T(t)p^*(t) \qquad (6.89)$$

Equation (6.89) yields immediately

$$\mathbf{u}^*(t) = -R^{-1}(t)B^T(t)\mathbf{p}^*(t) \tag{6.90}$$

Inserting Eq. (6.90) into Eq. (6.87), we can combine the state and costate equations into

$$\begin{bmatrix} \dot{\mathbf{x}}^*(t) \\ \dot{\mathbf{p}}^*(t) \end{bmatrix} = \begin{bmatrix} A(t) & -B(t)R^{-1}(t)B^T(t) \\ -Q(t) & -A^T(t) \end{bmatrix} \begin{bmatrix} \mathbf{x}^*(t) \\ \mathbf{p}^*(t) \end{bmatrix} + \begin{bmatrix} 0 \\ Q(t)\mathbf{r}(t) \end{bmatrix} \tag{6.91}$$

The solution of Eq. (6.91) can be obtained by means of the transition matrix (Section 3.9). Denoting the transition matrix by Φ, the solution of Eq. (6.91) is

$$\begin{bmatrix} \mathbf{x}^*(t_f) \\ \mathbf{p}^*(t_f) \end{bmatrix} = \Phi(t_f, t) \begin{bmatrix} \mathbf{x}^*(t) \\ \mathbf{p}^*(t) \end{bmatrix} + \int_t^{t_f} \Phi(t_f, \tau) \begin{bmatrix} 0 \\ Q(\tau)\mathbf{r}(\tau) \end{bmatrix} d\tau \tag{6.92}$$

which can be rewritten in the partitioned form

$$\mathbf{x}^*(t_f) = \Phi_{11}(t_f, t)\mathbf{x}^*(t) + \Phi_{12}(t_f, t)\mathbf{p}^*(t) + \mathbf{X}(t) \tag{6.93a}$$

$$\mathbf{p}^*(t_f) = \Phi_{21}(t_f, t)\mathbf{x}^*(t) + \Phi_{22}(t_f, t)\mathbf{p}^*(t) + \mathbf{P}(t) \tag{6.93b}$$

where

$$\mathbf{X}(t) = \int_t^{t_f} \Phi_{12}(t_f, \tau)Q(\tau)\mathbf{r}(\tau)\,d\tau, \quad \mathbf{P}(t) = \int_t^{t_f} \Phi_{22}(t_f, \tau)Q(\tau)\mathbf{r}(\tau)\,d\tau \tag{6.94a, b}$$

Equations (6.93) are subject to boundary conditions which, from Eq. (6.45d), can be shown to be

$$\mathbf{p}^*(t_f) = \frac{\partial}{\partial \mathbf{x}} \left\{ \frac{1}{2} [\mathbf{x}^*(t_f) - \mathbf{r}(t_f)]^T H [\mathbf{x}^*(t_f) - \mathbf{r}(t_f)] \right\} = H[\mathbf{x}^*(t_f) - \mathbf{r}(t_f)] \tag{6.95}$$

Inserting Eqs. (6.93a) and (6.95) into (6.93b), we have

$$H[\Phi_{11}(t_f, t)\mathbf{x}^*(t) + \Phi_{12}(t_f, t)\mathbf{p}^*(t) + \mathbf{X}(t) - \mathbf{r}(t_f)]$$
$$= \Phi_{21}(t_f, t)\mathbf{x}^*(t) + \Phi_{22}(t_f, t)\mathbf{p}^*(t) + \mathbf{P}(t) \tag{6.96}$$

so that, solving for $\mathbf{p}^*(t)$, we obtain the optimal costate

$$\mathbf{p}^*(t) = [\Phi_{22}(t_f, t) - H\Phi_{12}(t_f, t)]^{-1} \{[H\Phi_{11}(t_f, t) - \Phi_{21}(t_f, t)]\mathbf{x}^*(t)$$
$$+ H[\mathbf{X}(t) - \mathbf{r}(t_f)] - \mathbf{P}(t)\} = K(t)\mathbf{x}^*(t) + \mathbf{s}(t) \tag{6.97}$$

where

$$K(t) = [\Phi_{22}(t_f, t) - H\Phi_{12}(t_f, t)]^{-1}[H\Phi_{11}(t_f, t) - \Phi_{21}(t_f, t)] \quad (6.98a)$$

$$\mathbf{s}(t) = [\Phi_{22}(t_f, t) - H\Phi_{12}(t_f, t)]^{-1}\{H[\mathbf{X}(t) - \mathbf{r}(t_f)] - \mathbf{P}(t)\} \quad (6.98b)$$

Finally, introducing Eq. (6.97) into Eq. (6.90), we obtain the optimal control law

$$\mathbf{u}^*(t) = -R^{-1}(t)B^T(t)K(t)\mathbf{x}^*(t) - R^{-1}(t)B^T(t)\mathbf{s}(t) \quad (6.99)$$

6.7 PONTRYAGIN'S MINIMUM PRINCIPLE

In our preceding discussions of optimal control, no limit was placed on the magnitude of the control forces. In practice, however, there are cases in which the actuators reach saturation. For example, in space applications, there is a limit to the thrust magnitude that a rocket engine can achieve. Indeed, the thrust magnitude is equal to the product of the rate of change of mass and the exhaust velocity of the hot gas, where the first cannot exceed a certain value and the latter depends on several factors, such as fuel type, nozzle design and atmospheric conditions. Similarly, an electric motor is limited in the torque it can produce. Clearly, when the actuators become saturated the admissible controls are no longer unbounded.

It was indicated in Section 6.3 that when the admissible controls are bounded Eq. (6.45c) is no longer valid. Indeed, it can be shown [P4] that when the admissible controls are constrained Eq. (6.45c) must be replaced by

$$\mathcal{H}(\mathbf{x}^*(t), \mathbf{u}^*(t), \mathbf{p}^*(t), t) \leq \mathcal{H}(\mathbf{x}^*(t), \mathbf{u}(t), \mathbf{p}^*(t), t) \quad (6.100)$$

Inequality (6.100) indicates that *an optimal control must minimize the Hamiltonian* and is known as *Pontryagin's minimum principle*. Again, this is a necessary condition for optimality, and in general is not a sufficient condition.

Hence, if $\mathbf{x}(t) = \mathbf{x}^*(t)$ is an admissible trajectory satisfying the equation

$$\dot{\mathbf{x}}(t) = \mathbf{a}(\mathbf{x}(t), \mathbf{u}(t), t) \quad (6.101)$$

then the necessary conditions for $\mathbf{u}(t) = \mathbf{u}^*(t)$ to be an optimal control for the performance measure

$$J(\mathbf{u}) = h(\mathbf{x}(t_f), t_f) + \int_{t_0}^{t_f} g(\mathbf{x}(t), \mathbf{u}(t), t)\, dt \quad (6.102)$$

are

$$\dot{\mathbf{x}}^*(t) = \frac{\partial \mathcal{H}(\mathbf{x}^*(t), \mathbf{u}^*(t), \mathbf{p}^*(t), t)}{\partial \mathbf{p}}, \quad t_0 < t < t_f \tag{6.103a}$$

$$\dot{\mathbf{p}}^*(t) = -\frac{\partial \mathcal{H}(\mathbf{x}^*(t), \mathbf{u}^*(t), \mathbf{p}^*(t), t)}{\partial \mathbf{x}}, \quad t_0 < t < t_f \tag{6.103b}$$

$$\mathcal{H}(\mathbf{x}^*(t), \mathbf{u}^*(t), \mathbf{p}^*(t), t) \le \mathcal{H}(\mathbf{x}^*(t), \mathbf{u}(t), \mathbf{p}^*(t), t), \quad t_0 < t < t_f \tag{6.103c}$$

and

$$\left[\frac{\partial h(\mathbf{x}^*(t_f), t_f)}{\partial \mathbf{x}} - \mathbf{p}^*(t_f) \right]^T \delta \mathbf{x}_f$$
$$+ \left[\mathcal{H}(\mathbf{x}^*(t_f), \mathbf{u}^*(t_f), \mathbf{p}^*(t_f), t_f) + \frac{\partial h(\mathbf{x}^*(t_f), t_f)}{\partial t} \right] \delta t_f = 0 \tag{6.103d}$$

where the Hamiltonian is given by

$$\mathcal{H}(\mathbf{x}(t), \mathbf{u}(t), \mathbf{p}(t), t) = g(\mathbf{x}(t), \mathbf{u}(t), t) + \mathbf{p}^T(t)[\mathbf{a}(\mathbf{x}(t), \mathbf{u}(t), t)] \tag{6.104}$$

As indicated earlier, conditions (6.103) are necessary for $\mathbf{u}(t) = \mathbf{u}^*(t)$ to be an optimal control, but they are generally not sufficient.

Although Pontryagin's minimum principle was derived for the case in which the admissible controls are bounded, the principle can be used also for unbounded admissible controls. In this case it is possible to establish not only a necessary but also a sufficient condition for optimality. Clearly, as the bounds of the region of admissible controls recede to infinity, the region becomes unbounded. Then, as for any other unconstrained function, the necessary condition for $\mathbf{u}^*(t)$ to render the Hamiltonian a minimum is

$$\frac{\partial \mathcal{H}(\mathbf{x}^*(t), \mathbf{u}^*(t), \mathbf{p}^*(t), t)}{\partial \mathbf{u}} = \mathbf{0} \tag{6.105}$$

which implies that $\mathcal{H}$ has a stationary value for $\mathbf{u}(t) = \mathbf{u}^*(t)$. If, in addition,

$$\frac{\partial^2 \mathcal{H}(\mathbf{x}^*(t), \mathbf{u}^*(t), \mathbf{p}^*(t), t)}{\partial \mathbf{u}^2} > 0 \tag{6.106}$$

where we recognize that $\partial^2 \mathcal{H}/\partial \mathbf{u}^2$ represents a matrix, then (6.106) is a sufficient condition for the stationary value to be an extremum, in this particular case a minimum. Hence, the positive definiteness of the matrix $\partial^2 \mathcal{H}/\partial \mathbf{u}^2$ represents a sufficient condition for $\mathbf{u}^*(t)$ to cause $\mathcal{H}$ to have a local minimum.

On occasions, the Hamiltonian has the special form

$$\mathcal{H}(\mathbf{x}(t), \mathbf{u}(t), \mathbf{p}(t), t) = f(\mathbf{x}(t), \mathbf{p}(t), t) + [\mathbf{c}^T(\mathbf{x}(t), \mathbf{p}(t), t)]\mathbf{u}(t)$$
$$+ \frac{1}{2} \mathbf{u}^T(t) R(t) \mathbf{u}(t) \quad (6.107)$$

so that

$$\frac{\partial \mathcal{H}(\mathbf{x}^*(t), \mathbf{u}^*(t), \mathbf{p}^*(t), t)}{\partial \mathbf{u}} = \mathbf{c}(\mathbf{x}^*(t), \mathbf{p}^*(t), t) + R(t)\mathbf{u}^*(t) = 0 \quad (6.108a)$$

$$\frac{\partial^2 \mathcal{H}(\mathbf{x}^*(t), \mathbf{u}^*(t), \mathbf{p}^*(t), t)}{\partial \mathbf{u}^2} = R(t) \quad (6.108b)$$

Hence, if the matrix $R(t)$ is positive definite, then the control

$$\mathbf{u}^*(t) = -R^{-1}(t)\mathbf{c}(\mathbf{x}^*(t), \mathbf{p}^*(t), t) \quad (6.109)$$

minimizes the Hamiltonian. Because Eq. (6.108a) is linear in $\mathbf{u}^*(t)$, the equation has only one solution, so that the minimum of $\mathcal{H}$ is *global*.

A case occurring very frequently is the one in which the vector **a** in Eq. (6.101) does not depend explicitly on time, or

$$\dot{\mathbf{x}}(t) = \mathbf{a}(\mathbf{x}(t), \mathbf{u}(t)) \quad (6.110)$$

Moreover, if we consider the case in which the final state $\mathbf{x}(t_f)$ is fixed, although the final time t_f is free, then the performance measure has the special form

$$J(\mathbf{u}) = \int_{t_0}^{t_f} g(\mathbf{x}(t), \mathbf{u}(t)) \, dt \quad (6.111)$$

It follows that the Hamiltonian does not depend explicitly on time either, i.e.,

$$\mathcal{H}(\mathbf{x}(t), \mathbf{u}(t), \mathbf{p}(t)) = g(\mathbf{x}(t), \mathbf{u}(t)) + \mathbf{p}^T(t)[\mathbf{a}(\mathbf{x}(t), \mathbf{u}(t))] \quad (6.112)$$

Then, it can be shown [K5] that the Hamiltonian must be identically zero when evaluated on an extremal trajectory, or

$$\mathcal{H}(\mathbf{x}^*(t), \mathbf{u}^*(t), \mathbf{p}^*(t)) = 0, \quad t_0 < t < t_f \quad (6.113)$$

6.8 MINIMUM-TIME PROBLEMS

One of the classical problems of the calculus of variations is the brachistochrone problem, formulated by Johann Bernoulli in 1696. The problem can be stated as follows: Two points A and B lying in a vertical plane are to be connected by a frictionless curved track such that a mass point moves from A to B under gravity in the shortest possible time. A modern version of the brachistochrone problem is the minimum-time control problem, in which the object is to minimize the transition time from an initial state to a target set in the state space. In seeking a solution to this optimal control problem, the necessary conditions provided by Pontryagin's minimum principle prove very useful in identifying the admissible controls.

Let us consider the system

$$\dot{\mathbf{x}}(t) = \mathbf{a}(\mathbf{x}(t), t) + B(\mathbf{x}(t), t)\mathbf{u}(t) \tag{6.114}$$

where B is an $n \times r$ matrix and $\mathbf{u}(t)$ is an r-vector. The components of the control vector $\mathbf{u}(t)$ are constrained in magnitude by the conditions

$$|u_j(t)| \le 1, \quad j = 1, 2, \ldots, r \tag{6.115}$$

The performance measure for time-optimal control is simply

$$J(\mathbf{u}) = \int_{t_0}^{t_f} dt \tag{6.116}$$

Hence, the problem is of determining the control $\mathbf{u}(t)$ whose components $u_j(t)$ satisfy the constraints (6.115) and one that drives the initial state $\mathbf{x}(t_0) = \mathbf{x}_0$ to the target set $S(t)$ and minimizes the performance measure $J(\mathbf{u})$ given by Eq. (6.116). Note that constraints of the form

$$|u_j(t)| \le m_j, \quad j = 1, 2, \ldots, r \tag{6.117}$$

can be reduced to the form (6.115) by merely dividing the columns $\mathbf{b}_j(\mathbf{x}(t), t)$ of $B(\mathbf{x}(t), t)$ by m_j ($j = 1, 2, \ldots, r$).

From Eqs. (6.102) and (6.116), it follows that $h = 0$ and

$$g(\mathbf{x}(t), \mathbf{u}(t), t) = 1 \tag{6.118}$$

Moreover, comparing Eqs. (6.42) and (6.114), we conclude that

$$\mathbf{a}(\mathbf{x}(t), \mathbf{u}(t), t) = \mathbf{a}(\mathbf{x}(t), t) + B(\mathbf{x}(t), t)\mathbf{u}(t) \tag{6.119}$$

so that, inserting Eqs. (6.118) and (6.119) into Eq. (6.44), the Hamiltonian takes the form

$$\mathcal{H}(\mathbf{x}(t), \mathbf{u}(t), \mathbf{p}(t), t) = 1 + \mathbf{p}^T(t)[\mathbf{a}(\mathbf{x}(t), t) + B(\mathbf{x}(t), t)\mathbf{u}(t)] \tag{6.120}$$

From Pontryagin's minimum principle, inequality (6.100), we conclude that when the Hamiltonian has the form given by Eq. (6.120) a necessary condition for optimality is

$$1 + \mathbf{p}^{*T}(t)[\mathbf{a}(\mathbf{x}^*(t), t) + B(\mathbf{x}^*(t), t)\mathbf{u}^*(t)]$$
$$\leq 1 + \mathbf{p}^{*T}(t)[\mathbf{a}(\mathbf{x}^*(t), t) + B(\mathbf{x}^*(t), t)\mathbf{u}(t)] \quad (6.121)$$

which reduces to

$$\mathbf{p}^{*T}(t)B(\mathbf{x}^*(t), t)\mathbf{u}^*(t) \leq \mathbf{p}^{*T}(t)B(\mathbf{x}^*(t), t)\mathbf{u}(t) \quad (6.122)$$

It follows that $\mathbf{p}^{*T}(t)B(\mathbf{x}^*(t), t)\mathbf{u}(t)$ attains its minimum value at $\mathbf{u}(t) = \mathbf{u}^*(t)$.

Next, let us consider

$$\min_{\mathbf{u}(t) \in \Omega} \mathbf{p}^{*T}(t)B(\mathbf{x}^*(t), t)\mathbf{u}(t) = \min_{\mathbf{u}(t) \in \Omega} \sum_{j=1}^{r} u_j(t)\mathbf{b}_j^T(\mathbf{x}^*(t), t)\mathbf{p}^*(t)$$

$$= \sum_{j=1}^{r} [\min_{|u_j(t)| \leq 1} u_j(t)\mathbf{b}_j^T(\mathbf{x}^*(t), t)\mathbf{p}^*(t)] \quad (6.123)$$

where Ω is a subset of the r-dimensional space of the control variables having the form of a hypercube. Moreover, we assumed that the minimization and summation processes are interchangeable on the basis that the functions $u_j(t)$ are independent of one another. But,

$$\min_{|u_j(t)| \leq 1} [u_j(t)\mathbf{b}_j^T(\mathbf{x}^*(t), t)\mathbf{p}^*(t)] = -|\mathbf{b}_j^T(\mathbf{x}^*(t), t)\mathbf{p}^*(t)| \quad (6.124)$$

It follows that the time-optimal control $u_j^*(t)$ minimizing the function $u_j(t)\mathbf{b}_j^T(\mathbf{x}^*(t), t)\mathbf{p}^*(t)$ is as follows:

$$u_j^*(t) = \begin{cases} +1 & \text{if } \mathbf{b}_j^T(\mathbf{x}^*(t), t)\mathbf{p}^*(t) < 0 \\ -1 & \text{if } \mathbf{b}_j^T(\mathbf{x}^*(t), t)\mathbf{p}^*(t) > 0 \\ \text{indeterminate} & \text{if } \mathbf{b}_j^T(\mathbf{x}^*(t), t)\mathbf{p}^*(t) = 0 \end{cases} \quad (6.125)$$

If the function $\mathbf{b}_j^T(\mathbf{x}^*(t), t)\mathbf{p}^*(t)$ is zero only at a finite number of discrete times and is nonzero otherwise, then the time-optimal control problem is said to be *normal*. If the function $\mathbf{b}_j^T(\mathbf{x}^*(t), t)\mathbf{p}^*(t)$ is zero over one, or more, subintervals of time, then the time-optimal control problem is said to be *singular*. We shall be concerned only with normal problems. The dependence of the optimal control $u_j^*(t)$ on the function $\mathbf{b}_j^T(\mathbf{x}^*(t), t)\mathbf{p}^*(t)$ for a typical normal time-optimal control problem is shown in Fig. 6.7.

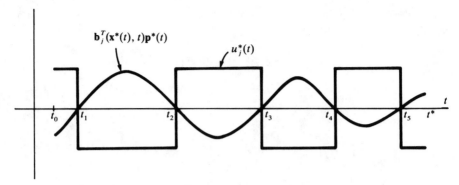

Figure 6.7

Relations (6.125) can be written in the compact form

$$u_j^*(t) = -\text{sgn}\,[\mathbf{b}_j^T(\mathbf{x}^*(t), t)\mathbf{p}^*(t)], \quad j = 1, 2, \ldots, r \qquad (6.126)$$

where sgn denotes the signum function, defined as $+1$ if the argument is positive and -1 if the argument is negative. With reference to Fig. 6.7 and Eqs. (6.126), we conclude that, *if the time-optimal control problem is normal, the components of the optimal control* $\mathbf{u}^*(t)$ *are piecewise constant functions of time.* Such functions are known as *bang-bang* and the preceding statement is referred to as the *bang-bang principle*. Equations (6.126) can be expressed in the vector form

$$\mathbf{u}^*(t) = -\text{SGN}\,[B^T(\mathbf{x}^*(t), t)\mathbf{p}^*(t)] \qquad (6.127)$$

where the symbol SGN indicates that the signum function applies to every component of the vector. The implication of the bang-bang principle is that the *time-optimal control is obtained by exerting maximum control force until the target set is reached.*

The questions of existence and uniqueness of time-optimal control for nonlinear systems of the type (6.114) remain. The situation is considerably better for linear time-invariant systems.

6.9 MINIMUM-TIME CONTROL OF LINEAR TIME-INVARIANT SYSTEMS

Let us consider a system described by the state equation

$$\dot{\mathbf{x}}(t) = A\mathbf{x}(t) + B\mathbf{u}(t) \qquad (6.128)$$

where A and B are constant $n \times n$ and $n \times r$ matrices, respectively. Moreover, the components of the control vector $\mathbf{u}(t)$ are constrained in magnitude by the relations

$$|u_j(t)| < 1, \quad j = 1, 2, \ldots, r \qquad (6.129)$$

It is assumed that the system is controllable and normal. Then, the *stationary, linear regulator, time-optimal control problem* can be defined as determining the control that transfers the system (6.128) from the initial state $\mathbf{x}(0) = \mathbf{x}_0$ to the final state $\mathbf{x}(t_f) = \mathbf{0}$ in minimum time.

The necessary conditions for optimality can be obtained from Eqs. (6.103). To this end, we first use Eq. (6.120) and write the Hamiltonian

$$\mathcal{H}(\mathbf{x}(t), \mathbf{u}(t), \mathbf{p}(t)) = 1 + \mathbf{p}^T(t)[A\mathbf{x}(t) + B\mathbf{u}(t)] \qquad (6.130)$$

Then, Eqs. (6.103a) and (6.103b) yield

$$\dot{\mathbf{x}}^*(t) = \frac{\partial \mathcal{H}(\mathbf{x}^*(t), \mathbf{u}^*(t), \mathbf{p}^*(t))}{\partial \mathbf{p}^*(t)} = A\mathbf{x}^*(t) + B\mathbf{u}^*(t) \qquad (6.131)$$

and

$$\dot{\mathbf{p}}^*(t) = -\frac{\partial \mathcal{H}(\mathbf{x}^*(t), \mathbf{u}^*(t), \mathbf{p}^*(t))}{\partial \mathbf{x}^*(t)} = -A^T \mathbf{p}^*(t) \qquad (6.132)$$

respectively, where the state vector is subject to the boundary conditions

$$\mathbf{x}^*(0) = \mathbf{x}_0, \quad \mathbf{x}^*(t_f) = \mathbf{0} \qquad (6.133)$$

Moreover, from (6.121), the inequality

$$1 + \mathbf{p}^{*T}(t)[A\mathbf{x}^*(t) + B\mathbf{u}^*(t)] \leq 1 + \mathbf{p}^{*T}(t)[A\mathbf{x}^*(t) + B\mathbf{u}(t)] \qquad (6.134)$$

must hold for every admissible control vector $\mathbf{u}(t)$, so that in this case the bang-bang principle, Eq. (6.126), yields the time-optimal control

$$\mathbf{u}^*(t) = -\text{SGN}[B^T \mathbf{p}^*(t)] \qquad (6.135)$$

Finally, because the Hamiltonian does not depend explicitly on time and the final state is fixed, Eq. (6.113) gives

$$\mathcal{H}(\mathbf{x}^*(t), \mathbf{u}^*(t), \mathbf{p}^*(t)) = 1 + \mathbf{p}^{*T}(t)[A\mathbf{x}^*(t) + B\mathbf{u}^*(t)] = 0 \qquad (6.136)$$

Equation (6.135) states that if an optimal control exists, then it is bang-bang. The question of existence is answered [A18] by the *Existence Theorem*: *If in the case of a stationary, linear regulator problem all the eigenvalues of the matrix A have nonpositive real part, then a time-optimal control exists.* Moreover, on the basis of the assumption that the system is controllable and normal, it is possible to prove [A18] the *Uniqueness Theorem*: *If an optimal control exists, then it is unique.*

Finally, a qualitative statement can be made concerning the shape of the control. In particular, *if the eigenvalues of the matrix A are all real, and a unique time-optimal control exists, then each component of the optimal control vector* $\mathbf{u}^*(t)$ *can switch from* $+1$ *to* -1 *or from* -1 *to* $+1$ *at most* $n - 1$ *times*. Hence, if reaching the origin of the state space requires more than $n - 1$ switchings, the control is not time-optimal.

Optimal-time problems of meaningful order admitting solutions are not very common and almost always they involve some special condition. The situation is considerably better in the case of low-order systems, and in particular for second-order systems. To develop an appreciation for the problem, we consider here two examples, each one concerned with a second-order system.

Example 6.5 Determine the time-optimal control designed to transfer a particle of mass m from a given initial state to the origin of the state space. The magnitude of the control force per unit mass must not exceed unity.

The equation of motion of the particle is

$$m\ddot{q}(t) = Q(t) \tag{a}$$

where $q(t)$ is the position and $Q(t)$ is the control force. Introducing the notation

$$q(t) = x_1(t), \quad \dot{q}(t) = x_2(t), \quad \frac{Q(t)}{m} = u(t) \tag{b}$$

we can write the state equations

$$\begin{aligned} \dot{x}_1(t) &= x_2(t) \\ \dot{x}_2(t) &= u(t) \end{aligned} \tag{c}$$

which are subject to $x_1(0) = q(0)$, $x_2(0) = \dot{q}(0)$. Moreover, the control $u(t)$ is subject to the constraint

$$|u(t)| \leq 1 \tag{d}$$

Equations (c) are of the type (6.128), in which

$$A = \begin{bmatrix} 0 & 1 \\ 0 & 0 \end{bmatrix}, \quad B = \begin{bmatrix} 0 \\ 1 \end{bmatrix} \tag{e}$$

Hence, inserting Eqs. (e) into Eq. (6.130), we obtain the Hamiltonian

$$\mathcal{H}(\mathbf{x}(t), u(t), \mathbf{p}(t)) = 1 + [p_1(t) \quad p_2(t)] \left(\begin{bmatrix} 0 & 1 \\ 0 & 0 \end{bmatrix} \begin{bmatrix} x_1(t) \\ x_2(t) \end{bmatrix} + \begin{bmatrix} 0 \\ 1 \end{bmatrix} u(t) \right)$$

$$= 1 + p_1(t) x_2(t) + p_2(t) u(t) \tag{f}$$

The bang-bang principle, Eq. (6.135), yields the optimal control

$$u^*(t) = -\operatorname{sgn}[B^T \mathbf{p}^*(t)] = -\operatorname{sgn}(p_2^*(t)) \tag{g}$$

or

$$u^*(t) = \begin{cases} -1 & \text{for } p_2^*(t) > 0 \\ +1 & \text{for } p_2^*(t) < 0 \end{cases} \tag{h}$$

Inserting Eq. (f) into Eqs. (6.132), we obtain the costate equations

$$\begin{aligned} \dot{p}_1^*(t) &= -\frac{\partial \mathcal{H}}{\partial x_1^*(t)} = 0 \\ \dot{p}_2^*(t) &= -\frac{\partial \mathcal{H}}{\partial x_2^*(t)} = -p_1^*(t) \end{aligned} \tag{i}$$

which have the solution

$$\begin{aligned} p_1^*(t) &= c_1 \\ p_2^*(t) &= -c_1 t + c_2 \end{aligned} \tag{j}$$

where c_1 and c_2 are constants of integration.

We observe from the second of Eqs. (j) that $p_2^*(t)$ can change sign at most once, and from Eq. (h) it follows that $u^*(t)$ also can change sign at most once. This is consistent with the statement on switchings, as this is a second-order system and both eigenvalues of A are zero and hence real. Because there is at most one switching, it follows that there are four possible control sequences:

$$\begin{aligned} u^*(t) &= +1, \quad 0 \leq t \leq t^* \\ u^*(t) &= -1, \quad 0 \leq t \leq t^* \\ u^*(t) &= \begin{cases} +1, & 0 < t < t_1 \\ -1, & t_1 \leq t < t^* \end{cases} \\ u^*(t) &= \begin{cases} -1, & 0 < t < t_1 \\ +1, & t_1 \leq t < t^* \end{cases} \end{aligned} \tag{k}$$

where t_1 is the switching time.

Next, we wish to determine the optimal trajectories. To this end, we consider Eqs. (c) in conjunction with $u^*(t) = \pm 1$ and with $x_1(t) = x_1^*(t)$, $x_2(t) = x_2^*(t)$. Integrating the second of Eqs. (c), we obtain

$$x_2^*(t) = \pm t + c_3 \tag{l}$$

so that the first of Eqs. (c) yields

$$x_1^*(t) = \pm \frac{1}{2} t^2 + c_3 t + c_4 \tag{m}$$

where c_3 and c_4 are constants of integration. To decide when to use the control $u^*(t) = +1$ and when to use $u^*(t) = -1$, we propose to plot the solution in the state plane, which amounts to eliminating the time t from the solution. The result is

$$\begin{aligned} x_1^*(t) &= \frac{1}{2} [x_2^*(t)]^2 + c_5 \quad \text{for } u^*(t) = +1 \\ x_1^*(t) &= -\frac{1}{2} [x_2^*(t)]^2 + c_6 \quad \text{for } u^*(t) = -1 \end{aligned} \tag{n}$$

where c_5 and c_6 are constants of integration. Equations (n) represent two families of parabolas, one family corresponding to $u^*(t) = +1$ and one corresponding to $u^*(t) = -1$. They are shown in Fig. 6.8 in solid and dashed lines, respectively, with the arrows indicating the direction of increasing time.

Because the objective is to drive the system to the origin of the state plane, two trajectories are of particular interest, namely those two leading

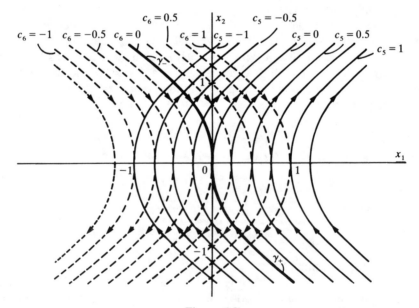

Figure 6.8

to the origin. The first is the trajectory $x_1^*(t) = \frac{1}{2}[x_2^*(t)]^2$ corresponding to $u^*(t) = +1$ and lying in the lower half of the plane and the second is the trajectory $x_1^*(t) = -\frac{1}{2}[x_2^*(t)]^2$ corresponding to $u^*(t) = -1$ and lying in the upper half of the plane. They are denoted by γ_+ and γ_-, respectively. The question remains as to the control sequence to be followed. The answer depends on the initial conditions. Figure 6.9a,b shows two typical cases. In the first case, shown in Fig. 6.9a, the initial state $\mathbf{x}(0) = \mathbf{x}_0$ lies to the left of the curve $\gamma = \gamma_+ + \gamma_-$. First, the optimal control is $u^*(t) = +1$ until the trajectory emanating from $\mathbf{x}_0$ reaches the curve γ_- and then the control switches to $u^*(t) = -1$ and the system follows the curve γ_- to the origin. This case corresponds to the third of Eqs. (k). In the second case, shown in Fig. 6.9b, the initial state lies to the right of γ and the control is $u^*(t) = -1$ until the system reaches γ_+ when the control switches to $u^*(t) = +1$, following the curve γ_+ to the origin. This case corresponds to the fourth of Eqs. (k). Because the switching in control takes place when the trajectory reaches the curve γ, the curve γ is referred to as the *switching curve*. Note that if the initial state $\mathbf{x}_0$ lies on γ_+ or on γ_-, then the control is $u^*(t) = +1$ or $u^*(t) = -1$, respectively, and no switching is involved. These two cases correspond to the first two of Eqs. (k). Figure 6.10 shows the totality of time-optimal trajectories, and there are no other time-optimal trajectories [A18].

The question remains as to how to implement the control as a feedback control, i.e., one depending on the state. To this end, we observe that the

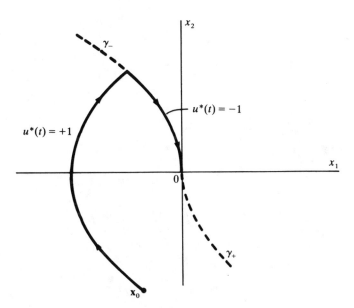

Figure 6.9a

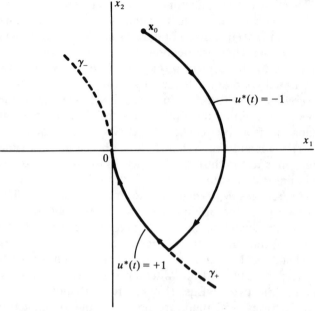

Figure 6.9b

switching curve $\gamma = \gamma_+ + \gamma_-$ divides the state plane into two regions, the first in which $u^*(t) = -1$ and the second in which $u^*(t) = +1$. They are denoted in Fig. 6.10 by R_- and R_+, respectively. Then, we introduce the switching function

$$\gamma(\mathbf{x}(t)) = x_1(t) + \frac{1}{2} x_2(t)|x_2(t)| \tag{o}$$

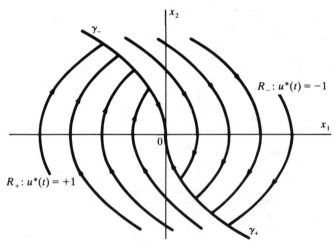

Figure 6.10

which can be used to define the control regions as follows: $\gamma(\mathbf{x}(t)) > 0$ implies that $\mathbf{x}(t)$ lies in R_-, $\gamma(\mathbf{x}(t)) < 0$ implies that $\mathbf{x}(t)$ lies in R_+ and $\gamma(\mathbf{x}(t)) = 0$ implies that $\mathbf{x}(t)$ lies on the switching curve γ. With this in mind, the optimal control law can be expressed in the form

$$u^*(t) = \begin{cases} -1 & \text{for } \mathbf{x}(t) \text{ such that } \gamma(\mathbf{x}(t)) > 0 \\ +1 & \text{for } \mathbf{x}(t) \text{ such that } \gamma(\mathbf{x}(t)) < 0 \\ -1 & \text{for } \mathbf{x}(t) \text{ such that } \gamma(\mathbf{x}(t)) = 0 \text{ and } x_2(t) > 0 \\ +1 & \text{for } \mathbf{x}(t) \text{ such that } \gamma(\mathbf{x}(t)) = 0 \text{ and } x_2(t) < 0 \\ 0 & \text{for } \mathbf{x}(t) = \mathbf{0} \end{cases} \quad (p)$$

Clearly, this is a nonlinear control law. Its implementation is shown in the block diagram of Fig. 6.11.

Example 6.6 Determine the time-optimal control designed to transfer a mass-spring system from a given initial state to the origin of the state space. The magnitude of the control force per unit mass must not exceed unity.

The equation of motion of a mass-spring system is

$$m\ddot{q}(t) + kq(t) = Q(t) \quad (a)$$

where m is the mass, k the spring constant, $q(t)$ the displacement and $Q(t)$ the control force. Introducing the notation

$$q(t) = x_1(t), \quad \dot{q}(t) = \omega_n x_2(t), \quad Q(t)/m\omega_n = u(t) \quad (b)$$

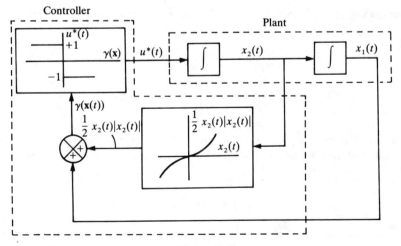

Figure 6.11

216 CONTROL OF LUMPED-PARAMETER SYSTEMS. MODERN APPROACH

where $\omega_n = \sqrt{(k/m)}$ is the natural frequency, we can write the state equations

$$\dot{x}_1(t) = \omega_n x_2(t)$$
$$\dot{x}_2(t) = -\omega_n x_1(t) + u(t) \qquad \text{(c)}$$

which are subject to the initial conditions $x_1(0) = q(0)$, $x_2(0) = \dot{q}(0)/\omega_n$. Moreover, the control $u(t)$ is subject to the constraint

$$|u(t)| \le 1 \qquad \text{(d)}$$

Equations (c) are of the form (6.128), so that the system is characterized by the matrices

$$A = \begin{bmatrix} 0 & \omega_n \\ -\omega_n & 0 \end{bmatrix}, \quad B = \begin{bmatrix} 0 \\ 1 \end{bmatrix} \qquad \text{(e)}$$

Using Eq. (6.130) in conjunction with Eqs. (e), we obtain the Hamiltonian

$$\mathcal{H}(\mathbf{x}(t), u(t), \mathbf{p}(t)) = 1 + [p_1(t) \quad p_2(t)]\left(\begin{bmatrix} 0 & \omega_n \\ -\omega_n & 0 \end{bmatrix}\begin{bmatrix} x_1(t) \\ x_2(t) \end{bmatrix} + \begin{bmatrix} 0 \\ 1 \end{bmatrix}u(t)\right)$$

$$= 1 + \omega_n p_1(t) x_2(t) + p_2(t)[-\omega_n x_1(t) + u(t)] \qquad \text{(f)}$$

As in Example 6.5, the bang-bang principle yields the optimal control

$$u^*(t) = \begin{cases} -1 & \text{for } p_2^*(t) > 0 \\ +1 & \text{for } p_2^*(t) < 0 \end{cases} \qquad \text{(g)}$$

Introducing Eq. (f) into Eq. (6.132), we obtain the costate equations

$$\dot{p}_1^*(t) = -\frac{\partial \mathcal{H}}{\partial x_1^*(t)} = \omega_n p_2^*(t)$$

$$\dot{p}_2^*(t) = -\frac{\partial \mathcal{H}}{\partial x_2^*(t)} = -\omega_n p_1^*(t) \qquad \text{(h)}$$

which are the equations of a harmonic oscillator. From Eq. (g), we note that our interest lies in $p_2^*(t)$ alone, so that from Eqs. (h) we can write the solution

$$p_2^*(t) = P \sin(\omega_n t + \theta) \qquad \text{(i)}$$

where the amplitude P and phase angle θ are constants of integration. Then, the time-optimal control can be determined according to Eq. (g). The

MINIMUM-TIME CONTROL OF LINEAR TIME-INVARIANT SYSTEMS   217

dependence of $u^*(t)$ on $p_2^*(t)$ is shown in Fig. 6.12. The figure permits us to make several observations:

1. The time-optimal control must be piecewise constant, switching periodically between the two values $u^*(t) = +1$ and $u^*(t) = -1$, with the exception of the initial and final stages of the control action which are generally shorter.
2. The time-optimal control remains constant for a maximum time interval of π/ω_n.
3. The number of switchings of the time-optimal control is unlimited.
4. The problem is normal.

The fact that the number of switchings is unlimited highlights the difference between the mass-spring system of this example and the unrestrained mass of Example 6.5, where in the latter the control could switch at most once. This, of course, is because here the eigenvalues are pure imaginary, $\pm i\omega_n$. As a consequence, the switching curve in this example is entirely different from the switching curve of Example 6.5, as we shall see shortly. Moreover, because $p_2^*(t)$ is harmonic, it is zero only at discrete points, so that the possibility that the problem is singular is excluded.

Next, we consider the optimal trajectories. To this end, we let $u(t) = \Delta =$ constant, where Δ can take the value $+1$ or -1, so that the state equations, Eqs. (c), become

$$\dot{x}_1(t) = \omega_n x_2(t)$$
$$\dot{x}_2(t) = -\omega_n x_1(t) + \Delta \tag{j}$$

The solution of Eqs. (j) can be obtained by means of the Laplace transformation. The results can be shown to be

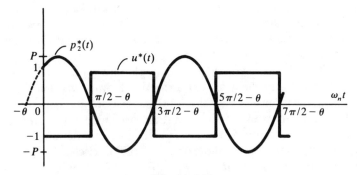

Figure 6.12

$$x_1(t) = x_1(0) \cos \omega_n t + x_2(0) \sin \omega_n t + \frac{\Delta}{\omega_n} (1 - \cos \omega_n t)$$
$$x_2(t) = \frac{1}{\omega_n} [\Delta - \omega_n x_1(0)] \sin \omega_n t + x_2(0) \cos \omega_n t \qquad (k)$$

To determine the trajectories, we must eliminate the time t from Eqs. (k). To this end, we rewrite Eqs. (k) in the form

$$\omega_n x_1(t) - \Delta = [\omega_n x_1(0) - \Delta] \cos \omega_n t + \omega_n x_2(0) \sin \omega_n t$$
$$\omega_n x_2(t) = -[\omega_n x_1(0) - \Delta] \sin \omega_n t + \omega_n x_2(0) \cos \omega_n t \qquad (l)$$

Squaring both Eqs. (l) and adding up the result, we obtain the equation of the trajectories

$$[\omega_n x_1(t) - \Delta]^2 + [\omega_n x_2(t)]^2 = [\omega_n x_1(0) - \Delta]^2 + [\omega_n x_2(0)]^2 \qquad (m)$$

In the case $u^*(t) = +1$, the trajectories equation is

$$[\omega_n x_1(t) - 1]^2 + [\omega_n x_2(t)]^2 = [\omega_n x_1(0) - 1]^2 + [\omega_n x_2(0)]^2 \qquad (n)$$

The trajectories represent circles in the plane $\omega_n x_1$, $\omega_n x_2$ with the center at $\omega_n x_1 = 1$ and with radii depending on the initial conditions $x_1(0) = q(0)$, $x_2(0) = \dot{q}(0)/\omega_n$. Similarly, the trajectories corresponding to $u^*(t) = -1$ have the equation

$$[\omega_n x_1(t) + 1]^2 + [\omega_n x_2(t)]^2 = [\omega_n x_1(0) + 1]^2 + [\omega_n x_2(0)]^2 \qquad (o)$$

and they represent circles with the center at $\omega_n x_1 = -1$. The two families of circles are plotted in Fig. 6.13.

The optimal trajectories are a combination of arcs of the circles plotted in Fig. 6.13. To narrow down the choice, we must eliminate trajectories that are not optimal. To this end, we note that a radius vector from the center of any given circle to a point on the circle rotates with the angular velocity ω_n, so that to go around the circle once requires the time $2\pi/\omega_n$. Observation 2, however, states that a time-optimal control remains constant for a maximum time interval of π/ω_n. It follows that an optimal trajectory cannot contain an arc of any one circle corresponding to an angle larger than 180°. From Fig. 6.13, it is clear that the final arc on an optimal trajectory must belong to either Γ_+ or Γ_-, because these are the only circles passing through the origin. Moreover, in view of the fact that no angle larger than 180° is admissible, it follows that only the lower half of Γ_+ and the upper half of Γ_- are suitable candidates. We denote these two semicircles by γ_+^0 and γ_-^0, respectively. The semicircles are shown in Fig. 6.14. Now, suppose that the preceding portion of the optimal trajectory intersects γ_+^0 at point A. Because this portion is not final, and assuming it is not the initial one either,

MINIMUM-TIME CONTROL OF LINEAR TIME-INVARIANT SYSTEMS   219

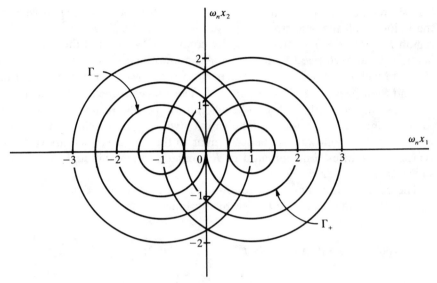

Figure 6.13

we conclude from the discussion above that this portion must be a semicircle with the center at $\omega_n x_1 = -1$. Hence, the portion must begin at point B on the semicircle γ_-^1 centered at $\omega_n x_1 = -3$. Using the same argument, the portion of the trajectory terminating at B must have initiated at point C on the semicircle γ_+^2 centered at $\omega_n x_1 = 5$. The process is obvious by now, so that we conclude the argument by assuming that the initial point $\omega_n \mathbf{x}_0$ is the starting point of the arc terminating at C, as can be seen in Fig. 6.14. Figure

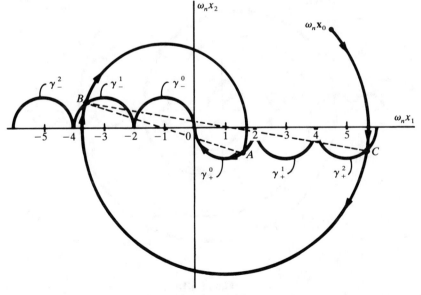

Figure 6.14

6.15 shows a complete picture, in which we identify the switching curve γ as the union of all the semicircles $\gamma_+^0, \gamma_+^1, \gamma_+^2 \cdots$ and $\gamma_-^0, \gamma_-^1, \gamma_-^2, \ldots$. The region R_- above γ is characterized by controls $u^* = -1$ and the region R_+ below γ is characterized by $u^* = +1$. A time-optimal trajectory initiated in R_- and reaching γ_+^j will switch the control from $u^*(t) = -1$ to $u^*(t) = +1$ at γ_+^j and then back to $u^*(t) = -1$ at γ_-^{j-1}. The process repeats itself until the trajectory reaches either γ_+^0 or γ_-^0, when it follows the respective semicircle to the origin. A trajectory initiated in R_+ will begin switching at γ_-^j and from then on it follows a sequence similar to the one described above. Two typical trajectories, one initiated in R_- and the other initiated in R_+, are shown in Fig. 6.15.

The control implementation follows the same pattern as that in Example 6.5. To this end, we define the switching function as

$$\gamma(\mathbf{x}(t)) = \gamma_+(\mathbf{x}(t)) \cup \gamma_-(\mathbf{x}(t)) = \left[\bigcup_{j=0}^{\infty} \gamma_+^j(\mathbf{x}(t))\right] \cup \left[\bigcup_{j=0}^{\infty} \gamma_-^j(\mathbf{x}(t))\right] \quad \text{(p)}$$

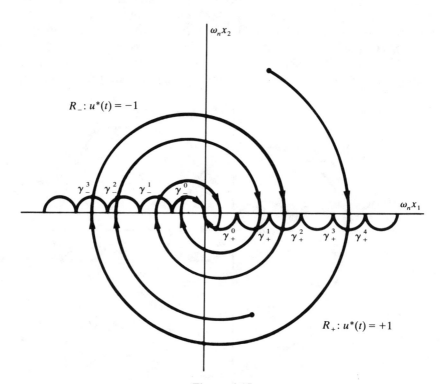

Figure 6.15

where

$$\gamma_+^j(\mathbf{x}(t)): [\omega_n x_1(t) - (2j+1)]^2 + [\omega_n x_2(t)]^2 = 1, \quad \omega_n x_2(t) < 0$$
$$\gamma_-^j(\mathbf{x}(t)): [\omega_n x_1(t) + (2j+1)]^2 + [\omega_n x_2(t)]^2 = 1, \quad \omega_n x_2(t) > 0$$
(q)

so that $\gamma(\mathbf{x}(t)) > 0$ implies that $\mathbf{x}(t)$ lies in R_-, $\gamma(\mathbf{x}(t)) < 0$ implies that $\mathbf{x}(t)$ lies in R_+, $\gamma_+^0(\mathbf{x}(t)) = 0$ implies that $\mathbf{x}(t)$ lies on the upper semicircle with the center at $\omega_n x_1 = -1$ and $\gamma_-^0(\mathbf{x}(t)) = 0$ implies that $\mathbf{x}(t)$ lies on the lower semicircle with the center at $\omega_n x_1 = +1$. In view of this, the optimal control law can be expressed in the form

$$u^*(t) = \begin{cases} -1 & \text{for } \mathbf{x}(t) \text{ such that } \gamma(\mathbf{x}(t)) > 0 \\ +1 & \text{for } \mathbf{x}(t) \text{ such that } \gamma(\mathbf{x}(t)) < 0 \\ -1 & \text{for } \mathbf{x}(t) \text{ such that } \gamma_-^0(\mathbf{x}(t)) = 0 \\ +1 & \text{for } \mathbf{x}(t) \text{ such that } \gamma_+^0(\mathbf{x}(t)) = 0 \\ 0 & \text{for } \mathbf{x}(t) = \mathbf{0} \end{cases}$$
(r)

The control implementation is shown in the block diagram of Fig. 6.16.

Examples 6.5 and 6.6 are both concerned with second-order systems for which the motion can be represented in the state plane. This graphical description provides a relatively easy way of recognizing the points at which the control switches from +1 to −1 and vice versa, because these points must lie on the switching curve. Although one can extend these concepts to higher-order systems, $n \geq 3$, by conceiving of trajectories in the state space and of switching hypersurfaces, such an approach is not practical. In Section 6.15, we examine an approach that can provide some solution to the time-optimal control of higher-order systems.

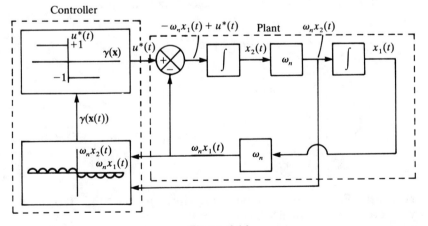

Figure 6.16

6.10 MINIMUM-FUEL PROBLEMS

In many engineering applications control is achieved by means of thrusters, which generate the thrust by expelling burned fuel. Hence, we can conceive of an optimal control problem in which the fuel consumed is minimized.

Let us consider once again the control problem defined by the state equations in the form

$$\dot{\mathbf{x}}(t) = \mathbf{a}(\mathbf{x}(t), t) + B(\mathbf{x}(t), t)\mathbf{u}(t) \tag{6.137}$$

where B is an $n \times r$ matrix and $\mathbf{u}(t)$ is an r-dimensional control vector whose components are constrained in magnitude by

$$|u_j(t)| \leq 1, \quad j = 1, 2, \ldots, r \tag{6.138}$$

We assume that the performance measure for the minimum-fuel problem is given by

$$J(\mathbf{u}) = \int_{t_0}^{t_f} \left[\sum_{j=1}^{r} c_j |u_j(t)| \right] dt \tag{6.139}$$

where c_j are positive constants. Then, the minimum-fuel problem is defined as the problem of determining the control $\mathbf{u}(t)$ whose components $u_j(t)$ satisfy the constraints (6.138) and one that drives the initial state $\mathbf{x}(t_0) = \mathbf{x}_0$ to the target set $S(t)$ and minimizes the performance measure $J(\mathbf{u})$ given by Eq. (6.139).

The approach to the minimum-fuel problem is similar to that to the minimum-time problem, the main difference lying in the details, which tend to be more complicated in the minimum-fuel problem. Recognizing from Eqs. (6.102) and (6.139) that in the case at hand $h = 0$ and

$$g(\mathbf{x}(t), \mathbf{u}(t), t) = \sum_{j=1}^{r} c_j |u_j(t)| \tag{6.140}$$

and using Eq. (6.44), the Hamiltonian becomes

$$\mathcal{H}(\mathbf{x}(t), \mathbf{u}(t), t) = \sum_{j=r}^{r} c_j |u_j(t)| + \mathbf{p}^T(t)[\mathbf{a}(\mathbf{x}(t), t) + B(\mathbf{x}(t), t)\mathbf{u}(t)] \tag{6.141}$$

Hence, using Pontryagin's minimum principle, inequality (6.100), a necessary condition for optimality has the form

$$\sum_{j=1}^{r} c_j |u_j^*(t)| + \mathbf{p}^{*T}(t)[\mathbf{a}(\mathbf{x}^*(t), t) + B(\mathbf{x}^*(t), t)\mathbf{u}^*(t)]$$

$$\leq \sum_{j=1}^{r} c_j |u_j(t)| + \mathbf{p}^{*T}(t)[\mathbf{a}(\mathbf{x}^*(t), t) + B(\mathbf{x}^*(t), t)\mathbf{u}(t)] \quad (6.142)$$

which reduces to

$$\sum_{j=1}^{r} c_j |u_j^*(t)| + \mathbf{p}^{*T}(t) B(\mathbf{x}^*(t), t)\mathbf{u}^*(t)$$

$$\leq \sum_{j=1}^{r} c_j |u_j(t)| + \mathbf{p}^{*T}(t) B(\mathbf{x}^*(t), t)\mathbf{u}(t) \quad (6.143)$$

The interpretation of inequality (6.143) is that $\sum_{j=1}^{r} c_j |u_j(t)| + \mathbf{p}^{*T}(t) B(\mathbf{x}^*(t), t)\mathbf{u}(t)$ attains its minimum value when $\mathbf{u}(t) = \mathbf{u}^*(t)$.

Next, let us consider

$$\min_{\mathbf{u}(t) \in \Omega} \left\{ \sum_{j=1}^{r} c_j |u_j(t)| + \mathbf{p}^{*T}(t) B(\mathbf{x}^*(t), t)\mathbf{u}(t) \right\}$$

$$= \min_{\mathbf{u}(t) \in \Omega} \left\{ \sum_{j=1}^{r} c_j \left[|u_j(t)| + u_j(t) \frac{1}{c_j} \mathbf{b}_j^T(\mathbf{x}^*(t), t)\mathbf{p}^*(t) \right] \right\}$$

$$= \sum_{j=1}^{r} c_j \left\{ \min_{|u_j(t)| \leq 1} \left[|u_j(t)| + u_j(t) \frac{1}{c_j} \mathbf{b}_j^T(\mathbf{x}^*(t), t)\mathbf{p}^*(t) \right] \right\} \quad (6.144)$$

where Ω is a subset of the r-dimensional space of control variables having the form of a hypercube and $\mathbf{b}_j$ are the columns of B. Moreover, the minimization and summation were interchanged on the basis of the assumption that the functions $u_j(t)$ can be constrained independently of one another. It can be verified that [A18]

$$\min_{|u_j(t)| \leq 1} \left[|u_j(t)| + u_j(t) \frac{1}{c_j} \mathbf{b}_j^T(\mathbf{x}^*(t), t)\mathbf{p}^*(t) \right]$$

$$= \begin{cases} 0 & \text{if } \left| \frac{1}{c_j} \mathbf{b}_j^T(\mathbf{x}^*(t), t)\mathbf{p}^*(t) \right| < 1 \\ 1 - \left| \frac{1}{c_j} \mathbf{b}_j^T(\mathbf{x}^*(t), t)\mathbf{p}^*(t) \right| & \text{if } \left| \frac{1}{c_j} \mathbf{b}_j^T(\mathbf{x}^*(t), t)\mathbf{p}^*(t) \right| \geq 1 \end{cases}$$

$$(6.145)$$

Then, recalling that the above minimum is obtained when $u_j(t) = u_j^*(t)$, we conclude that

$$u_j^*(t) = 0 \qquad \text{if } \left|\frac{1}{c_j} \mathbf{b}_j^T(\mathbf{x}^*(t), t)\mathbf{p}^*(t)\right| < 1$$

$$u_j^*(t) = -\text{sgn}\left[\frac{1}{c_j} \mathbf{b}_j^T(\mathbf{x}^*(t), t)\mathbf{p}^*(t)\right] \quad \text{if } \left|\frac{1}{c_j} \mathbf{b}_j^T(\mathbf{x}^*(t), t)\mathbf{p}^*(t)\right| > 1$$
(6.146)

$$0 \le u_j^*(t) \le 1 \qquad \text{if } \frac{1}{c_j} \mathbf{b}_j^T(\mathbf{x}^*(t), t)\mathbf{p}^*(t) = -1$$

$$-1 \le u_j^*(t) \le 0 \qquad \text{if } \frac{1}{c_j} \mathbf{b}_j^T(\mathbf{x}^*(t), t)\mathbf{p}^*(t) = +1$$

The dependence of $u_j^*(t)$ on $(1/c_j)\mathbf{b}_j^T(\mathbf{x}^*(t), t)\mathbf{p}^*(t)$ is displayed in Fig. 6.17. Implied in Fig. 6.17 is the assumption that *the problem is normal*, i.e., there is no finite time interval over which $|(1/c_j)\mathbf{b}_j^T(\mathbf{x}^*(t), t)\mathbf{p}^*(t)| = 1$. The functional dependence expressed by Eqs. (6.146) can be written in the compact form

$$u_j^*(t) = -\text{dez}\left[\frac{1}{c_j} \mathbf{b}_j^T(\mathbf{x}^*(t), t)\mathbf{p}^*(t)\right], \quad j = 1, 2, \ldots, r \quad (6.147)$$

where dez [ ] is known as the *dead-zone function*, a three-level piecewise constant function of time defined as

$$\begin{aligned}
f &= \text{dez}(x) \\
f &= 0 & &\text{if } |x| < 1 \\
f &= \text{sgn}(x) & &\text{if } |x| > 1 \\
0 &\le f \le 1 & &\text{if } x = +1 \\
-1 &\le f \le 0 & &\text{if } x = -1
\end{aligned} \qquad (6.148)$$

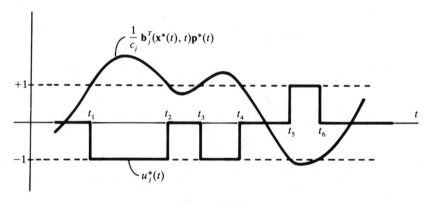

Figure 6.17

and we note that the levels are $+1$, 0 and -1. The dead-zone function is also known as *bang-off-bang* and as *on-off*. The above results can be summarized as follows: *If the fuel-optimal problem is normal, then the components of the fuel-optimal control are piecewise constant functions of time satisfying Eqs. (6.147)*. This statement is known as the *on-off principle*. Equations (6.147) can be expressed in the vector form

$$\mathbf{u}^*(t) = -\text{DEZ}[C^{-1}B^T(\mathbf{x}^*(t), t)\mathbf{p}^*(t)] \tag{6.149}$$

where $C = \text{diag}(c_j)$.

Example 6.7 Design a fuel-optimal control for the harmonic oscillator of Example 6.6.

Equations (a)–(e) of Example 6.6 remain the same in this example. On the other hand, here the performance measure is

$$J(\mathbf{u}) = \int_0^{t_f} |u|\, dt \tag{a}$$

so that the Hamiltonian is

$$\mathcal{H}(\mathbf{x}(t), u(t), \mathbf{p}(t)) = |u(t)| + \omega_n p_1(t)x_2(t) + p_2(t)[-\omega_n x_1(t) + u(t)] \tag{b}$$

Using the on-off principle, we obtain the optimal control

$$u^*(t) = \begin{cases} 0 & \text{if } |p_2^*(t)| < 1 \\ -\text{sgn}\,|p_2^*(t)| & \text{if } |p_2^*(t)| > 1 \end{cases} \tag{c}$$

The costate equations are the same as in Example 6.6, so that $p_2^*(t)$ has the expression

$$p_2^*(t) = P \sin(\omega_n t + \theta) \tag{d}$$

where P is the amplitude and θ the phase angle. Unlike Example 6.6, here P and θ play a much larger role in the design of a fuel-optimal control. This is so because the control is activated only when $|p_2^*(t)| > 1$, and not merely when $|p_2^*(t)| > 0$ as in time-optimal problems. Moreover, it is clear that the phase angle determines the amount of shift of $p_2^*(t) = P \sin(\omega_n t + \theta)$ along the time axis, and hence the timing of the control.

Because the state equations are the same, the general equation of the trajectories remains in the same form as for the circles of Example 6.6, or

$$[\omega_n x_1(t) - \Delta]^2 + [\omega_n x_2(t)]^2 = [\omega_n x_1(0) - \Delta]^2 + [\omega_n x_2(0)]^2 \tag{e}$$

Here, however, there are three levels for the control $u^*(t)$, namely $+1$, 0

and -1, so that Δ can take these three values. Hence, here there are three families of circles. The first family corresponds to $\Delta = 1$ and has the equation

$$[\omega_n x_1(t) - 1]^2 + [\omega_n x_2(t)]^2 = [\omega_n x_1(0) - 1]^2 + [\omega_n x_2(0)]^2 \tag{f}$$

and we note that the circles have the center at $\omega_n x_1 = 1$ in the state plane $\omega_n x_1$, $\omega_n x_2$. The second family corresponds to $\Delta = 0$ and has the equation

$$[\omega_n x_1(t)]^2 + [\omega_n x_2(t)]^2 = [\omega_n x_1(0)]^2 + [\omega_n x_2(0)]^2 \tag{g}$$

so that the center of the circles is at the origin of the state plane. Finally, the third family corresponds to $\Delta = -1$ and has the equation

$$[\omega_n x_1(t) + 1]^2 + [\omega_n x_2(t)]^2 = [\omega_n x_1(0) + 1]^2 + [\omega_n x_2(0)]^2 \tag{h}$$

The center of these circles is at $\omega_n x_1 = -1$. The three families of circles are shown in Fig. 6.18.

Next, let us turn our attention to the costate variable $p_2^*(t)$ and examine the role of the amplitude P and phase angle θ in the level of the control. To this end, we consider a typical plot $p_2^*(t)$ versus t, as shown in Fig. 6.19. This figure permits us to conclude that the control sequence is as follows: $u^*(t) = 0$ for $0 < t < t_1$, $u^*(t) = -1$ for $t_1 < t < t_2$, $u^*(t) = 0$ for $t_2 < t < t_3$, $u^*(t) = 1$ for $t_3 < t < t_4$, etc. We also observe that the time intervals during which the control is activated are all equal to one another and the time intervals during which the control is inactive are also equal to one another.

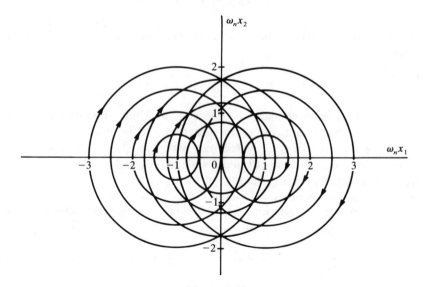

Figure 6.18

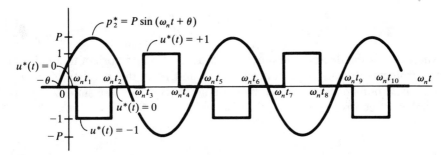

Figure 6.19

It can be shown [S26] that during the control time intervals the trajectory sweeps the same angle in the state plane, where the angle is bisected by the axis $\omega_n x_2$. It follows that the state plane can be divided into four sectors, two control sectors and two inactive sectors, where the control sectors sweep an angle α and are bisected by the $\omega_n x_2$ axis and the inactive sectors sweep an angle $\pi - \alpha$ and are bisected by the $\omega_n x_1$ axis, as shown in Fig. 6.20. Any trajectory represents the union of arcs of circles of the type shown in Fig. 6.18. The inactive sectors are denoted by R_0 and the control sectors by R_+ and R_- corresponding to the controls $u^*(t) = +1$ and $u^*(t) = -1$, respectively. Denoting by $r_0 = |\mathbf{x}_0|$ the magnitude of the initial state vector, it can be shown that when the motion is initiated in an inactive sector the following relations hold [S26]

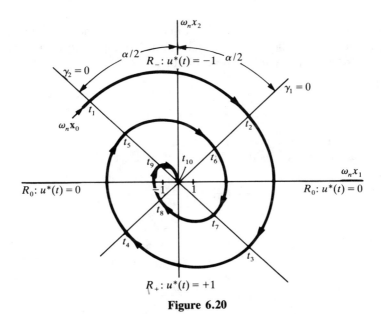

Figure 6.20

$$\Delta r_0 = 2 \sin \frac{\alpha}{2} = \frac{r_0}{n_c} \qquad \text{(i)}$$

where Δr_0 is the decrease in the radial distance from the origin to a point on the trajectory during one control action and n_c is the number of control actions, a number to be chosen by the analyst. But, because $\sin \alpha/2 \le 1$, the number of control actions must satisfy

$$n_c > \frac{r_0}{2} \qquad \text{(j)}$$

Moreover, the amplitude and phase angle of the costate variable $p_2^*(t)$ are given by

$$P = \left(\cos \frac{\alpha}{2}\right)^{-1}, \quad \theta = \pi - \tan^{-1}\left[\frac{x_2(0)}{x_1(0)}\right] \qquad \text{(k)}$$

Figure 6.20 shows a typical optimal trajectory for the case in which the initial state lies in an inactive sector. Consistent with Fig. 6.19, the number of control actions was chosen as $n_c = 5$, so that the trajectory approaches the origin of the state plane at the rate of $\Delta r_0 = r_0/5$ per control action.

When the initial state lies inside a control sector, then the first control action is of shorter duration than α/ω_n. In this case, it can be shown that [S26]

$$\Delta r_0 = \frac{r_1}{n_c - 1} \qquad \text{(l)}$$

where r_1 is the radial distance to the trajectory at the end of the first control action and $n_c - 1$ is the number of full control actions. Moreover, the angle defining the control sectors can be obtained from

$$\sin \frac{\alpha}{2} = \frac{r_0^2 - 2r_0 \sin \phi}{4(n_c^2 - n_c)} \qquad \text{(m)}$$

where ϕ is the angle between $\mathbf{x}_0$ and axis $\omega_n x_2$. Finally, the amplitude P of the costate $p_2^*(t)$ is defined by the first of Eqs. (k) and the phase angle by

$$\theta = \pi - \tan^{-1}\left[\frac{x_2(0)}{x_1(0)}\right] + \delta_1 - \delta_2 \qquad \text{(n)}$$

where

$$\begin{aligned}\delta_1 &= \sin^{-1} \frac{\cos \phi}{(r_0^2 + 1 - 2r_0 \sin \phi)^{1/2}} \\ \delta_2 &= \sin^{-1} \frac{\cos (\alpha/2)}{(r_0^2 + 1 - 2r_0 \sin \phi)^{1/2}}\end{aligned} \qquad \text{(o)}$$

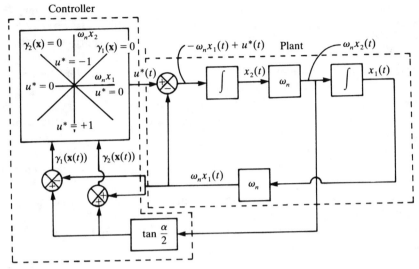

Figure 6.21

To implement the minimum-fuel control as a feedback control, we introduce the switching functions

$$\gamma_1(\mathbf{x}(t)) = \tan(\alpha/2)\omega_n x_2(t) - \omega_n x_1(t)$$
$$\gamma_2(\mathbf{x}(t)) = \tan(\alpha/2)\omega_n x_2(t) + \omega_n x_1(t) \qquad (p)$$

so that $\gamma_1(\mathbf{x}(t)) > 0$ and $\gamma_2(\mathbf{x}(t)) > 0$ implies that $\mathbf{x}(t)$ lies in R_-, $\gamma_1(\mathbf{x}(t)) < 0$ and $\gamma_2(\mathbf{x}(t)) < 0$ implies that $\mathbf{x}(t)$ lies in R_+ and $\gamma_1(\mathbf{x}(t)) > 0$ and $\gamma_2(\mathbf{x}(t)) < 0$ or $\gamma_1(\mathbf{x}(t)) < 0$ and $\gamma_2(\mathbf{x}(t)) > 0$ implies that $\mathbf{x}(t)$ lies in R_0. Note that $\gamma_1(\mathbf{x}(t)) = 0$ and $\gamma_2(\mathbf{x}(t)) = 0$ represent the equations of the switching lines shown in Fig. 6.20. In view of this, the control law can be expressed in the form

$$u^*(t) = \begin{cases} -1 & \text{for } \mathbf{x}(t) \text{ such that } \gamma_1(\mathbf{x}(t)) > 0 \text{ and } \gamma_2(\mathbf{x}(t)) > 0 \\ 0 & \text{for } \mathbf{x}(t) \text{ such that } \gamma_1(\mathbf{x}(t)) < 0 \text{ and } \gamma_2(\mathbf{x}(t)) > 0 \\ +1 & \text{for } \mathbf{x}(t) \text{ such that } \gamma_1(\mathbf{x}(t)) < 0 \text{ and } \gamma_2(\mathbf{x}(t)) < 0 \\ 0 & \text{for } \mathbf{x}(t) \text{ such that } \gamma_1(\mathbf{x}(t)) > 0 \text{ and } \gamma_2(\mathbf{x}(t)) < 0 \end{cases} \qquad (q)$$

The control implementation is displayed in the block diagram of Fig. 6.21.

6.11 A SIMPLIFIED ON-OFF CONTROL

In the time-optimal and fuel-optimal control problems, the control depends explicitly on the costate variables and not on the state variables, so that in this sense they are not feedback controls. In some cases, however, the control can be made to depend on the state variables, as witnessed by Examples 6.5, 6.6 and 6.7, so that in such cases it does represent feedback control. Still, the controls can be complicated functions of the state variables, causing difficulties in implementation, so that a simpler control law appears desirable. One such control law is the on-off control with a dead zone shown in Fig. 6.22, and we note that here the control is related to the velocity directly and not indirectly through the costate. In fact, the control law has the simple form

$$u_j(t) = -c_j \, \text{dez}\,(\dot{q}_j(t)), \quad j = 1, 2, \ldots, r \quad (6.150)$$

where dez( ) denotes the dead-zone function defined in Section 6.10, except that here the amplitude is c_j and the width of the dead zone is $2d_j$. Equations (6.150) can be generalized somewhat by writing the control law in the vector form

$$\mathbf{u}(t) = -C \, \text{DEZ}\,[\mathbf{x}(t)] \quad (6.151)$$

where C is an $r \times n$ matrix with entries multiplying displacements equal to zero and entries multiplying velocities having a diagonal form with c_j on the diagonal. Then, using Eq. (6.137), the state equations become

$$\dot{\mathbf{x}}(t) = \mathbf{a}(\mathbf{x}(t), t) - B(\mathbf{x}(t), t) C \, \text{DEZ}\,[\mathbf{x}(t)] \quad (6.152)$$

For time-invariant linear plants, Eq. (6.152) reduces to

$$\dot{\mathbf{x}}(t) = A\mathbf{x}(t) - BC \, \text{DEZ}\,[\mathbf{x}(t)] \quad (6.153)$$

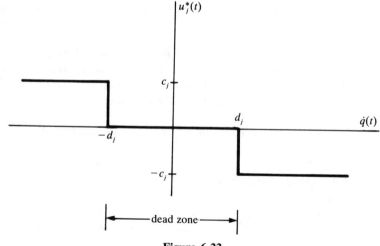

Figure 6.22

A SIMPLIFIED ON-OFF CONTROL

When the system enters ultimately the dead zone and stays there, there will be some residual motion. For an undamped system, this residual motion has the form of oscillation, but the implication is that the dead zone is relatively narrow, so that the amplitude of oscillation is sufficiently small that it can be tolerated.

For multi-input or for high-order systems, the above control scheme can still encounter difficulties, in the same way that the time-optimal and fuel-optimal controls do. But, for second-order systems with a single input, the scheme provides an attractive alternative to the time-optimal and fuel-optimal controls. Some degree of optimality may be achieved by changing the parameters c_j and d_j, particularly the latter.

Example 6.8 Consider the harmonic oscillator of Examples 6.6 and 6.7 and use an on-off control of the type shown in Fig. 6.22 to drive an arbitrary initial state to the origin of the state space. Let $c_j = 1$ and $d_j = d$.

The state equations are the same as in Example 6.6 and the equation of the trajectories remains in the form of Eq. (e) of Example 6.7. It follows that the three families of circles shown in Fig. 6.18 still apply. The only question remaining is when to switch from one circle to another. Because the dead zone is defined by

$$u(t) = 0, \quad |\dot{q}(t)| = |\omega_n x_2(t)| < d \tag{a}$$

it follows that the switching curve reduces to two lines parallel to the $\omega_n x_1$ axis and at distances $\pm d$ from it, as shown in Fig. 6.23. The two lines divide

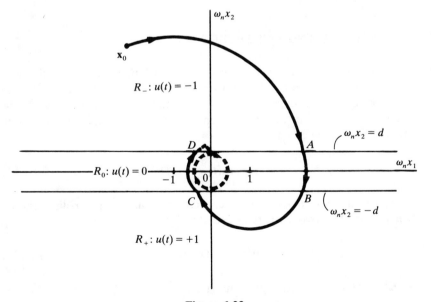

Figure 6.23

the state plane into three regions. The first region, denoted by R_-, is defined by $\omega_n x_2 > d$ and is characterized by trajectories in the form of circles with the center at $\omega_n x_1 = -1$. In this region the control is $u(t) = -1$. The second region is the dead zone, denoted by R_0 and characterized by Eq. (a). The third region is denoted by R_+ and defined by $\omega_n x_2 < -d$. The trajectories in R_+ are circles centered at $\omega_n x_1 = +1$ and the control is $u(t) = +1$. A trajectory initiated in R_- will move clockwise on a circle centered at $\omega_n x_1 = -1$ until it reaches $\omega_n x_2 = d$ at point A. If at this point $\omega_n x_1 > 0$, then the trajectory will enter R_0 and move clockwise on a circle with the center at the origin until it reaches $\omega_n x_2 = -d$ at point B. At this point it enters R_+ and continues on a circle centered at $\omega_n x_1 = +1$ until it reaches once again $\omega_n x_2 = -d$, this time at point C. If point C is such that $\omega_n x_1 > 0$, upon entering R_0, clockwise motion on a circle with the center at the origin will tend to push the trajectory back into R_+, giving rise to a phenomenon known as chattering. On the other hand, if point C is such that $\omega_n x_1 < 0$, the trajectory enters R_0 and it follows a circle centered at the origin until it reaches $\omega_n x_2 = d$ at point D. Upon entering R_-, the control will push the trajectory back into R_0, so that in this case chattering occurs along $\omega_n x_2 = d$. This case is shown in Fig. 6.23.

To prevent chattering, it is advisable to continue without control a little longer while in R_-. Then, the control $u(t) = -1$ should be delayed a little while in R_0. In this manner, the next portion of the trajectory will be a circle centered at the origin and of radius $\omega_n (x_1^2 + x_2^2)^{1/2} < d$. At this time, the control task is terminated, so that the system will continue with a small-amplitude oscillation. A trajectory in which the removal of the control was delayed twice is shown in dashed line in Fig. 6.23.

6.12 CONTROL USING OBSERVERS

All the feedback control designs discussed until now have one thing in common, namely they are all based on the assumption that the full state vector is available for measurement. Then, the current control vector is assumed as a function of the current state vector. Indeed, the state equations have the vector form

$$\dot{\mathbf{x}}(t) = A\mathbf{x}(t) + B\mathbf{u}(t) \qquad (6.154)$$

where $\mathbf{x}(t)$ is the n-dimensional state vector and $\mathbf{u}(t)$ is the r-dimensional control vector. The two vectors are related by

$$\mathbf{u}(t) = -G\mathbf{x}(t) \qquad (6.155)$$

where G is the $n \times r$ control gain matrix.

Quite often, it is not practical to measure the full state vector, but only certain linear combinations of the state. Assuming that there are q measured quantities and denoting the measurement vector by $\mathbf{y}(t)$, the relation between the measurement vector and the state vector can be written in the form

$$\mathbf{y}(t) = C\mathbf{x}(t) \tag{6.156}$$

where $\mathbf{y}(t)$ is the output vector. The assumption is made here that the system is observable (see Section 3.13), which implies that the initial state can be deduced from the outputs within a finite time period.

The object of this section is to introduce a device permitting an estimate of the full state vector from the output vector. Such a state estimator is known as an *observer* [K16]. An observer for (6.154) is assumed to have the form

$$\dot{\hat{\mathbf{x}}}(t) = A\hat{\mathbf{x}}(t) + b\mathbf{u}(t) + K[\mathbf{y}(t) - C\hat{\mathbf{x}}(t)] \tag{6.157}$$

where $\hat{\mathbf{x}}(t)$ is the n-dimensional observer state vector and K is an $n \times q$ matrix, and we note that there are two types of inputs to the observer, the first is identical to the input to the original system and the second involves the output of the original system. Because the dimension of the observer state vector $\hat{\mathbf{x}}(t)$ is the same as the dimension of the state vector $\mathbf{x}(t)$ of the original system, an observer of the type (6.157) is said to be a *full-order observer*. In view of Eq. (6.156), when $\hat{\mathbf{x}}(t) = \mathbf{x}(t)$ the observer reduces to the original system. A block diagram of the observer is shown in Fig. 6.24.

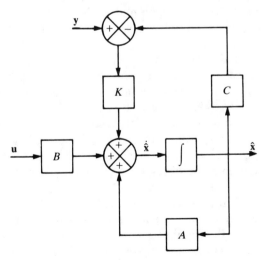

Figure 6.24

Introducing Eq. (6.156) into Eq. (6.157) and subtracting Eq. (6.154), we obtain

$$\dot{\mathbf{x}}(t) - \dot{\hat{\mathbf{x}}}(t) = [A - KC][\mathbf{x}(t) - \hat{\mathbf{x}}(t)] \qquad (6.158)$$

From Eq. (6.158), we conclude that if the observer initial state is the same as the initial state of the original system, $\hat{\mathbf{x}}(0) = \mathbf{x}(0)$, then $\hat{\mathbf{x}}(t) = \mathbf{x}(t)$ for all $t > 0$ and for all $\mathbf{u}(t)$. In this case, the observer state tracks the state of the original system. Note that the full-order observer of the type (6.157) is also known as an *identity observer*.

Next, we consider the case in which $\hat{\mathbf{x}}(0) \neq \mathbf{x}(0)$ and introduce the *observer error vector*

$$\mathbf{e}(t) = \mathbf{x}(t) - \hat{\mathbf{x}}(t) \qquad (6.159)$$

so that Eq. (6.158) can be rewritten in the form

$$\dot{\mathbf{e}}(t) = [A - KC]\mathbf{e}(t) \qquad (6.160)$$

Because Eq. (6.160) represents a linear homogeneous system of equations with constant coefficients, the nature of the solution vector $\mathbf{e}(t)$ depends on the eigenvalues of $A - KC$, where the eigenvalues are known as the *observer poles*. Hence, if all the observer poles lie in the left half of the complex plane, the solution $\mathbf{e}(t)$ is asymptotically stable, so that the error vector approaches zero asymptotically as $t \to \infty$. This implies that the observer state $\hat{\mathbf{x}}(t)$ approaches the original system state $\mathbf{x}(t)$ asymptotically. Clearly, the observer poles depend on the choice of the matrix K, where K is known as the *observer gain matrix*. If the system is completely observable, all the observer poles can be placed arbitrarily in the complex plane by a suitable choice of the observer gain matrix. Note that, because the solution $\mathbf{e}(t)$ must be real, any complex poles must occur in pairs of complex conjugates.

The question remains as to how to determine the observer gains. Clearly, for fast convergence, the observer poles should be deep in the left half of the complex plane, which implies a large gain matrix K. But, a large gain matrix makes the observer sensitive to sensor noise, which is added to the the original system output vector $\mathbf{y}(t)$. Optimal gains can be computed by adopting a stochastic approach to the problem, leading to the so-called *Kalman–Bucy filter*. Note that, in contrast, the observer (6.157) is deterministic. A deterministic observer is commonly known as a *Luenberger observer* [L16].

One question of interest is whether the choice of the observer poles is affected by the choice of the closed-loop poles of the original system. In this regard, we recognize that in implementing the feedback control we must use the estimated state $\hat{\mathbf{x}}(t)$, as the actual state $\mathbf{x}(t)$ is not available. Hence, Eq. (6.155) must be replaced by

CONTROL USING OBSERVERS   235

$$\mathbf{u}(t) = -G\hat{\mathbf{x}}(t) \tag{6.161}$$

Introducing Eq. (6.161) into Eq. (6.154) and considering Eq. (6.159), we obtain

$$\dot{\mathbf{x}}(t) = [A - BG]\mathbf{x}(t) + BG\mathbf{e}(t) \tag{6.162}$$

Equations (6.160) and (6.162) can be combined into

$$\begin{bmatrix} \dot{\mathbf{x}}(t) \\ \dot{\mathbf{e}}(t) \end{bmatrix} = \begin{bmatrix} A - BG & BG \\ 0 & A - KC \end{bmatrix} \begin{bmatrix} \mathbf{x}(t) \\ \mathbf{e}(t) \end{bmatrix} \tag{6.163}$$

The matrix of coefficients in Eq. (6.163) is in block-triangular form, so that the characteristic polynomial of the combined system is the product of the characteristic polynomials of $A - BG$ and $A - KC$. The preceding statement is known as the *deterministic separation principle*. It implies that the observer poles can be chosen independently of the closed-loop poles of the original system (6.154).

Before we discuss the subject of stochastic observers, we wish to examine the possibility of using observers of lower order than the full-order observer described by Eq. (6.157). To this end, we consider once again Eqs. (6.154) and (6.156), and note that Eq. (6.156) provides q equations in the n unknowns $x_1(t), x_2(t), \ldots, x_n(t)$, where the latter are the n components of the state vector $\mathbf{x}(t)$. It follows that, to reconstruct the state, it is only necessary to derive $n - q$ linear combinations of the state. Hence, let us introduce the $(n - q)$-dimensional vector

$$\mathbf{p}(t) = C'\mathbf{x}(t) \tag{6.164}$$

where C' is an $(n - q) \times n$ matrix. Equations (6.156) and (6.164) can be combined into

$$\begin{bmatrix} \mathbf{y}(t) \\ \mathbf{p}(t) \end{bmatrix} = \begin{bmatrix} C \\ C' \end{bmatrix} \mathbf{x}(t) \tag{6.165}$$

Then, assuming that the matrix of coefficients on the right side of Eq. (6.165) is nonsingular, we can obtain the state vector by writing

$$\mathbf{x}(t) = \begin{bmatrix} C \\ C' \end{bmatrix}^{-1} \begin{bmatrix} \mathbf{y}(t) \\ \mathbf{p}(t) \end{bmatrix} = L_1 \mathbf{y}(t) + L_2 \mathbf{p}(t) \tag{6.166}$$

where we introduced the notation

$$\left[\frac{C}{C'}\right]^{-1} = [L_1 \mid L_2] \qquad (6.167)$$

in which L_1 and L_2 are $n \times q$ and $n \times (n - q)$ matrices, respectively. Because the vector $\mathbf{y}(t)$ is known, through measurements, it follows from Eq. (6.166) that to estimate the state it is only necessary to estimate $\mathbf{p}(t)$. Hence, denoting the estimate of $\mathbf{p}(t)$ by $\hat{\mathbf{p}}(t)$, the state estimate is simply

$$\hat{\mathbf{x}}(t) = L_1 \mathbf{y}(t) + L_2 \hat{\mathbf{p}}(t) \qquad (6.168)$$

To construct an observer for $\mathbf{p}(t)$, let us differentiate Eq. (6.164) with respect to time, consider Eqs. (6.154) and (6.166) and write

$$\dot{\mathbf{p}}(t) = C' \dot{\mathbf{x}}(t) = C'[A\mathbf{x}(t) + B\mathbf{u}(t)]$$
$$= C'AL_2 \mathbf{p}(t) + C'AL_1 \mathbf{y}(t) + C'B\mathbf{u}(t) \qquad (6.169)$$

Following the same procedure, but starting with Eq. (6.156) instead of Eq. (6.164), we obtain directly

$$\dot{\mathbf{y}}(t) = CAL_2 \mathbf{p}(t) + CAL_1 \mathbf{y}(t) + CB\mathbf{u}(t) \qquad (6.170)$$

Replacing $\mathbf{p}(t)$ by $\hat{\mathbf{p}}(t)$ in Eqs. (6.169) and (6.170), we can conceive of an observer for $\mathbf{p}(t)$ in the form

$$\dot{\hat{\mathbf{p}}}(t) = C'AL_2 \hat{\mathbf{p}}(t) + C'AL_1 \mathbf{y}(t) + C'B\mathbf{u}(t)$$
$$+ K[\dot{\mathbf{y}}(t) - CAL_1 \mathbf{y}(t) - CB\mathbf{u}(t) - CAL_2 \hat{\mathbf{p}}(t)] \qquad (6.171)$$

The solution of Eq. (6.171) is then introduced into Eq. (6.168), thus obtaining an estimate $\hat{\mathbf{x}}(t)$ of the state.

Equation (6.171) involves the time derivative of the output $\mathbf{y}(t)$, so that the question arises whether this added operation is necessary. To address this question, we consider

$$\hat{\mathbf{r}}(t) = \hat{\mathbf{p}}(t) - K\mathbf{y}(t) \qquad (6.172)$$

Differentiating Eq. (6.172) with respect to time and using Eqs. (6.171) and (6.172), we obtain

$$\dot{\hat{\mathbf{r}}}(t) = [C'AL_2 - KCAL_2]\hat{\mathbf{r}}(t) + [C'AL_2 K + C'AL_1$$
$$- KCAL_1 - KCAL_2 K]\mathbf{y}(t) + [C'B - KCB]\mathbf{u}(t) \qquad (6.173)$$

which is free of $\dot{\mathbf{y}}(t)$. The estimated state is obtained by introducing Eq. (6.172) into Eq. (6.168) with the result

$$\hat{\mathbf{x}}(t) = L_2\hat{\mathbf{r}}(t) + [L_1 + L_2 K]\mathbf{y}(t) \tag{6.174}$$

so that the state estimation problem reduces to solving a differential equation for the $(n - q)$-dimensional vector $\hat{\mathbf{r}}(t)$. The lower-order observer is commonly known as a *reduced-order observer* [K16].

Example 6.9 The harmonic oscillator of example 6.6 has a sensor measuring displacements. Design a full-order Luenberger observer to estimate the state. Let the initial state of the system and of the observer be

$$\mathbf{x}(0) = \begin{bmatrix} 1 \\ 0 \end{bmatrix}, \quad \hat{\mathbf{x}}(0) = \mathbf{0} \tag{a}$$

The natural frequency of the oscillator is $\omega_n = 10$ rad/s. Let the observer poles have the values

$$\begin{matrix} s_1 \\ s_2 \end{matrix} = -10 \pm i20 \tag{b}$$

Plot the components of the system state vector and of the observer state vector as functions of time and compare results.

In the first place, we wish to compute the observer gains. Because this is a second-order system using one measurement, the observer gain matrix is actually a two-dimensional vector. Moreover, from Eq. (6.156), we conclude that

$$C = [1 \quad 0] \tag{c}$$

Hence, considering Eq. (6.157) and recalling the matrix A from Example 6.6, the characteristic equation is

$$\det[sI - A + KC] = \det\left[s\begin{bmatrix} 1 & 0 \\ 0 & 1 \end{bmatrix} - \begin{bmatrix} 0 & \omega_n \\ -\omega_n & 0 \end{bmatrix} + \begin{bmatrix} k_1 \\ k_2 \end{bmatrix}[1 \quad 0]\right]$$

$$= \begin{bmatrix} s + k_1 & -\omega_n \\ \omega_n + k_2 & s \end{bmatrix} = s^2 + k_1 s + \omega_n(\omega_n + k_2) = s^2 + k_1 s + 10(10 + k_2) = 0 \tag{d}$$

Introducing Eqs. (b) into Eq. (d) and solving for k_1 and k_2, we obtain the observer gain matrix

$$K = \begin{bmatrix} k_1 \\ k_2 \end{bmatrix} = \begin{bmatrix} 20 \\ 40 \end{bmatrix} \tag{e}$$

and we note that the gains k_1 and k_2 have units of frequency. Then, using Eq. (6.157), the observer equation is

$$\begin{bmatrix} \dot{\hat{x}}_1(t) \\ \dot{\hat{x}}_2(t) \end{bmatrix} = \begin{bmatrix} 0 & 10 \\ -10 & 0 \end{bmatrix} \begin{bmatrix} \hat{x}_1(t) \\ \hat{x}_2(t) \end{bmatrix} + \begin{bmatrix} 0 \\ 1 \end{bmatrix} u(t) + \begin{bmatrix} 20 \\ 40 \end{bmatrix} [x_1(t) - \hat{x}_1(t)] \quad \text{(f)}$$

Of course, the system equation is obtained from Eq. (f) by letting $\hat{x}_j = x_j(t)$ ($j = 1, 2$).

Plots of $x_j(t)$ vs. t and $\hat{x}_j(t)$ vs. t ($j = 1, 2$) are shown in Fig. 6.25 for the case in which the function $u(t)$ represents the unit step function. As can be concluded from the plots, convergence of the observer state to the actual state is relatively fast.

Example 6.10 Design a reduced-order observer for the system of Example 6.9.

Because this is a second-order system and there is one measurement, namely the displacement $x_1(t)$, the reduced-order observer is of order one

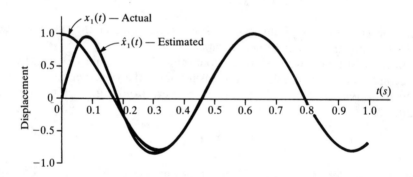

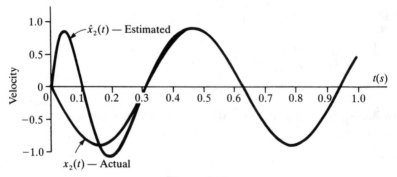

Figure 6.25

and is designed to estimate the velocity $x_2(t)$. Hence, from Eq. (6.164), we have

$$\mathbf{p}(t) = p(t) = x_2(t) = C'\mathbf{x}(t) \tag{a}$$

from which we conclude that

$$C' = [0 \quad 1] \tag{b}$$

so that, recalling Eq. (c) of Example 6.9, we can write

$$\left[\frac{C}{C'}\right] = \begin{bmatrix} 1 & 0 \\ 0 & 1 \end{bmatrix} \tag{c}$$

It follows from Eq. (6.167) that

$$L_1 = \begin{bmatrix} 1 \\ 0 \end{bmatrix}, \quad L_2 = \begin{bmatrix} 0 \\ 1 \end{bmatrix} \tag{d}$$

Inserting Eqs. (b) and (d) into Eq. (6.173) and recalling the matrices A, B and C from Examples 6.6 and 6.9, the reduced-order observer becomes

$$\dot{\hat{r}}(t) = -k\omega_n \hat{r}(t) - \omega_n(1 + k^2)x_1(t) + u(t) \tag{e}$$

where, from Eq. (6.172), we recognized that the gain matrix is really a scalar, $K = k$, and we note that k is dimensionless. Finally, from Eq. (6.174), we obtain the estimated state

$$\hat{\mathbf{x}}(t) = \begin{bmatrix} 0 \\ 1 \end{bmatrix} \hat{r}(t) + \begin{bmatrix} 1 \\ k \end{bmatrix} x_1(t) \tag{f}$$

The actual velocity and the estimated velocity corresponding to $k = 20$ are plotted in Fig. 6.26 for the same initial conditions as in Example 6.9.

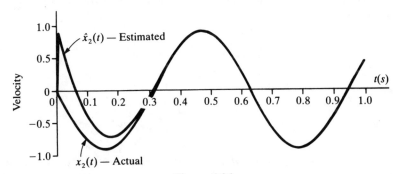

Figure 6.26

6.13 OPTIMAL OBSERVERS. THE KALMAN–BUCY FILTER

In section 6.12, we pointed out that the presence of sensor noise tends to affect the convergence of a Luenberger observer adversely. This leads naturally to a stochastic observer that not only can handle noise better but also is characterized by observer gains that are optimal in some sense. The optimal observer, or stochastic observer, is known as a *Kalman–Bucy Filter* [K16].

In the presence of noise, the state equations, Eq. (6.154), can be rewritten as

$$\dot{\mathbf{x}}(t) = A\mathbf{x}(t) + B\mathbf{u}(t) + \mathbf{v}(t) \qquad (6.175)$$

where $\mathbf{v}(t)$ is known as the *state excitation noise*. Note that $\mathbf{v}(t)$ can be regarded as including actuator noise and/or any external random disturbances. Similarly, the output equations, Eq. (6.156), can be rewritten as

$$\mathbf{y}(t) = C\mathbf{x} + \mathbf{w}(t) \qquad (6.176)$$

where $\mathbf{w}(t)$ is the *observation*, or *sensor noise*. Noise is stochastic in nature and its characteristics are generally described in terms of statistical quantities. Hence, before discussing the stochastic observer, we wish to introduce a number of definitions.

A function whose value at a given time cannot be predicted is said to be *random* and the function itself is called a *random* or *stochastic process*. The probability that the value of a random variable $x(t)$ lies in the interval defined by x and $x + dx$ is $p(x)\,dx$, where $p(x)$ is known as the *probability density function*. Then, the *mathematical expectation*, or the *expected value* of a given function $f(x)$ is defined as

$$G\{f(x)\} = \int_{-\infty}^{\infty} f(x)p(x)\,dx \qquad (6.177)$$

In the special case in which $f(x) = x$, the expected value is known as the *mean value*, or

$$E\{x\} = \bar{x} = \int_{-\infty}^{\infty} xp(x)\,dx \qquad (6.178)$$

When $f(x) = x^2$, we obtain

$$E\{x^2\} = \overline{x^2} = \int_{-\infty}^{\infty} x^2 p(x)\,dx \qquad (6.179)$$

which is known as the *mean square value* of x; its square root is called the *root mean square value*, or *rms value*. Similarly, the *variance* of x is defined as

$$\sigma_x^2 = E\{(x - \bar{x})^2\} = \int_{-\infty}^{\infty} (x - \bar{x})^2 p(x)\, dx = \overline{x^2} - (\bar{x})^2 \quad (6.180)$$

Equation (6.180) takes into consideration the fact that $\int_{-\infty}^{\infty} p(x)\, dx = 1$, which implies that the probability that x will fall in the interval $-\infty < x < \infty$ is a certainty. The square root of the variance is known as the *standard deviation* σ_x. A *Gaussian stochastic process* is characterized by the probability density function

$$p(x) = \frac{1}{\sigma_x \sqrt{2\pi}} \exp\left[\frac{-(x - \bar{x})^2}{2\sigma_x^2}\right] \quad (6.181)$$

and it is clear that a Gaussian process is fully defined by two quantities only, the mean value and the standard deviation.

In our study, we are concerned with quantities involving more than one stochastic process. In particular, let $x(t)$ and $y(t)$ be two random variables. Then, the *covariance function* between x and y is defined as

$$C_{xy} = E\{(x - \bar{x})(y - \bar{y})\}$$
$$= \int_{-\infty}^{\infty} \int_{-\infty}^{\infty} (x - \bar{x})(y - \bar{y}) p(x, y)\, dx\, dy \quad (6.182)$$

where $p(x, y)$ is the *joint probability density function*. In the case of Gaussian processes, the joint probability density function has the expression

$$p(x, y) = \frac{1}{2\pi \sigma_x \sigma_y (1 - \rho_{xy}^2)^{1/2}} \exp\left\{-\frac{1}{2(1 - \rho_{xy}^2)^{1/2}} \left[\left(\frac{x - \bar{x}}{\sigma_x}\right)^2 \right.\right.$$
$$\left.\left. - 2\rho_{xy} \frac{x - \bar{x}}{\sigma_x} \frac{y - \bar{y}}{\sigma_y} + \left(\frac{y - \bar{y}}{\sigma_y}\right)^2\right]\right\} \quad (6.183)$$

in which

$$\rho_{xy} = \frac{C_{xy}}{\sigma_x \sigma_y} \quad (6.184)$$

is known as the *correlation coefficient*. Its value lies between -1 and $+1$. Note that when $C_{xy} = 0$, which implies that $\rho_{xy} = 0$, the random variables x and y are uncorrelated. This does not mean that x and y are statistically independent. However, in the important case of Gaussian processes, the fact that $C_{xy} = 0$ does imply statistical independence, as can be concluded from Eq. (6.183). Another function of interest is the *correlation function* between x and y, given by

$$R_{xy} = E\{xy\} \quad (6.185)$$

The correlation function is also known as the *second-order joint moment*.

In the above discussion, it is implied that the various statistics are computed for a given time t. Covariance functions, however, can be defined also for the same function but at different times. Indeed, the covariance function between $x(t_1)$ and $x(t_2)$ is

$$C_x(t_1, t_2) = E\{[x(t_1) - \bar{x}(t_1)][x(t_2) - \bar{x}(t_2)]\} \tag{6.186}$$

Similarly, the correlation function between $x(t_1)$ and $x(t_2)$ is

$$R_x(t_1, t_2) = E\{x(t_1)x(t_2)\} \tag{6.187}$$

and is known as the *autocorrelation function*. For two random variables evaluated at different times, the function

$$C_{xy}(t_1, t_2) = E\{[x(t_1) - \bar{x}(t_1)][y(t_2) - \bar{y}(t_2)]\} \tag{6.188}$$

defines a *cross-covariance function* and the function

$$R_{xy}(t_1, t_2) = E\{x(t_1)y(t_2)\} \tag{6.189}$$

represents a *cross-correlation function*.

The above definitions can be extended to *vector-valued random variables*, or simply *random vectors*, defined as vectors with components representing stochastic processes. In particular, the *mean value* of an n-dimensional random vector $\mathbf{x}(t)$ is defined as

$$E\{\mathbf{x}(t)\} = \bar{\mathbf{x}}(t) = \int_{-\infty}^{\infty} \cdots \int_{-\infty}^{\infty} \mathbf{x}(t) p(x_1, x_2, \ldots, x_n) \, dx_1 \, dx_2 \cdots dx_n \tag{6.190}$$

Moreover, the *covariance matrix* is given by

$$C_\mathbf{x}(t_1, t_2) = E\{[\mathbf{x}(t_1) - \bar{\mathbf{x}}(t_1)][\mathbf{x}(t_2) - \bar{\mathbf{x}}(t_2)]^T\} \tag{6.191}$$

and the *correlation matrix*, or the *second-order joint moment matrix*, has the expression

$$R_\mathbf{x}(t_1, t_2) = E\{\mathbf{x}(t_1)\mathbf{x}^T(t_2)\} \tag{6.192}$$

Similarly, for two random vectors

$$C_{\mathbf{xy}}(t_1, t_2) = E\{[\mathbf{x}(t_1) - \bar{\mathbf{x}}(t_1)][\mathbf{y}(t_2) - \bar{\mathbf{y}}(t_2)]^T\} \tag{6.193}$$

defines the *cross-covariance matrix* and

$$R_{xy}(t_1, t_2) = E\{\mathbf{x}(t_1)\mathbf{y}^T(t_2)\} \tag{6.194}$$

defines the *cross-correlation matrix*.

The covariance matrix and the correlation matrix have the following properties

$$C_x(t_2, t_1) = C_x^T(t_1, t_2) \tag{6.195a}$$

$$R_x(t_2, t_1) = R_x^T(t_1, t_2) \tag{6.195b}$$

$$C_x(t_1, t_2) = R_x(t_1, t_2) - \bar{\mathbf{x}}(t_1)\bar{\mathbf{x}}^T(t_2) \tag{6.195c}$$

$$C_x(t, t) \geq 0 \tag{6.195d}$$

$$R_x(t, t) \geq 0 \tag{6.195e}$$

where $C_x(t, t) \geq 0$ implies that the covariance matrix of $\mathbf{x}(t)$ is positive semidefinite, with a similar statement for the correlation matrix.

An important class of stochastic processes is the one in which statistical properties do not change with time. Such processes are known as *stationary*. If for a random process $\mathbf{x}(t)$ the mean value $\bar{\mathbf{x}}(t)$ is constant and the covariance matrix $C_x(t_1, t_2)$ depends on $t_2 - t_1$ alone, and not on t_1, then the random process is said to be *weakly stationary*. If in addition the second-order moment matrix $R_x(t, t)$ is finite for all t, then the random process is said to be *wide-sense stationary*.

Next, we consider a Gaussian stochastic process $\mathbf{x}(t)$, where $\mathbf{x}(t)$ is an n-dimensional random vector. Associated with the times $t_1, t_2, \ldots, t_m$ there are m vectors $\mathbf{x}(t_1), \mathbf{x}(t_2), \ldots, \mathbf{x}(t_m)$. Then, we can define the compound covariance matrix as

$$C_x = \begin{bmatrix} C_x(t_1, t_1) & C_x(t_1, t_2) & \cdots & C_x(t_1, t_m) \\ C_x(t_2, t_1) & C_x(t_2, t_2) & \cdots & C_x(t_2, t_m) \\ \vdots & \vdots & & \vdots \\ C_x(t_m, t_1) & C_x(t_m, t_2) & \cdots & C_x(t_m, t_m) \end{bmatrix} \tag{6.196}$$

which is a matrix containing m^2 submatrices $C_x(t_i, t_j)$ of dimensions $n \times n$. Assuming that the matrix C_x is nonsingular, the joint probability density function has the form

$$p(\mathbf{x}(t_1), \mathbf{x}(t_2), \ldots, \mathbf{x}(t_m)) =$$

$$\frac{1}{[(2\pi)^{mn} \det C_x]^{1/2}} \exp\left\{-\frac{1}{2} \sum_{i=1}^{m} \sum_{j=1}^{m} [\mathbf{x}(t_i) - \bar{\mathbf{x}}(t_i)]^T D_{ij} [\mathbf{x}(t_j) - \bar{\mathbf{x}}(t_j)]\right\} \tag{6.197}$$

where D_{ij} are $n \times n$ submatrices of C_x^{-1}, or

$$D = C_x^{-1} = \begin{bmatrix} D_{11} & D_{12} & \cdots & D_{1m} \\ D_{21} & D_{22} & \cdots & D_{2m} \\ \vdots & \vdots & & \vdots \\ D_{m1} & D_{m2} & \cdots & D_{mm} \end{bmatrix} \quad (6.198)$$

From Eq. (6.197), we conclude that the process is characterized completely by the mean values $\bar{x}(t_i)$ and covariances $C_x(t_i, t_j)$, so that, if the Gaussian process is weakly stationary, then it is wide-sense stationary as well.

It was pointed out earlier that if $x(t)$ is a wide-sense stationary vector stochastic process, then the covariance matrix $C_x(t_1, t_2)$ depends on $t_2 - t_1$ alone, and not on t_1. Hence, letting $t_2 - t_1 = \tau$, we can denote the covariance matrix as follows:

$$C_x(t_1, t_2) = C_x(\tau) \quad (6.199)$$

The covariance matrix $C_x(\tau)$ provides information concerning properties of the random vector $x(t)$ in the time domain. At times it is more convenient to present the information not in the time domain but in the frequency domain. Such information can be provided by introducing the *power spectral density matrix* defined as the Fourier transform of the correlation matrix, or

$$S_x(\omega) = \int_{-\infty}^{\infty} R_x(\tau) e^{-i\omega \tau} \, d\tau \quad (6.200)$$

where $i = \sqrt{-1}$, and we note that $S_x(\omega)$ is a complex matrix. Consistent with the properties of $C_x(t_1, t_2)$ described by Eqs. (6.195), the matrix $S_x(\omega)$ posseses the properties

$$S_x(-\omega) = S_x(\omega) \quad (6.201a)$$

$$S_x^H(\omega) = S_x(\omega) \quad (6.201b)$$

$$S_x(\omega) \geq 0 \quad (6.201c)$$

where $S_x^H(\omega)$ represents the complex conjugate transpose of $S_x(\omega)$. Stationary random processes are often described in terms of the power spectral density.

Next, let us consider a scalar zero-mean stochastic process $x(t)$ such that $x(t_1)$ and $x(t_2)$ are uncorrelated even when t_1 and t_2 are close to each other. Because $R_x(t_1, t_2) = 0$ for $t_1 \neq t_2$, the autocorrelation function can be expressed in the idealized form

$$R_x(t_1, t_2) = X(t_1) \delta(t_2 - t_1) \quad (6.202)$$

where the function $X(t_1)$ is known as the *intensity* of the process at time t_1 and $\delta(t_2 - t_1)$ is the Dirac delta function. A process of this type is known as

OPTIMAL OBSERVERS. THE KALMAN–BUCY FILTER   245

white noise, a term explained shortly. When the intensity X of the white noise is constant, the process is wide-sense stationary, so that Eq. (6.202) can be replaced by

$$R_x(\tau) = X\delta(\tau) \qquad (6.203)$$

Introducing Eq. (6.203) into the scalar version of Eq. (6.200), we conclude that

$$S_x(\omega) = X \qquad (6.204)$$

or the power spectral density is the same for all frequencies. A random process $x(t)$ with constant power spectral density is called white noise by analogy with the white light, which has a flat spectrum over the visible range. Of course, the concept represents a physical impossibility because it implies infinite power, but judicious use of the concept can lead to meaningful results.

The above development can be extended to vector-valued processes. Indeed, if $\mathbf{x}(t)$ represents a zero-mean random vector with the correlation matrix

$$R_{\mathbf{x}}(\tau) = X\delta(\tau) \qquad (6.205)$$

where X is a constant positive semidefinite intensity matrix, then $\mathbf{x}(t)$ is a white noise process with the power spectral density matrix

$$S_{\mathbf{x}}(\omega) = X \qquad (6.206)$$

Now let us return to the problem of designing an optimal observer for the system described by Eqs. (6.175) and (6.176). We assume that the processes $\mathbf{v}(t)$ and $\mathbf{w}(t)$ are white noise with intensities $V(t)$ and $W(t)$, respectively, so that the correlation matrices have the form

$$E\{\mathbf{v}(t_1)\mathbf{v}^T(t_2)\} = V(t_1)\delta(t_2 - t_1) \qquad (6.207\text{a})$$

$$E\{\mathbf{w}(t_1)\mathbf{w}^T(t_2)\} = W(t_1)\delta(t_2 - t_1) \qquad (6.207\text{b})$$

Moreover, we assume that the processes $\mathbf{v}(t)$ and $\mathbf{w}(t)$ are uncorrelated, so that

$$E\{\mathbf{v}(t_1)\mathbf{w}^T(t_2)\} = E\{\mathbf{w}(t_1)\mathbf{v}^T(t_2)\} = 0 \qquad (6.208)$$

In the special case in which $W(t)$ is positive definite, or

$$W(t) > 0 \qquad (6.209)$$

the problem of state estimation is said to be *nonsingular*. This implies that all the components of the output vector are contaminated by white noise and that no information not containing white noise can be extracted from $\mathbf{y}(t)$. The initial state vector is such that

$$E\{\mathbf{x}(t_0)\} = \bar{\mathbf{x}}_0 \qquad (6.210\text{a})$$

$$E\{[\mathbf{x}(t_0) - \bar{\mathbf{x}}_0][\mathbf{x}(t_0) - \bar{\mathbf{x}}_0]^T\} = Q_0 \qquad (6.210\text{b})$$

and is assumed to be uncorrelated with the state excitation noise and observation noise, or

$$E\{\mathbf{x}(t_0)\mathbf{v}^T(t)\} = 0 \qquad (6.211\text{a})$$

$$E\{\mathbf{x}(t_0)\mathbf{w}^T(t)\} = 0 \qquad (6.211\text{b})$$

We consider a full-order observer of the form

$$\dot{\hat{\mathbf{x}}}(t) = A\hat{\mathbf{x}}(t) + B\mathbf{u}(t) + K(t)[\mathbf{y}(t) - C\hat{\mathbf{x}}(t)] \qquad (6.212)$$

where $\hat{\mathbf{x}}(t)$ is the estimated state, so that the problem is that of determining the optimal gain matrix and initial observer state $\hat{\mathbf{x}}(t_0)$. As performance measure, we consider the quadratic form

$$J = E\{\mathbf{e}^T(t)U(t)\mathbf{e}(t)\} \qquad (6.213)$$

where

$$\mathbf{e}(t) = \mathbf{x}(t) - \hat{\mathbf{x}}(t) \qquad (6.214)$$

is the observer error vector and $U(t)$ is a symmetric positive definite weighting matrix. Subtracting Eq. (6.212) from Eq. (6.175) and considering Eqs. (6.176) and (6.214), we obtain

$$\dot{\mathbf{e}}(t) = [A - K(t)C]\mathbf{e}(t) + \mathbf{v}(t) - K(t)\mathbf{w}(t) \qquad (6.215)$$

where the solution $\mathbf{e}(t)$ is subject to the initial condition

$$\mathbf{e}(t_0) = \mathbf{x}(t_0) - \hat{\mathbf{x}}(t_0) \qquad (6.216)$$

Next, let us introduce the notation

$$E\{\mathbf{e}(t)\} = \bar{\mathbf{e}}(t) \qquad (6.217\text{a})$$

$$E\{[\mathbf{e}(t) - \bar{\mathbf{e}}(t)][\mathbf{e}(t) - \bar{\mathbf{e}}(t)]^T\} = \tilde{Q}(t) \qquad (6.217\text{b})$$

where $\bar{e}(t)$ is the mean value and $\tilde{Q}(t)$ the variance matrix of $e(t)$. It follows from Eqs. (6.217) that

$$E\{e^T(t)e^T(t)\} = \tilde{Q}(t) + \bar{e}(t)\bar{e}^T(t) \tag{6.218}$$

It can be shown [K16], however, that

$$E\{e^T(t)U(t)e(t)\} = \bar{e}^T(t)U(t)\bar{e}(t) + \text{tr}\,[\tilde{Q}(t)U(t)] \tag{6.219}$$

Because $U(t)$ is positive definite, the quadratic form achieves its minimum value of zero for $\bar{e}(t) = 0$. Moreover, it can be shown that the error mean value satisfies the homogeneous differential equation [K16]

$$\dot{\bar{e}}(t) = [A - K(t)C]\bar{e}(t) \tag{6.220}$$

so that $\bar{e}(t) = 0$ if $\bar{e}(t_0) = 0$, which can be achieved by choosing

$$\hat{x}(t_0) = \bar{x}_0 \tag{6.221}$$

Because the second term on the right side of Eq. (6.219) does not depend on $\bar{e}(t)$, it can be minimized independently. It is further shown by Kwakernaak and Sivan [K16] that the variance $\tilde{Q}(t)$ satisfies the differential equation

$$\dot{\tilde{Q}}(t) = [A - K(t)C]\tilde{Q}(t) + \tilde{Q}^T(t)[A - K(t)C]^T + V(t)$$
$$+ K(t)W(t)K^T(t) \tag{6.222}$$

where $\tilde{Q}(t)$ is subject to the initial condition

$$\tilde{Q}(t_0) = Q_0 \tag{6.223}$$

Two transformations involving time reversals lead to the conclusion that the optimal observer is characterized by the gain matrix [K16]

$$K^*(t) = Q(t)C^T W^{-1}(t) \tag{6.224}$$

where $Q(t)$ is the variance matrix of $e(t)$ satisfying the Riccati equation

$$\dot{Q}(t) = AQ(t) + Q(t)A^T + V(t) - Q(t)C^T W^{-1}(t)CQ(t) \tag{6.225}$$

subject to

$$Q(t_0) = Q_0 \tag{6.226}$$

Implied in this is that for any symmetric positive definite matrix $U(t)$

$$\operatorname{tr}[Q(t)U(t)] \le \operatorname{tr}[\tilde{Q}(t)U(t)] \qquad (6.227)$$

so that the mean square weighted error for the optimal observer has the value

$$E\{\mathbf{e}^T(t)U(t)\mathbf{e}(t)\} = \operatorname{tr}[Q(t)U(t)] \qquad (6.228)$$

The above result is independent of the weighting matrix $U(t)$. It must be pointed out that, although in the above developments the implication was that the coefficient matrices A, B and C are constant, all the results remain the same for time-dependent matrices A, B and C.

The optimal observer characterized by the gain matrix given by Eq. (6.224) is known as the *Kalman–Bucy filter*. We observe that the gains in both the optimal regulator and optimal estimator are computed by solving a matrix Riccati equation order n. But, whereas in the optimal regulator the Riccati matrix is obtained by integrating backward in time from some final condition, in the optimal estimator the Riccati matrix is obtained by integrating forward in time from some initial condition. Hence, the process is simpler for the optimal estimator. The fact that both the optimal regulator and optimal observer involve the solution of a matrix Riccati equation points toward a *duality* that exists between the two, permitting conclusions to be drawn about observers from knowledge gained from regulators.

As in the case of the optimal regulator, the solution of a nonlinear $n \times n$ matrix differential equation can be avoided by solving a linear $2n \times 2n$ matrix differential equation.

For time-invariant nonsingular observers and white noise processes $\mathbf{v}(t)$ and $\mathbf{w}(t)$ with positive definite intensities V and W, respectively, the Riccati matrix approaches a constant value as the time t increases, independently of the value of Q_0. This case is known as the *steady-state case*, in which the Riccati matrix satisfies the *algebraic matrix Riccati equation*

$$AQ + QA^T + V - QC^TW^{-1}CQ = 0 \qquad (6.229)$$

yielding the *steady-state optimal observer gain matrix*

$$K^* = QC^TW^{-1} \qquad (6.230)$$

In this case, the steady-state observer is minimal in the sense that the mean square weighted error $\mathbf{e}^T(t)U\mathbf{e}(t)$ is minimized, where U is constant.

The main problem in implementing the Kalman–Bucy filter lies in the selection of the noise intensities V and W.

A Kalman filter for second-order systems, such as those associated with the lumped-parameter structures discussed in Chapter 4, was developed recently by Hashemipour and Laub (H12).

OPTIMAL OBSERVERS. THE KALMAN–BUCY FILTER   249

Example 6.11   Design a Kalman–Bucy filter for the harmonic oscillator of Example 6.9. Assume that the oscillator is subjected to actuator noise, where the noise is white and has the intensity $v_a = 24$ rad/s. The displacement sensor also contains white noise and the intensity is $w_s = 10^{-2}$ (rad/s)$^{-1}$. First solve the differential matrix Riccati equation and plot the gains $k_1(t)$ and $k_2(t)$ as functions of time for $Q_0 = 0$. Then solve the steady-state matrix Riccati equation, solve for the constant gains k_1 and k_2 and determine the observer poles. Plot the components of the system state vector and of the observer state vector as functions of time and compare results, both for time-dependent and constant observer gains.

Equations (6.175) and (6.176) have the explicit form

$$\begin{bmatrix} \dot{x}_1(t) \\ \dot{x}_2(t) \end{bmatrix} = \begin{bmatrix} 0 & \omega_n \\ -\omega_n & 0 \end{bmatrix} \begin{bmatrix} x_1(t) \\ x_2(t) \end{bmatrix} + \begin{bmatrix} 0 \\ 1 \end{bmatrix} u(t) + \begin{bmatrix} 0 \\ 1 \end{bmatrix} v(t) \qquad \text{(a)}$$

and

$$y(t) = \begin{bmatrix} 1 & 0 \end{bmatrix} \begin{bmatrix} x_1(t) \\ x_2(t) \end{bmatrix} + w(t) \qquad \text{(b)}$$

so that

$$A = \begin{bmatrix} 0 & \omega_n \\ -\omega_n & 0 \end{bmatrix}, \quad B = \begin{bmatrix} 0 \\ 1 \end{bmatrix}, \quad C = \begin{bmatrix} 1 & 0 \end{bmatrix} \qquad \text{(c)}$$

Moreover, the noise intensity matrices are

$$V(t) = \begin{bmatrix} 0 & 0 \\ 0 & 1 \end{bmatrix} v_a = \text{constant}, \quad W(t) = w_s = \text{constant} \qquad \text{(d)}$$

where the second is really a scalar.

Inserting from Eqs. (c) and (d) into Eq. (6.224), we can write the gain vector

$$\begin{bmatrix} k_1(t) \\ k_2(t) \end{bmatrix} = \begin{bmatrix} q_{11}(t) & q_{12}(t) \\ q_{12}(t) & q_{22}(t) \end{bmatrix} \begin{bmatrix} 1 \\ 0 \end{bmatrix} w_s^{-1} = \frac{1}{w_s} \begin{bmatrix} q_{11}(t) \\ q_{12}(t) \end{bmatrix} \qquad \text{(e)}$$

so that q_{22} is not really needed. To obtain $q_{11}(t)$ and $q_{12}(t)$, we turn to the Riccati equation, Eq. (6.225), and write

$$\begin{bmatrix} \dot{q}_{11}(t) & \dot{q}_{12}(t) \\ \dot{q}_{12}(t) & \dot{q}_{22}(t) \end{bmatrix} = \begin{bmatrix} 0 & \omega_n \\ -\omega_n & 0 \end{bmatrix} \begin{bmatrix} q_{11}(t) & q_{12}(t) \\ q_{12}(t) & q_{22}(t) \end{bmatrix}$$
$$+ \begin{bmatrix} q_{11}(t) & q_{12}(t) \\ q_{12}(t) & q_{22}(t) \end{bmatrix} \begin{bmatrix} 0 & -\omega_n \\ \omega_n & 0 \end{bmatrix} + v_a \begin{bmatrix} 0 & 0 \\ 0 & 1 \end{bmatrix}$$
$$- \frac{1}{w_s} \begin{bmatrix} q_{11}(t) & q_{12}(t) \\ q_{12}(t) & q_{22}(t) \end{bmatrix} \begin{bmatrix} 1 \\ 0 \end{bmatrix} \begin{bmatrix} 1 & 0 \end{bmatrix} \begin{bmatrix} q_{11}(t) & q_{12}(t) \\ q_{12}(t) & q_{22}(t) \end{bmatrix} \qquad \text{(f)}$$

Equation (f) represents three scalar simultaneous differential equations having the explicit form

$$\dot{q}_{11}(t) = 2\omega_n q_{12}(t) - \frac{1}{w_s} q_{11}^2(t)$$

$$\dot{q}_{12}(t) = \omega_n[q_{22}(t) - q_{11}(t)] - \frac{1}{w_s} q_{11}(t)q_{12}(t) \qquad (g)$$

$$\dot{q}_{22}(t) = -2\omega_n q_{12}(t) + v_a - \frac{1}{w_s} q_{12}^2(t)$$

Recalling that $\omega_n = 10$ rad/s and that $q_{11}(0) = q_{12}(0) = q_{22}(0) = 0$, using the values of v_a and w_s given above, solving for $q_{11}(t)$, $q_{12}(t)$ and $q_{22}(t)$ and inserting $q_{11}(t)$ and $q_{12}(t)$ into Eq. (e), we obtain the observer gains $k_1(t)$ and $k_2(t)$. They are plotted in Fig. 6.27 as functions of time.

The steady-state gains can be determined by letting $\dot{q}_{11} = \dot{q}_{12} = \dot{q}_{22} = 0$ in Eqs. (g). Then, solving the resulting algebraic equations, we obtain

$$q_{11} = w_s \left[2\omega_n \left(\sqrt{\omega_n^2 + \frac{v_a}{w_s}} - \omega_n \right) \right]^{1/2} = w_s \omega_n [2(\sqrt{1+\alpha} - 1)]^{1/2}$$

$$q_{12} = w_s \left(\sqrt{\omega_n^2 + \frac{v_a}{w_s}} - \omega_n \right) = w_s \omega_n (\sqrt{1+\alpha} - 1) \qquad (h)$$

$$q_{22} = w_s \left[2\omega_n^{-1} \left(\omega_n^2 + \frac{v_a}{w_s} \right)^{3/2} - 2 \left(\omega_n^2 + \frac{v_a}{w_s} \right) \right]^{1/2}$$

$$= w_s \omega_n [2(1+\alpha)^{3/2} - 2(1+\alpha)]^{1/2}$$

where we used the notation

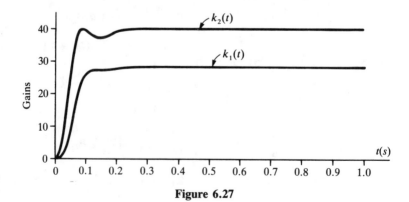

Figure 6.27

OPTIMAL OBSERVERS. THE KALMAN–BUCY FILTER

$$\frac{v_a}{w_s} = \alpha \omega_n^2 \tag{i}$$

Using the values given, we obtain

$$\alpha = 24 \tag{j}$$

so that, using Eq. (e), we compute the steady-state gains

$$k_1 = \frac{1}{w_s} q_{11} = \omega_n [2(\sqrt{1+\alpha} - 1)]^{1/2} = 20\sqrt{2} \text{ rad/s}$$

$$k_2 = \frac{1}{w_s} q_{12} = \omega_n (\sqrt{1+\alpha} - 1) = 40 \text{ rad/s} \tag{k}$$

From Example 6.9, and using the gains in Eqs. (k), the observer characteristic equation is

$$s^2 + k_1 s + 10(10 + k_2) = s^2 + 20\sqrt{2}s + 500 = 0 \tag{l}$$

which has the roots

$$\begin{matrix} s_1 \\ s_2 \end{matrix} = -10\sqrt{2} \pm \sqrt{(10\sqrt{2})^2 - 500} = -10\sqrt{2} \pm i10\sqrt{3} \tag{m}$$

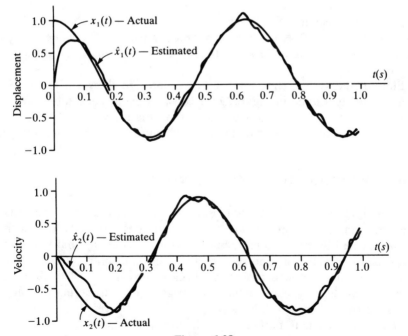

Figure 6.28

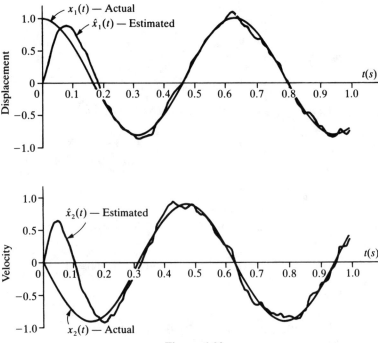

Figure 6.29

The roots are deeper in the left side of the complex plane than the observer poles in Example 6.9. Note that, to permit comparison, the values of v_a and w_s were chosen so as to produce the same steady-state value for k_2 as in Example 6.9.

Plots of the system state and observer state are shown in Fig. 6.28 for time-varying gains and in Fig. 6.29 for constant gains. As can be concluded from these figures, for larger values of time there is very little difference between the results using time-varying gains and those using constant gains.

6.14 DIRECT OUTPUT FEEDBACK CONTROL

One method of control, known as *output feedback control*, consists of generating the control inputs directly from the sensor outputs, i.e., without the use of an observer reconstructing the state. Assuming that the number and location of actuators and sensors are given, the control design problem amounts to choosing a matrix of feedback control gains. In this regard, it may be possible to select optimal control gains by considering the method described by Levine and Athans [L4].

Output feedback is particularly effective when there are as many actuators as sensors and the sensors and actuators are collocated. A given actuator force is first obtained by multiplying the signal from the sensor at the same location by the corresponding control gain and then fed back directly to the structure. For this reason, we refer to this approach as *direct output feedback control*.

To introduce the idea, it is not necessary to work with the state equations and the configuration equations suffice. Moreover, we assume that the gyroscopic and circulatory forces are zero. Under these circumstances, the equations of motion, Eq. (4.16), reduce to

$$M\ddot{\mathbf{q}}(t) + C\dot{\mathbf{q}}(t) + K\mathbf{q}(t) = \mathbf{Q}(t) \qquad (6.231)$$

where M, C and K are the $n \times n$ mass, damping and stiffness matrices, respectively, $\mathbf{q}(t)$ is the n-dimensional configuration vector and $\mathbf{Q}(t)$ is the n-dimensional force vector, which in this case is the control vector. If fewer than n inputs are desired, then the appropriate number of components of $\mathbf{Q}(t)$ will be equal to zero.

Next let us assume that the control law has the form

$$\mathbf{Q}(t) = -G_1 \mathbf{q}(t) - G_2 \dot{\mathbf{q}}(t) \qquad (6.232)$$

where G_1 and G_2 are control gain matrices. Because in direct output feedback control, the control force at a given location depends only on the state at the same location, it follows that *the gain matrices G_1 and G_2 are diagonal*. Inserting Eq. (6.232) into Eq. (6.231), we obtain the closed-loop equations

$$M\ddot{\mathbf{q}}(t) + (C + G_2)\dot{\mathbf{q}}(t) + (K + G_1)\mathbf{q}(t) = \mathbf{0} \qquad (6.233)$$

The main question remaining is the nature of the gain matrices G_1 and G_2. Displacement feedback is not really necessary, except in the case of semidefinite structures characterized by a positive semidefinite stiffness matrix K. Semidefinite structures admit rigid-body modes, and velocity feedback alone cannot stabilize the rigid-body modes. Hence, in the case of semidefinite structures the matrix G_1 should be such that $K + G_1$ is positive-definite. To examine the requirement on G_2, we resort to energy considerations. Following the approach of Section 4.2, we multiply Eq. (6.233) on the left by $\dot{\mathbf{q}}^T$ and rearrange to obtain

$$\frac{d}{dt}\left[\frac{1}{2}\dot{\mathbf{q}}^T M \dot{\mathbf{q}} + \frac{1}{2}\mathbf{q}^T(K + G_1)\mathbf{q}\right] = -\dot{\mathbf{q}}^T(C + G_2)\dot{\mathbf{q}} \qquad (6.234)$$

The expression inside brackets on the left side of Eq. (6.234) can be identified as the total energy of the system, including the effect of displacement feedback. The objective of the feedback control is to drive the total energy to zero. If the damping matrix C is positive definite, then passive damping alone can achieve this objective. If C is only positive semidefinite, then G_2 should be such that $C + G_2$ is positive definite. This assures that the right side of Eq. (6.234) is negative, except at the origin of the state space. Hence, if $C + G_2$ is positive definite, energy is being dissipated at all times until the state is driven to zero.

In view of the above discussion, it is clear that the diagonal elements of G_1 and G_2 must be nonnegative. Of course, in the case of full state feedback all the components of G_1 and G_2 must be positive. This implies that G_1 and G_2 are positive definite, which guarantees asymptotic stability of the structure. Full state feedback is not necessary, however, as for asymptotic stability only $K + G_1$ and $C + G_2$ must be positive definite and not G_1 and G_2.

Example 6.12 Consider the system shown in Fig. 6.30 and show how direct output feedback control can be used to induce asymptotic stability.

The equations of motion are given by Eq. (6.231), in which the coefficient matrices have the form

$$M = \begin{bmatrix} m_1 & 0 \\ 0 & m_2 \end{bmatrix}, \quad C = \begin{bmatrix} c & -c \\ -c & c \end{bmatrix}, \quad K = \begin{bmatrix} k & -k \\ -k & k \end{bmatrix} \quad (a)$$

Of course, the configuration and control vector are simply

$$\mathbf{q}(t) = [q_1(t) \quad q_2(t)]^T, \quad \mathbf{Q}(t) = [Q_1(t) \quad Q_2(t)]^T \quad (b)$$

We propose to control the system by means of a single sensor measuring the state of m_1 and a single actuator exerting a control force on m_1. Accordingly, we use the control law

$$Q_1(t) = -g_1 q_1(t) - g_2 \dot{q}_1(t), \quad Q_2(t) = 0 \quad (c)$$

where g_1 and g_2 are positive real scalars, so that the gain matrices are

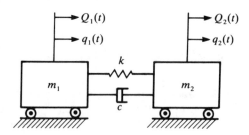

Figure 6.30

$$G_1 = \begin{bmatrix} g_1 & 0 \\ 0 & 0 \end{bmatrix}, \quad G_2 = \begin{bmatrix} g_2 & 0 \\ 0 & 0 \end{bmatrix} \tag{d}$$

It is not difficult to verify that the matrices

$$C + G_2 = \begin{bmatrix} c + g_2 & -c \\ -c & c \end{bmatrix}, \quad K + G_1 = \begin{bmatrix} k + g_1 & -k \\ -k & k \end{bmatrix} \tag{e}$$

are positive definite, so that the closed-loop system is asymptotically stable.

6.15 MODAL CONTROL

Modal control is a generic term for a class of techniques based on the idea that control of a system can be achieved by controlling its modes. One such technique is the pole allocation method discussed in Section 6.2. The pole allocation method is general in nature and, although it can be applied to lumped structures, it was not developed with structures in mind. As a result, in applying it to structures, it does not take into account the special features of structures. In particular, we recall from Chapter 4 that the motion of n-degree-of-freedom lumped structures is governed by n second-order ordinary differential equations. These n second-order configuration equations can be converted into $2n$ first-order state equations by the addition of n identities. Consistent with this, n components of the state vector represent displacements and n components represent velocities. Moreover, because n of the state equations are nothing but identities, they are not subject to control, so that there are at most n inputs. In contrast, the approach presented in this section is designed specifically for structures and does take into account the structural characteristics pointed out above.

The pole allocation method represents modal control in the sense that the open-loop poles are altered so as to induce asymptotic stability in some or in all the modes. In the process, the eigenvectors are also altered, and the question can be raised whether this is not wasted effort, as the eigenvectors do not have direct bearing on the system stability. In this section, we present a technique permitting design of controls for each mode independently of the other modes and without altering the system eigenvectors in the case of linear control. Moreover, unlike many of the techniques discussed earlier in this chapter, the technique permits nonlinear control for multi-degree-of-freedom structures.

We recall from Section 4.1 that the equations of motion of a general dynamical system have the form

$$M\ddot{q}(t) + (C + G)\dot{q}(t) + (K + H)q(t) = \mathbf{Q}(t) \tag{6.235}$$

where G and H are $n \times n$ gyroscopic and circulatory matrices, respectively.

CONTROL OF LUMPED-PARAMETER SYSTEMS. MODERN APPROACH

The remaining quantities are as defined in Section 6.14. Adjoining the identity $\dot{\mathbf{q}}(t) \equiv \dot{\mathbf{q}}(t)$, Eq. (6.235) can be rewritten in the state form

$$\dot{\mathbf{x}}(t) = A\mathbf{x}(t) + B\mathbf{Q}(t) \tag{6.236}$$

where $\mathbf{x}(t) = [\mathbf{q}^T(t) \mid \dot{\mathbf{q}}^T(t)]^T$ is the state vector and

$$A = \left[\begin{array}{c|c} 0 & I \\ \hline -M^{-1}(K+H) & -M^{-1}(C+G) \end{array}\right], \quad B = \left[\begin{array}{c} 0 \\ \hline M^{-1} \end{array}\right] \tag{6.237a, b}$$

are coefficient matrices. Note that we retained the notation $\mathbf{Q}(t)$ for the control vector, instead of the customary $\mathbf{u}(t)$, for reasons that will become obvious shortly.

Our object is to transform Eq. (6.236) into a set of modal equations. To this end, we recall the eigenvalue problems (3.105) and (3.106), where λ_i are the eigenvalues, $\mathbf{u}_i$ the right eigenvectors and $\mathbf{v}_j$ the left eigenvectors of A (Section 3.11), which explains our reluctance to use the notation $\mathbf{u}$ for the control vector. The right and left eigenvectors are biorthogonal and they are assumed to be normalized so as to satisfy the biorthonormality relations (3.108). The two sets of eigenvectors can be arranged conveniently in the modal matrices $U = [\mathbf{u}_1 \ \mathbf{u}_2 \ \cdots \ \mathbf{u}_{2n}]$, $V = [\mathbf{v}_1 \ \mathbf{v}_2 \ \cdots \ \mathbf{v}_{2n}]$, so that the biorthonormality relations can be rewritten in the compact form

$$V^T U = I, \quad V^T A U = \Lambda \tag{6.238a, b}$$

where $\Lambda = \text{diag } \lambda_i$ is the diagonal matrix of the eigenvalues.

Next, let us use the expansion theorem and express the solution of Eq. (6.236) as a linear combination of the right eigenvectors multiplied by time-dependent modal coordinates as follows:

$$\mathbf{x}(t) = \sum_{i=1}^{2n} \zeta_i(t)\mathbf{u}_i = U\boldsymbol{\zeta}(t) \tag{6.239}$$

where $\zeta_i(t)$ $(i = 1, 2, \ldots, 2n)$ are the modal coordinates and $\boldsymbol{\zeta}(t)$ is the corresponding vector. Inserting Eq. (6.239) into Eq. (6.236), multiplying on the left by V^T and considering Eqs. (6.238), we obtain the modal equations

$$\dot{\boldsymbol{\zeta}}(t) = \Lambda\boldsymbol{\zeta}(t) + \mathbf{Z}(t) \tag{6.240}$$

where

$$\mathbf{Z}(t) = V^T B \mathbf{Q}(t) = V_L^T M^{-1} \mathbf{Q}(t) \tag{6.241}$$

is the modal control vector, in which V_L is an $n \times 2n$ matrix representing the lower half of the matrix V.

Equation (6.240) can be written in the scalar form

$$\dot{\zeta}_i(t) = \lambda_i \zeta_i(t) + Z_i(t), \quad i = 1, 2, \ldots, 2n \tag{6.242}$$

In general, in feedback control the modal control force Z_i does not depend explicitly on t but on the modal coordinates $\zeta_1, \zeta_2, \ldots, \zeta_{2n}$, or

$$Z_i = Z_i(\zeta_1, \zeta_2, \ldots, \zeta_{2n}), \quad i = 1, 2, \ldots, 2n \tag{6.243}$$

In this case, Eqs. (6.242) represent a set of simultaneous equations, as the feedback control forces recouple the modal equations. For this reason, we refer to feedback control of the type (6.243) as *coupled control*. On the other hand, in the special case in which Z_i depends on ζ_i alone, or

$$Z_i = Z_i(\zeta_i), \quad i = 1, 2, \ldots, 2n \tag{6.244}$$

Eqs. (6.242) represent a set of independent equations. For this reason, we refer to feedback control of the type (6.244) as *independent control*. The method based on the use of Eqs. (6.242) and (6.244) is known as *independent modal-space control* (IMSC), because the independent design of the controls is carried out in the modal space rather than in the actual space [M20, M24, O3, M28]. We note that Eqs. (6.244) do not restrict the functions Z_i to being linear in ζ_i and indeed *this dependence can be linear or nonlinear*.

Questions of control implementation remain. In particular, the question is how to generate the actual control vector $\mathbf{Q}(t)$ from the modal control vector $\mathbf{Z}(t)$. Before addressing this question it will prove convenient to separate Eqs. (6.242) into the real and imaginary parts. To this end, we introduce the notation

$$\zeta_j(t) = \xi_j(t) + i\eta_j(t), \quad \lambda_j = \alpha_j + i\beta_j, \quad Z_j(t) = X_j(t) + iY_j(t),$$

$$i = 1, 2, \ldots, 2n \quad (6.245\text{a, b, c})$$

Inserting Eqs. (6.245) into Eqs. (6.242) and separating the real and imaginary parts, we obtain

$$\dot{\xi}_j(t) = \alpha_j \xi_j(t) - \beta_j \eta_j(t) + X_j(t),$$
$$\dot{\eta}_j(t) = \beta_j \xi_j(t) + \alpha_j \eta_j(t) + Y_j(t), \quad j = 1, 2, \ldots, 2n \tag{6.246}$$

where

$$X_j(t) = \operatorname{Re} \mathbf{v}_{jL}^T M^{-1} \mathbf{Q}(t), \quad Y_j(t) = \operatorname{Im} \mathbf{v}_{jL}^T M^{-1} \mathbf{Q}(t), \quad j = 1, 2, \ldots, 2n$$
(6.247a, b)

in which $\mathbf{v}_{jL}$ is the bottom half of the eigenvector $\mathbf{v}_j$. Actually, one half of Eqs. (6.246) and (6.247) are redundant. Indeed, because the problem (6.232) is real, the eigenvalues and eigenvectors appear in pairs of complex conjugates. Hence, letting

$$\zeta_{j+n}(t) = \bar{\zeta}_j(t), \quad \lambda_{j+n} = \bar{\lambda}_j, \quad Z_{j+n}(t) = \bar{Z}_j(t), \quad j = 1, 2, \ldots, n$$
(6.248a, b, c)

be the associated complex conjugates, we conclude that the equations corresponding to $n+1 \le j \le 2n$ contain precisely the same information as the equations corresponding to $1 \le j \le n$. Then, introducing the notation

$$\mathbf{w}_j(t) = [\xi_j(t) \ \ \eta_j(t)]^T, \quad \mathbf{W}_j(t) = [X_j(t) \ \ Y_j(t)]^T, \quad \Lambda_j = \begin{bmatrix} \alpha_j & -\beta_j \\ \beta_j & \alpha_j \end{bmatrix},$$
$$j = 1, 2, \ldots, n \quad (6.249\text{a, b, c})$$

where $\mathbf{w}_j(t)$ is the jth modal state vector, $\mathbf{W}_j(t)$ is the jth modal control vector and Λ_j is the jth coefficient matrix, Eqs. (6.246) can be written in the form

$$\dot{\mathbf{w}}_j(t) = \Lambda_j \mathbf{w}_j(t) + \mathbf{W}_j(t), \quad j = 1, 2, \ldots, n \tag{6.250}$$

Equations (6.250) can be combined into the single equation

$$\dot{\mathbf{w}}(t) = \Lambda \mathbf{w}(t) + \mathbf{W}(t) \tag{6.251}$$

where

$$\mathbf{w}(t) = [\mathbf{w}_1^T(t) \ \ \mathbf{w}_2^T(t) \ \ \cdots \ \ \mathbf{w}_n^T(t)]^T \tag{6.252}$$

is the overall modal vector

$$\Lambda = \text{block-diag } \Lambda_j \tag{6.253}$$

is the coefficient matrix and

$$\mathbf{W}(t) = [\mathbf{W}_1^T(t) \ \ \mathbf{W}_2^T(t) \ \ \cdots \ \ \mathbf{W}_n^T(t)]^T \tag{6.254}$$

is the overall modal control vector.

To replace Eqs. (6.247) by a single relation between the modal control vector $\mathbf{W}(t)$ and the actual control vector $\mathbf{Q}(t)$, we let

$$V_{jRI} = [\text{Re } \mathbf{v}_{jL} \quad \text{Im } \mathbf{v}_{jL}], \quad j = 1, 2, \ldots, n \tag{6.255}$$

be an $n \times 2$ matrix and combine Eqs. (6.247) into

$$\mathbf{W}_j(t) = V_{jRI}^T M^{-1} \mathbf{Q}(t), \quad j = 1, 2, \ldots, n \tag{6.256}$$

Then, letting

$$V_{RI} = [V_{1RI} \quad V_{2RI} \quad \cdots \quad V_{nRI}] \tag{6.257}$$

be an $n \times 2n$ matrix obtained by arranging Eqs. (6.255) in a single matrix, Eqs. (6.256) can be written in the compact form

$$\mathbf{W}(t) = V_{RI}^T M^{-1} \mathbf{Q}(t) \tag{6.258}$$

which is the desired relation.

In the IMSC method, the modal control vector $\mathbf{W}(t)$ is designed first and the actual control vector $\mathbf{Q}(t)$ is then determined from Eq. (6.258). This presents a problem, however, as $\mathbf{W}(t)$ is a $2n$-vector and $\mathbf{Q}(t)$ is an n-vector. One alternative is to determine $\mathbf{Q}(t)$ by writing

$$\mathbf{Q}(t) = M(V_{RI}^T)^\dagger \mathbf{W}(t) \tag{6.259}$$

where

$$(V_{RI}^T)^\dagger = (V_{RI} V_{RI}^T)^{-1} V_{RI} \tag{6.260}$$

is the pseudo-inverse of V_{RI}^T. Another alternative is to control only one half of the complex modes and leave the other half uncontrolled. As a rule of thumb, the higher modes are the ones to be left uncontrolled, as damping tends to have a larger effect on the higher modes.

It should be pointed out that in the absence of feedback control the modal equations are independent. In general, the feedback forces recouple the modal equations. But, in the special case of IMSC the modal controls remain uncoupled, as the linear transformation decoupling the plant also decouples the feedback control vector. Because this linear transformation is the modal matrix U (see Eq. (6.239)), it follows that the closed-loop eigenvectors are the same as the open-loop eigenvectors. This statement implies that the modal control vectors $\mathbf{W}_j(t)$ depend only on $\xi_j(t)$ and $\eta_j(t)$. The statement is true only when one half of the modes are controlled. It is only approximately true when Eq. (6.259) is used to compute the actual controls. The statement is definitely true in the important case of undamped, nongyroscopic structures, as shown later in this section.

In actual feedback, we do not use the modal coordinates $\xi_j(t)$ and $\eta_j(t)$ but the estimated modal coordinates $\hat{\xi}_j(t)$ and $\hat{\eta}_j(t)$, where the latter can be obtained from the system output by means of a modal Luenberger observer or a modal Kalman filter. To this end, we assume that the output is given by

$$\mathbf{y}(t) = C\mathbf{w}(t) \tag{6.261}$$

Then, we construct a modal observer in the form

$$\dot{\hat{\mathbf{w}}}(t) = \Lambda\hat{\mathbf{w}}(t) + \mathbf{W}(t) + K_E[\mathbf{y}(t) - C\hat{\mathbf{w}}(t)] \tag{6.262}$$

where $\hat{\mathbf{w}}(t)$ is the observer state and K_E is the matrix of observer gains. The question is how to tie the output to the modal state. To this end, we recall that the modal state vector $\mathbf{w}(t)$ is related to the actual state vector $\mathbf{x}(t)$ by means of Eq. (6.239). By analogy with Eq. (6.257), we can introduce the $2n \times 2n$ matrix

$$U_{RI} = [U_{1Ri} \quad U_{2RI} \quad \cdots \quad U_{nRI}] \tag{6.263}$$

where

$$U_{jRI} = 2[\operatorname{Re}\mathbf{u}_j \quad -\operatorname{Im}\mathbf{u}_j], \quad j = 1, 2, \ldots, n \tag{6.264}$$

are $2n \times 2$ matrices. Then, Eq. (6.239) can be rewritten in the form

$$\mathbf{x}(t) = U_{RI}\mathbf{w}(t) \tag{6.265}$$

Of course, the output vector $\mathbf{y}(t)$ is only a subset of the state vector $\mathbf{x}(t)$, which implies that the matrix C can be obtained from U_{RI} by retaining the rows corresponding to the components of $\mathbf{y}(t)$.

The problem of designing modal controls is considerably simpler in the case of *undamped structures with no gyroscopic and circulatory effects present*. In this case, the eigenvalues are pure imaginary, $\lambda_j = i\omega_j$, and the eigenvectors $\mathbf{u}_j$ and $\mathbf{v}_j$ are such that the upper half of the components is real and the lower half imaginary, or vice-versa [M24]. It follows that every second row in the matrix V_{RI}^T is zero. If the lower half of $\mathbf{v}_j$ is imaginary, then every odd-numbered row is zero, which makes all $X_j(t)$ automatically equal to zero. If one ignores the zero rows in V_{RI}^T, then the $2n \times n$ matrix collapses into an $n \times n$ matrix V_I^T. Moreover, introducing the n-vector

$$\mathbf{Y}(t) = [Y_1(t) \quad Y_2(t) \quad \cdots \quad Y_n(t)]^T \tag{6.266}$$

Eq. (6.259) reduces to

$$\mathbf{Q}(t) = M(V_I^T)^{-1}\mathbf{Y}(t) \tag{6.267}$$

so that the process of determining the actual control vector from the modal control vector no longer involves the computation of a pseudo-inverse but of a mere inverse.

Next, recognizing that in the case at hand

$$\Lambda_j = \begin{bmatrix} 0 & -\omega_j \\ \omega_j & 0 \end{bmatrix}, \quad j = 1, 2, \ldots, n \qquad (6.268)$$

and recalling that the top component of $\mathbf{W}_j(t)$ is zero, we can rewrite Eqs. (6.250) in the form

$$\dot{\xi}_j(t) = -\omega_j \eta_j(t),$$
$$\dot{\eta}_j(t) = \omega_j \xi_j(t) + Y_j(t), \qquad j = 1, 2, \ldots, n \quad (6.269)$$

which can be reduced to the single second-order modal equation

$$\ddot{\xi}_j(t) + \omega_j^2 \xi_j(t) = \omega_j Y_j(t), \quad j = 1, 2, \ldots, n \qquad (6.270)$$

As expected, Eqs. (6.270) represents a set of independently controlled harmonic oscillators, and modal control can be designed for every one of them by any of the techniques described earlier in this chapter, including the nonlinear controls. Of course, to determine the actual controls for implementation, we must use Eq. (6.267). As a result, if the modal controls are on-off, then the actual controls are a linear combination of on-off controls, which is likely to resemble a staircase in time.

In the undamped case, the steady-state solution of the optimal IMSC problem involving the quadratic performance measure can be obtained in closed form. Because in IMSC the modal controls have the form

$$Y_j(t) = Y_j(\xi_j(t), \eta_j(t)), \quad j = 1, 2, \ldots, n \qquad (6.271)$$

so that the jth modal equation is independent of any other modal equation, the performance measure can be written as

$$J = \sum_{j=1}^{n} J_j \qquad (6.272)$$

where

$$J_j = \int_0^\infty [\mathbf{w}_j^T(t) Q_j \mathbf{w}_j(t) + \mathbf{W}_j^T(t) R_j \mathbf{W}_j(t)] \, dt, \quad j = 1, 2, \ldots, n \qquad (6.273)$$

are modal performance measures, each of which can be minimized independently. Hence, the optimal modal controls are given by

$$\mathbf{W}_j^*(t) = -R_j^{-1}K_j(t)\mathbf{w}_j^*(t), \quad j = 1, 2, \ldots, n \qquad (6.274)$$

where $K_j(t)$ satisfy the matrix Riccati equations

$$\dot{K}_j(t) = -Q_j - K_j(t)\Lambda_j - \Lambda_j^T K_j(t) + K_j(t)R_j^{-1}K_j(t), \quad K_j(\infty) = 0,$$
$$j = 1, 2, \ldots, n \qquad (6.275)$$

The weighting matrices can be taken as diagonal. Moreover, because $X_j(t) = 0$, the matrices R_j must be such that

$$R_j^{-1} = \begin{bmatrix} 0 & 0 \\ 0 & R_{j2}^{-1} \end{bmatrix}, \quad j = 1, 2, \ldots, n \qquad (6.276)$$

Choosing arbitrarily $Q_j = I$, where I is the 2×2 identity matrix, the steady-state matrix Riccati equation yields the component equations

$$\dot{K}_{j11} = -1 - 2\omega_j K_{j12} + R_{j2}^{-1} K_{j12}^2,$$
$$\dot{K}_{j12} = \omega_j K_{j11} - \omega_j K_{j22} + R_{j2}^{-1} K_{j12} K_{j22}, \quad j = 1, 2, \ldots, n \qquad (6.277)$$
$$\dot{K}_{j22} = -1 + 2\omega_j K_{j12} + R_{j2}^{-1} K_{j22}^2,$$

where the symmetry of the matrix K_j was taken into account. The steady-state solution is obtained by letting $\dot{K}_j = 0$. Solving the resulting algebraic equations, we obtain [M24]

$$K_{j11} = [\omega_j^{-2} - R_{j2} - 2\omega_j^2 R_{j2}^2 + 2\omega_j^{-1} R_{j2}^{-1}(\omega_j^2 R_{j2}^2 + R_{j2})^{3/2}]^{1/2},$$
$$K_{j12} = \omega_j R_{j2} - (\omega_j^2 R_{j2}^2 + R_{j2})^{1/2}, \quad j = 1, 2, \ldots, n \qquad (6.278)$$
$$K_{j22} = [R_{j2} - 2\omega_j^2 R_{j2}^2 + 2\omega_j R_{j2}(\omega_j^2 R_{j2}^2 + R_{j2})^{1/2}]^{1/2},$$

Hence, using Eqs. (6.274), we obtain the optimal modal feedback controls

$$X_j^*(t) = 0,$$

$$Y_j^*(t) = -R_{j2}^{-1}[K_{j12}\xi_j(t) + K_{j22}\eta_j(t)]$$
$$= -[\omega_j - (\omega_j^2 + R_{j2}^{-1})^{1/2}]\xi_j(t) - [R_{j2}^{-1} \qquad j = 1, 2, \ldots, n \quad (6.279)$$
$$- 2\omega_j^2 + 2\omega_j(\omega_j^2 + R_{j2}^{-1})^{1/2}]^{1/2}\eta_j(t),$$

Inserting the second half of Eqs. (6.279) into Eqs. (6.269), we obtain the closed-loop modal equations

$$\dot{\xi}_j(t) = -\omega_j \eta_j(t),$$
$$\dot{\eta}_j(t) = (\omega_j^2 + R_{j2}^{-1})^{1/2}\xi_j(t) - [R_{j2}^{-1} - 2\omega_j^2 \qquad j = 1, 2, \ldots, n \quad (6.280)$$
$$+ 2\omega_j(\omega_j^2 + R_{j2}^{-1})^{1/2}]^{1/2}\eta_j(t),$$

The characteristic equation for the jth mode is

$$\det\begin{bmatrix} p_j & \omega_j \\ -(\omega_j^2 + R_{j2}^{-1})^{1/2} & p_j + [R_{j2}^{-1} - 2\omega_j^2 + 2\omega_j(\omega_j^2 + R_{j2}^{-1})^{1/2}]^{1/2} \end{bmatrix}$$
$$= p_j^2 + [R_{j2}^{-1} - 2\omega_j^2 + 2\omega_j(\omega_j^2 + R_{j2}^{-1})^{1/2}]^{1/2}p_j + \omega_j(\omega_j^2 + R_{j2}^{-1})^{1/2} = 0 \quad (6.281)$$

so that the closed-loop poles are

$$\begin{aligned}
\genfrac{}{}{0pt}{}{p_j}{p_{j+n}} &= -\frac{1}{2}[R_{j2}^{-1} - 2\omega_j^2 + 2\omega_j(\omega_j^2 + R_{j2}^{-1})^{1/2}]^{1/2} \\
&\quad \pm \frac{1}{2}[R_{j2}^{-1} - 2\omega_j^2 + 2\omega_j(\omega_j^2 + R_{j2}^{-1})^{1/2} - 4\omega_j(\omega_j^2 + R_{j2}^{-1})^{1/2}]^{1/2} \\
&= -\frac{1}{2}[R_{j2}^{-1} - 2\omega_j^2 + 2\omega_j(\omega_j^2 + R_{j2}^{-1})^{1/2}]^{1/2} \\
&\quad \pm \frac{1}{2}[R_{j2}^{-1} - 2\omega_j^2 - 2\omega_j(\omega_j^2 + R_{j2}^{-1})^{1/2}]^{1/2}, \quad j = 1, 2, \ldots, n \quad (6.282)
\end{aligned}$$

Of course, the forces producing the above closed-loop poles are obtained by inserting the modal forces $Y_j^*(t)$ from Eqs. (6.279) into Eq. (6.267).

It will prove of interest to examine the case in which $R_{j2}\omega_j^2 \gg 1$. Considering the binomial expansion

$$(\omega_j^2 + R_{j2}^{-1})^{1/2} = \omega_j\left[1 + \frac{1}{2}\omega_j^{-2}R_{j2}^{-1} - \frac{1}{8}(\omega_j^{-2}R_{j2}^{-1})^2 + \cdots\right] \quad (6.283)$$

and using two or three terms, as the case may be, the closed-loop poles can be approximated by

$$\genfrac{}{}{0pt}{}{p_j}{p_{j+n}} \cong -\left(\frac{1}{2}R_j^{-1}\right)^{1/2} \pm i\omega_j\left[1 - \frac{1}{16}(\omega_j^{-2}R_j^{-1})^2\right]^{1/2} = -\zeta_j\omega_j \pm i\omega_{dj},$$
$$j = 1, 2, \ldots, n \quad (6.284)$$

where

$$\zeta_j = \left(\frac{1}{2}\omega_j^{-2}R_j^{-1}\right)^{1/2}, \quad \omega_{dj} = \left(1 - \frac{\zeta_j^4}{4}\right)^{1/2}, \quad \omega_j \quad j = 1, 2, \ldots, n \quad (6.285\text{a, b})$$

so that the control provides artificial viscous damping with the modal viscous damping factors given by Eqs. (6.285a) and the frequencies of damped oscillation given by Eqs. (6.285b).

The fact that for undamped, nongyroscopic structures the modal equations remain decoupled in the presence of feedback control implies that in the linear case the closed-loop eigenvectors are the same as the open-loop eigenvectors. In the case of nonlinear control, the closed-loop equations are nonlinear, so that the concept of eigenvalues and eigenvectors no longer applies.

The undamped, nongyroscopic case is more important than it may appear. In the first place, damping in structures is very often sufficiently small that it can be ignored. If damping is light, but not negligible, then the control problem can still be solved as if the structure were undamped and the effect of damping on the closed-loop eigenvalues and eigenvectors can be treated by the perturbation technique discussed in Section 3.14.

Example 6.13 Assume that the fourth-order system of Fig. 6.4 has the parameters $m_1 = 1$, $m_2 = 2$, $k_1 = 1$ and $k_2 = 4$ and design an optimal IMSC. Use as performance measure

$$J = \sum_{j=1}^{2} J_j = \sum_{j=1}^{2} \int_0^\infty (\mathbf{w}_j^T \mathbf{w}_j + R_j Y_j^2) \, dt \tag{a}$$

and consider the two cases: (1) $R_1 = R_2 = 1$ and (2) $R_1 = R_2 = 10$. Derive expressions for the actual controls and compute the closed-loop poles in each case and draw conclusions as to the effect of the weighting factors $R_j (j = 1, 2)$.

The mass and stiffness matrices have the values

$$M = \begin{bmatrix} 1 & 0 \\ 0 & 2 \end{bmatrix}, \quad K = \begin{bmatrix} 5 & -4 \\ -4 & 4 \end{bmatrix} \tag{b}$$

while the damping matrix, the gyroscopic matrix and the circulatory matrix are all zero. Inserting Eqs. (b) into Eq. (6.237a), we obtain

$$A = \begin{bmatrix} 0 & 0 & 1 & 0 \\ 0 & 0 & 0 & 2 \\ -5 & 4 & 0 & 0 \\ 2 & -2 & 0 & 0 \end{bmatrix} \tag{c}$$

so that, solving the eigenvalue problem for A, we obtain the eigenvalues

$$\lambda_1 = i0.5463, \quad \lambda_2 = i2.5887, \quad \lambda_3 = -i0.5463, \quad \lambda_4 = -i2.5887 \tag{d}$$

as well as the matrices of right and left eigenvectors

$$U = \begin{bmatrix} 0.8585 & 0.2573 & 0.8585 & 0.2573 \\ 1.0091 & -0.1094 & 1.0091 & -0.1094 \\ i0.4690 & i0.6660 & -i0.4690 & -i0.6660 \\ i0.5512 & -i0.2833 & -i0.5512 & i0.2833 \end{bmatrix}$$

$$V = \begin{bmatrix} -0.1548 & 1.4270 & -0.1548 & 1.4270 \\ -0.3638 & -1.2141 & -0.3638 & -1.2141 \\ i0.2833 & -i0.5512 & -i0.2833 & i0.5512 \\ i0.6660 & -i0.4690 & -i0.6660 & i0.4690 \end{bmatrix}$$ (e)

where the eigenvectors have been normalized so as to satisfy the biothonormality relations (6.238). Hence, from the second of Eqs. (e), the matrix V_I^T to be used in Eq. (6.267) is simply

$$V_I^T = \begin{bmatrix} 0.2833 & 0.6660 \\ -0.5512 & -0.4690 \end{bmatrix}$$ (f)

so that

$$(V_I^T)^{-1} = \begin{bmatrix} -2.0026 & -2.8437 \\ 2.3536 & 1.2097 \end{bmatrix}$$ (g)

Before we can compute the actual controls, we must compute the modal controls. These are given by the second half of Eqs. (6.279), where in the case at hand the natural frequencies are

$$\omega_1 = 05463, \quad \omega_2 = 2.5887$$ (h)

Recognizing that $R_{j2} = R_j$, we consider the cases:

1. $R_1 = R_2 = 1$
Using the second half of Eqs. (6.279), the optimal modal forces are

$$\begin{aligned} Y_1^*(t) &= -[\omega_1 - (\omega_1^2 + R_1^{-1})^{1/2}]\xi_1(t) - [R_1^{-1} - 2\omega_1^2 + 2\omega_1(\omega_1^2 \\ &+ R_1^{-1})^{1/2}]^{1/2}\eta_1(t) = -[0.5463 - (0.5463^2 + 1)^{1/2}]\xi_1(t) \\ &- [1 - 2 \times 0.5463^2 + 2 \times 0.5463(0.5463^2 + 1)^{1/2}]^{1/2}\eta_1(t) \\ &= 0.5932\xi_1(t) - 1.2838\eta_1(t) \\ Y_2^*(t) &= -[\omega_2 - (\omega_2^2 + R_2^{-1})^{1/2}]\xi_2(t) - [R_2^{-1} - 2\omega_2^2 + 2\omega_2(\omega_2^2 \\ &+ R_2^{-1})^{1/2}]^{1/2}\eta_2(t) = -[2.5887 - (2.5887^2 + 1)^{1/2}]\xi_2(t) \\ &- [1 - 2 \times 2.5887^2 + 2 \times 2.5887(2.5887^2 + 1)^{1/2}]^{1/2}\eta_2(t) \\ &= 0.1864\xi_2(t) - 1.4019\eta_2(t) \end{aligned}$$ (i)

Hence, inserting the first of Eqs. (b), Eq. (g) and Eqs. (i) into Eq. (6.267), we obtain the actual controls

$$Q(t) = M(V_I^T)^{-1}Y(t) = \begin{bmatrix} 1 & 0 \\ 0 & 2 \end{bmatrix} \begin{bmatrix} -2.0026 & -2.8437 \\ 2.3536 & 1.2097 \end{bmatrix} \begin{bmatrix} Y_1^*(t) \\ Y_2^*(t) \end{bmatrix}$$

$$= \begin{bmatrix} -1.1879\xi_1(t) + 2.5709\eta_1(t) - 0.5301\xi_2(t) + 3.9866\eta_2(t) \\ 2.7923\xi_1(t) - 6.0431\eta_1(t) + 0.4510\xi_2(t) - 3.3918\eta_2(t) \end{bmatrix} \quad (j)$$

Moreover, using Eqs. (6.282), the closed-loop poles are

$$\begin{aligned} \frac{p_1}{p_3} &= -\frac{1}{2}[R_1^{-1} - 2\omega_1^2 + 2\omega_1(\omega_1^2 + R_1^{-1})^{1/2}]^{1/2} \pm \frac{1}{2}[R_1^{-1} - 2\omega_1^2 - 2\omega_1(\omega_1^2 \\ &\quad + R_1^{-1})^{1/2}]^{1/2} = -\frac{1}{2}[1 - 2 \times 0.5463^2 + 2 \times 0.5463(0.5463^2 + 1)^{1/2}]^{1/2} \\ &\quad \pm \frac{1}{2}[1 - 2 \times 0.5463^2 - 2 \times 0.5463(0.5463^2 + 1)^{1/2}]^{1/2} \\ &= -0.6419 \pm i0.4588 \end{aligned}$$

(k)

$$\begin{aligned} \frac{p_2}{p_4} &= -\frac{1}{2}[R_2^{-1} - 2\omega_2^2 + 2\omega_2(\omega_2^2 + R_2^{-1})^{1/2}]^{1/2} \pm \frac{1}{2}[R_2^{-1} - 2\omega_2^2 - 2\omega_2(\omega_2^2 \\ &\quad + R_2^{-1})^{1/2}]^{1/2} = -\frac{1}{2}[1 - 2 \times 2.5887^2 + 2 \times 2.5887(2.5887^2 \\ &\quad + 1)^{1/2}]^{1/2} \pm \frac{1}{2}[1 - 2 \times 2.5887^2 - 2 \times 2.5887(2.5887^2 + 1)^{1/2}]^{1/2} \\ &= -0.7009 \pm i2.5870 \end{aligned}$$

2. $R_1 = R_2 = 10$

Using the second half of Eqs. (6.279), the optimal modal forces are

$$\begin{aligned} Y_1^*(t) &= -[0.5463 - (0.5463^2 + 0.1)^{1/2}]\xi_1(t) - [0.1 - 2 \times 0.5463^2 \\ &\quad + 2 \times 0.5463(0.5463^2 + 0.1)^{1/2}]^{1/2}\eta_1(t) \\ &= 0.0849\xi_1(t) - 0.1928\eta_1(t) \\ Y_2^*(t) &= -[2.5887 - (2.5887^2 + 0.1)^{1/2}]\xi_2(t) - [0.1 - 2 \times 2.5887^2 \\ &\quad + 2 \times 2.5887(2.5887^2 + 0.1)^{1/2}]^{1/2}\eta_2(t) \\ &= 0.0192\xi_2(t) - 0.4468\eta_2(t) \end{aligned}$$

(l)

and, using Eqs. (6.282), the closed-loop poles are

$$\begin{aligned}\rho_1 \\ \rho_3\end{aligned} = -\frac{1}{2}[0.1 - 2 \times 0.5463^2 + 2 \times 0.5463(0.5463^2 + 0.1)^{1/2}]^{1/2}$$

$$\pm \frac{1}{2}[0.1 - 2 \times 0.5463^2 - 2 \times 0.5463(0.5463^2 + 0.1)^{1/2}]^{1/2}$$

$$= -0.0914 \pm i0.5447 \qquad (m)$$

$$\begin{aligned}\rho_2 \\ \rho_4\end{aligned} = -\frac{1}{2}[0.1 - 2 \times 2.5887^2 + 2 \times 2.5887(2.5887^2 + 0.1)^{1/2}]^{1/2}$$

$$\pm \frac{1}{2}[0.1 - 2 \times 2.5887^2 - 2 \times 2.5887(2.5887^2 + 0.1)^{1/2}]^{1/2}$$

$$= -0.2234 \pm i2.5887$$

It is clear from the above results that large weighting factors R_j imply small control forces and closed-loop poles with relatively small negative real parts and imaginary parts close in values to the natural frequencies. Hence, the effect of the weighting factors decreases as their magnitude increases.

CHAPTER 7

DISTRIBUTED-PARAMETER STRUCTURES. EXACT AND APPROXIMATE METHODS

In general, flexible structures are characterized by parameters, such as stiffness and mass, that are functions of spatial variables. Hence, structures represent distributed-parameter systems. The motion of a given structure is governed by partial differential equations to be satisfied over the domain of extension of the structure and boundary conditions to be satisfied at the boundaries of the domain. The response of a flexible structure is ordinarily obtained by modal analysis, which requires the solution of the eigenvalue problem. For distributed systems, this is a differential eigenvalue problem and its solution consists of an infinite set of eigenvalues and corresponding eigenfunctions. Then, the displacement is assumed in the form of a series of space-dependent eigenfunctions multiplied by time-dependent generalized coordinates. This helps transform the partial differential equations into a set of independent ordinary differential equations, where the latter can be solved by the methods of Chapters 3 and 4.

For the most part, closed-form solutions to the partial differential equations describing distributed structures are not possible or feasible. The alternative is to produce approximate solutions. This invariably implies discretization in space of the distributed structure. The simplest discretization process is lumping, but this is often unsatisfactory. Widely used discretization procedures in structural dynamics are the classical Rayleigh–Ritz method, the finite element method, Galerkin's method and the collocation method. For structures consisting of a number of given substructures, substructure synthesis is a powerful modeling technique.

7.1 BOUNDARY-VALUE PROBLEMS

Let us consider a distributed system defined over the domain $D: 0 \le x \le L$, where x is the spatial variable, and denote by $w(x, t)$ the displacement from an equilibrium position. The kinetic energy can be written in the general form

$$T(t) = \int_0^L \hat{T}(x, t)\, dx \qquad (7.1)$$

where

$$\hat{T}(x, t) = \hat{T}(\dot{w}) \qquad (7.2)$$

is the kinetic energy density, in which dots designate partial derivatives with respect to time. Similarly, the potential energy can be written as

$$V(t) = \int_0^L \hat{V}(x, t)\, dx \qquad (7.3)$$

where

$$\hat{V}(x, t) = \hat{V}(w, w', w'') \qquad (7.4)$$

is the potential energy density, in which primes denote partial derivatives with respect to x. Moreover, the virtual work associated with the nonconservative forces has the form

$$\delta W(t) = \int_0^L \delta \hat{W}(x, t)\, dx \qquad (7.5)$$

where

$$\delta \hat{W}(x, t) = f(x, t)\delta w(x, t) \qquad (7.6)$$

is the virtual work density, in which $f(x, t)$ is the distributed nonconservative force and $\delta w(x, t)$ is the virtual displacement.

The differential equation governing the motion of the system can be derived in general form by means of the extended Hamilton's principle (Section 2.4). The principle can be stated as

$$\int_{t_1}^{t_2} (\delta L + \delta W)\, dt = 0, \quad \delta w = 0, \quad 0 \le x \le L, \quad \text{at } t = t_1, t_2 \qquad (7.7)$$

where $L = T - V$ is the Lagrangian. Introducing Eqs. (7.1), (7.3), (7.5) and (7.6) into Eq. (7.7), we obtain

$$\int_{t_1}^{t_2} \int_0^L (\delta \hat{L} + f \delta w) \, dx \, dt = 0, \quad \delta w = 0, \quad 0 \le x \le L, \quad \text{at } t = t_1, t_2 \tag{7.8}$$

where

$$\hat{L} = \hat{T} - \hat{V} = \hat{L}(w, w', w'', \dot{w}) \tag{7.9}$$

is the Lagrangian density. From Eq. (7.9), we can write the variation in the Lagrangian density

$$\delta \hat{L} = \frac{\partial \hat{L}}{\partial w} \delta w + \frac{\partial \hat{L}}{\partial w'} \delta w' + \frac{\partial \hat{L}}{\partial w''} \delta w'' + \frac{\partial \hat{L}}{\partial \dot{w}} \delta \dot{w} \tag{7.10}$$

Inserting Eq. (7.10) into Eq. (7.8), integrating by parts both with respect to x and t and recalling that $\delta w = 0$ at $t = t_1, t_2$, we obtain

$$\int_{t_1}^{t_2} \left\{ \int_0^L \left[\frac{\partial \hat{L}}{\partial w} - \frac{\partial}{\partial x}\left(\frac{\partial \hat{L}}{\partial w'}\right) + \frac{\partial^2}{\partial x^2}\left(\frac{\partial \hat{L}}{\partial w''}\right) - \frac{\partial}{\partial t}\left(\frac{\partial \hat{L}}{\partial \dot{w}}\right) + f \right] \delta w \, dx \right.$$
$$\left. + \left[\frac{\partial \hat{L}}{\partial w'} - \frac{\partial}{\partial x}\left(\frac{\partial \hat{L}}{\partial w''}\right) \right] \delta w \bigg|_0^L + \frac{\partial \hat{L}}{\partial w''} \delta w' \bigg|_0^L \right\} dt = 0 \tag{7.11}$$

Because δw is arbitrary, Eq. (7.11) can be satisfied for all values of δw if and only if

$$\frac{\partial \hat{L}}{\partial w} - \frac{\partial}{\partial x}\left(\frac{\partial \hat{L}}{\partial w'}\right) + \frac{\partial^2}{\partial x^2}\left(\frac{\partial \hat{L}}{\partial w''}\right) - \frac{\partial}{\partial t}\left(\frac{\partial \hat{L}}{\partial \dot{w}}\right) + f = 0, \quad 0 < x < L \tag{7.12}$$

and

$$\left[\frac{\partial \hat{L}}{\partial w'} - \frac{\partial}{\partial x}\left(\frac{\partial \hat{L}}{\partial w''}\right)\right] \delta w \bigg|_0^L = 0, \quad \frac{\partial \hat{L}}{\partial w''} \delta w' \bigg|_0^L = 0 \tag{7.13a, b}$$

Equations (7.13) can be satisfied in a variety of ways. The various possibilities are

$$\frac{\partial \hat{L}}{\partial w'} - \frac{\partial}{\partial x}\left(\frac{\partial \hat{L}}{\partial w''}\right) = 0 \quad \text{at } x = 0, L \tag{7.14a}$$

or

$$w = 0 \quad \text{at } x = 0, L \tag{7.14b}$$

and

$$\frac{\partial \hat{L}}{\partial w''} = 0 \quad \text{at } x = 0, L \qquad (7.15a)$$

or

$$w' = 0 \quad \text{at } x = 0, L \qquad (7.15b)$$

Equation (7.12) is the *Lagrange differential equation of motion* for this distributed system and Eqs. (7.14) and (7.15) represent *boundary conditions*. Only two boundary conditions need be satisfied at each end. The differential equation and the boundary conditions constitute a *boundary-value problem*. Of course, the complete problem also involves the initial conditions $w(x, 0)$ and $\dot{w}(x, 0)$, so that it is a boundary-value and *initial-value problem* at the same time.

As a simple illustration, let us consider the bending vibration of a cantilever beam, as shown in Fig. 7.1. In this case, the kinetic energy is

$$T = \frac{1}{2} \int_0^L m(x)\dot{w}^2(x, t)\, dx \qquad (7.16)$$

in which $m(x)$ is the mass density, and the potential energy is

$$V = \frac{1}{2} \int_0^L EI(x)[w''(x, t)]^2\, dx \qquad (7.17)$$

where $EI(x)$ is the bending stiffness, in which E is the modulus of elasticity and $I(x)$ is the cross-sectional area moment of inertia, so that

$$\hat{L} = \frac{1}{2} m(x)\dot{w}^2(x, t) - \frac{1}{2} EI(x)[w''(x, t)]^2 \qquad (7.18)$$

Introducing Eq. (7.18) into Eq. (7.12), we obtain

$$-\frac{\partial^2}{\partial x^2}\left(EI\frac{\partial^2 w}{\partial x^2}\right) + f = m\frac{\partial^2 w}{\partial t^2}, \quad 0 < x < L \qquad (7.19)$$

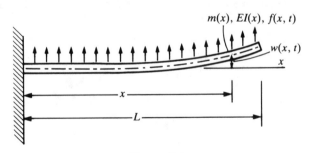

Figure 7.1

Moreover, from Eqs. (7.14) and (7.15), we must choose the boundary conditions

$$w = 0, \quad w' = 0 \quad \text{at } x = 0 \qquad (7.20a, b)$$

and

$$-\frac{\partial}{\partial x}\left(EI \frac{\partial^2 w}{\partial x^2}\right) = 0, \quad EI \frac{\partial^2 w}{\partial x^2} = 0 \quad \text{at } x = L \qquad (7.21a, b)$$

as the other possibilities do not conform to the physical situation.

The formulation given by Eqs. (7.12) and (7.13) is not the most general. In particular, it does not make allowance for lumped masses and springs at boundaries. The presence of such factors introduces terms in Eqs. (7.13), as shown in the following example.

In the case of a flexible body in space, such as a flexible spacecraft, the motion includes right-body translations and rotations, in addition to the elastic displacements. The equations governing the motion of such a body can be obtained by adjoining to partial differential equations of the type (7.12) the Lagrange's equations of motion in terms of quasi-coordinates, Eqs. (2.45). Of course, in this case the Lagrangian has the functional dependence $L = L(R_X, R_Y, R_Z, \theta_1, \theta_2, \theta_3, V_x, V_y, V_z, \omega_x, \omega_y, \omega_z, w, \dot{w}, w', w'')$, and so does the Lagrangian density $\hat{L}$. The complete set of equations can be found in Meirovitch [M51].

Example 7.1 Let us consider the rod in axial vibration shown in Fig. 7.2 and derive the boundary-value problem by means of Hamilton's principle.

The kinetic energy is given by

$$T(t) = \frac{1}{2}\int_0^L m(x)\dot{w}^2(x, t)\, dx \qquad (a)$$

and the potential energy by

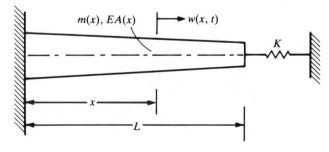

Figure 7.2

where $A(x)$ is the cross-sectional area and K is the spring constant. The Hamilton principle, Eq. (2.23), can be written as

$$\delta \int_{t_1}^{t_2} L \, dt = \delta \int_{t_1}^{t_2} (T - V) \, dt = \delta \int_{t_1}^{t_2} \left\{ \frac{1}{2} \int_0^L [m\dot{w}^2 - EA(w')^2] \, dx \right.$$
$$\left. - \frac{1}{2} Kw^2(L, t) \right\} dt = 0, \quad \delta w = 0 \quad \text{at } t = t_1, t_2 \tag{c}$$

$$V(t) = \frac{1}{2} \int_0^L EA(x)[w'(x, t)]^2 \, dx + \frac{1}{2} Kw^2(L, t) \tag{b}$$

Taking the variation in Eq. (c), integrating by parts with respect to t and x and recalling that the virtual displacement is zero at $t = t_1, t_2$, we obtain

$$\int_{t_1}^{t_2} \left\{ -\int_0^L [m\ddot{w} - (EAw')'] \, \delta w \, dx - (EAw' + Kw) \, \delta w \bigg|_{x=L} \right.$$
$$\left. + Kw \, \delta w \bigg|_{x=0} \right\} dt = 0 \tag{d}$$

Due to the arbitrariness of δw, Eq. (d) is satisfied for all possible δw provided that

$$(EAw')' = m\ddot{w}, \quad 0 < x < L \tag{e}$$

$$(EAw' + Kw) \, \delta w \bigg|_{x=L} = 0, \quad Kw \, \delta w \bigg|_{x=0} = 0 \tag{f}$$

Geometry requires that w be zero at $x = 0$. Moreover, w is in general not zero at $x = L$. Hence, the boundary conditions are

$$w = 0 \quad \text{at } x = 0, \quad EAw' + Kw = 0 \text{ at } x = L \tag{g}$$

The boundary-value problem consists of the differential equation (e) and the boundary conditions (g).

7.2 THE DIFFERENTIAL EIGENVALUE PROBLEM

Let us consider the free vibration problem, $f = 0$, and assume a solution of Eq. (7.19) in the form

$$w(x, t) = W(x)F(t) \tag{7.22}$$

The function $F(t)$ can be shown to have the exponential form $F_0 e^{i\omega t}$ [M26]. Hence, letting $\ddot{F}(t) = -\omega^2 F(t)$, Eq. (7.19) can be reduced to

THE DIFFERENTIAL EIGENVALUE PROBLEM

$$\frac{d^2}{dx^2}\left(EI\frac{d^2W}{dx^2}\right) = \lambda mW, \quad \lambda = \omega^2, \quad 0 < x < L \quad (7.23)$$

where W is subject to the boundary conditions

$$W = 0, \quad W' = 0 \quad \text{at } x = 0 \quad (7.24\text{a, b})$$

and

$$-\frac{d}{dx}\left(EI\frac{d^2W}{dx^2}\right) = 0, \quad EI\frac{d^2W}{dx^2} = 0 \quad \text{at } x = L \quad (7.25\text{a, b})$$

The problem of determining the values of the parameter λ for which Eq. (7.23) has nontrivial solutions satisfying Eqs. (7.24) and (7.25) is known as the *differential eigenvalue problem* for the distributed system under consideration. Before discussing the eigenvalue problem and its solution, it will prove advantageous to generalize it, so that the discussion will not appear to be restricted to the particular problem described by Eqs. (7.23)–(7.25).

Let us consider the more general eigenvalue problem consisting of the differential equation

$$\mathcal{L}W(x) = \lambda m(x)W(x), \quad 0 < x < L \quad (7.26)$$

where $\mathcal{L}$ is a linear homogeneous differential stiffness operator of order $2p$, in which p is an integer, and the boundary conditions

$$B_i W = 0, \quad i = 1, 2, \ldots, p, \quad x = 0, L \quad (7.27)$$

where B_i are linear homogeneous differential operators of maximum order $2p - 1$. To establish the relevance of the above formulation, we note that for the eigenvalue problem (7.23)–(7.25)

$$\mathcal{L} = \frac{d^2}{dx^2}\left(EI\frac{d^2}{dx^2}\right) \quad (7.28\text{a})$$

$$B_1 = 1, \quad B_2 = \frac{d}{dx} \quad \text{at } x = 0 \quad (7.28\text{b})$$

$$B_1 = -\frac{d}{dx}\left(EI\frac{d^2}{dx^2}\right), \quad B_2 = EI\frac{d^2}{dx^2} \quad \text{at } x = L \quad (7.28\text{c})$$

The solution of Eq. (7.26) contains $2p$ constants of integration, in addition to the undetermined parameter λ. Use of the $2p$ boundary conditions (7.27) determines $2p - 1$ of the constants and at the same time generates an equation, generally transcendental, known as the *characteristic equation*. The solution of the characteristic equation consists of *a denumerably infinite* set of eigenvalues λ_r $(r = 1, 2, \ldots)$. Upon substituting these eigenvalues into the general solution, we obtain a denumerably infinite set of associated *eigenfunctions* W_r $(r = 1, 2, \ldots)$. The eigenfunctions are unique, except for the amplitude, which is consistent with the fact that for homogeneous problems the solutions are unique within a multiplicative constant. The amplitudes can be rendered unique by means of a process known as *normalization*. The eigensolutions possess certain useful properties. Before discussing these properties, it is necessary to introduce additional definitions.

Let us consider two functions f and g that are piecewise continuous in the domain $0 \le x \le L$ and define the *inner product*

$$(f, g) = \int_0^L fg \, dx \tag{7.29}$$

If the inner product vanishes, the functions f and g are said to be *orthogonal* over the domain $0 \le x \le L$. Next, let us consider a set of linearly independent functions $\phi_1, \phi_2, \ldots$ and define

$$f_n = \sum_{r=1}^n c_r \phi_r \tag{7.30}$$

where c_r $(r = 1, 2, \ldots, n)$ are constant coefficients. Then, if by choosing n large enough the mean square error

$$M = \|f - f_n\|^2 = (f - f_n, f - f_n) = \int_0^L (f - f_n)^2 \, dx \tag{7.31}$$

can be made less than any arbitrarily small positive number ε, the set of functions $\phi_1, \phi_2, \ldots$ is said to be *complete*.

There are three classes of functions of special interest: *eigenfunctions*, *comparison functions* and *admissible functions*. The eigenfunctions must satisfy both the differential equation (7.26) and the boundary conditions (7.27). This is a relatively small class of functions, and there is no guarantee that it is possible to solve the eigenvalue problem for a given system exactly and obtain the eigenfunctions. In such cases, we must be content with approximate solutions of the eigenvalue problem, which leads naturally to the subject of the other two classes of functions. The class of *comparison functions* comprises functions that are $2p$ times differentiable and *satisfy all the boundary conditions* (7.27), but not necessarily the differential equation (7.26). Clearly, the class of comparison functions is considerably larger than the class of eigenfunctions. Quite often, satisfaction of all the boundary conditions can create difficulties. Hence, a closer look at the nature of the boundary conditions is in order. There are two types of boundary conditions, namely, *geometric boundary conditions* and *natural boundary condi-*

tions. As the name indicates, the first type reflects a geometric condition at the boundary, such as zero displacement or zero slope. In the case of geometric boundary conditions, the operator B_i is of maximum order $p - 1$. On the other hand, the second type reflects force or bending moment balance at the boundary. In the case of natural boundary conditions, the operator B_i is of maximum order $2p - 1$. Note that the first of the boundary conditions (g) in Example 7.1 is geometric and the second one is natural. The class of *admissible functions* comprises functions that are p times differentiable and *satisfy only the geometric boundary conditions*, which is a far larger class of functions than the class of comparison functions. Approximate solutions are generated from either comparison or admissible functions, depending on the problem formulation. We return to this subject later in this chapter.

Next, let us consider two comparison functions u and v and define the inner product

$$(u, \mathcal{L}v) = \int_0^L u\mathcal{L}v \, dx \tag{7.32}$$

Then, the differential operator $\mathcal{L}$ is said to be *self-adjoint* if

$$(u, \mathcal{L}v) = (v, \mathcal{L}u) \tag{7.33}$$

Whether Eq. (7.33) holds or not can be ascertained through integrations by parts, taking into consideration the boundary conditions. The mass density is a mere function, so that m is self-adjoint by definition. It follows that, if $\mathcal{L}$ is self-adjoint, then the system is self-adjoint. Self-adjointness implies a certain symmetry of the eigenvalue problem. Indeed, if the system is self-adjoint, then integrations by parts yield

$$\int_0^L u\mathcal{L}v \, dx = \int_0^L \sum_{k=0}^{p} a_k \frac{d^k u}{dx^k} \frac{d^k v}{dx^k} dx + \sum_{l=0}^{p-1} b_l \frac{d^l u}{dx^l} \frac{d^l v}{dx^l} \Big|_0^L \tag{7.34}$$

where a_k ($k = 0, 1, \ldots, p$) and b_l ($l = 0, 1, \ldots, p - 1$) are in general functions of x, and we note that the right side of Eq. (7.34) is symmetric in u and v. Clearly, an inner product involving the function m instead of the operator $\mathcal{L}$ is symmetric in u and v.

It will prove convenient to introduce the *energy inner product* defined by

$$[u, v] = \int_0^L \sum_{k=0}^{p} a_k \frac{d^k u}{dx^k} \frac{d^k v}{dx^k} dx + \sum_{l=0}^{p-1} b_l \frac{d^l u}{dx^l} \frac{d^l v}{dx^l} \Big|_0^L \tag{7.35}$$

and we observe that the energy inner product is defined even when u and v are admissible functions rather than comparison functions. To explain the term energy inner product, let $v = u$ in Eq. (7.35) and write

$$[u, u] = \int_0^L \sum_{k=0}^{P} a_k \left(\frac{d^k u}{dx^k}\right)^2 dx + \sum_{l=0}^{p-1} b_l \left(\frac{d^l u}{dx^l}\right)^2 \Big|_0^L \quad (7.36)$$

which can be interpreted as twice the maximum potential energy of the system whose eigenvalue problem is described by Eqs. (7.26) and (7.27). The inner product $[u, u]$ can be used to define another type of norm, known as the *energy norm* and denoted by

$$|u| = [u, u]^{1/2} \quad (7.37)$$

Next, let

$$u_n = \sum_{r=1}^{n} c_r \phi_r, \quad r = 1, 2, \ldots \quad (7.38)$$

Then, if by choosing n sufficiently large the quantity $|u - u_n|^2$ can be made less than any arbitrarily small positive number ε, the set of functions $\phi_1, \phi_2, \ldots$ is said to be *complete in energy*.

Now, let us examine the properties of the eigensolution. In earlier discussions, nothing was said about the nature of λ. As a consequence of the self-adjointness of the operator $\mathcal{L}$, and hence the self-adjointness of the system, it can be shown that λ is real. The question remains as to the sign of λ, which depends on the nature of the operator $\mathcal{L}$. If for any comparison function u

$$(u, \mathcal{L}u) = \int_0^L u\mathcal{L}u \, dx \geq 0 \quad (7.39)$$

and the equality sign holds if and only if u is identically zero, then the operator $\mathcal{L}$ is *positive definite*. Because the mass density is positive by definition, an inner product involving the function m instead of the operator $\mathcal{L}$ is positive definite. Hence, if $\mathcal{L}$ is positive definite, then the system is positive definite. In this case, *all the system eigenvalues λ_r ($r = 1, 2, \ldots$) are positive*. Then, the square roots of the eigenvalues, $\omega_r = \sqrt{\lambda_r}$, are the system *natural frequencies*. Consistent with this, the eigenfunctions are called *natural modes* of vibration. On the other hand, if the equality sign in (7.39) can be realized even when u is not identically zero, then the operator $\mathcal{L}$, and hence the system, is only *positive semidefinite* and *there exists at least one eigenvalue equal to zero*. The eigenfunctions corresponding to zero eigenvalues represent *rigid-body modes*. Such rigid-body modes, characterized by zero natural frequencies, occur when the structure is either partly or fully unrestrained.

The self-adjointness of a system has another very important implication. Indeed, *if a system is self-adjoint, then the eigenfunctions are orthogonal* [M26]. If the eigenfunctions are normalized so that they satisfy

$\int_0^L mW_r^2 \, dx = 1$ $(r = 1, 2, \ldots)$, then they become *orthonormal*, where the orthonormality property is expressed by

$$\int_0^L mW_r W_s \, dx = \delta_{rs}, \quad r, s = 1, 2, \ldots \quad (7.40)$$

Moreover, because the eigenfunctions satisfy Eq. (7.26), we also have

$$\int_0^L W_r \mathcal{L} W_s \, dx = \int_0^L W_s \mathcal{L} W_r \, dx = \lambda_r \delta_{rs}, \quad r, s = 1, 2, \ldots \quad (7.41)$$

Normalized eigenfunctions are known as *normal modes*.

Because the eigenfunctions are orthogonal over $0 \le x \le L$, they comprise a complete set. This fact permits the formulation of the following *expansion theorem*: Every function w with continuous $\mathcal{L}w$ and satisfying the boundary conditions of the system can be expanded in an absolutely and uniformly convergent series in the eigenfunctions W_r $(r = 1, 2, \ldots)$ in the form

$$w = \sum_{r=1}^{\infty} c_r W_r \quad (7.42)$$

where the coefficients c_r are given by

$$c_r = \int_0^L mw W_r \, dx, \quad r = 1, 2, \ldots \quad (7.43)$$

The expansion theorem is of fundamental importance in the vibration of distributed systems, as it is the basis of the modal analysis for the determination of the response.

As an illustration, let us consider the eigenvalue problem for a cantilever beam described by Eqs. (7.23)–(7.25). Assuming that the beam is *uniform*, so that $EI(x) = EI = $ constant, $m(x) = m = $ constant, Eq. (7.23) reduces to

$$W'''' - \beta^4 W = 0, \quad \beta^4 = \omega^2 m/EI, \quad 0 < x < L \quad (7.44)$$

Boundary conditions (7.24) retain the same form. On the other hand, boundary conditions (7.25) simplify to

$$W'' = 0, \quad W''' = 0 \quad \text{at } x = L \quad (7.45)$$

The solution of Eq. (7.44) is

$$W(x) = C_1 \sin \beta x + C_2 \cos \beta x + C_3 \sinh \beta x + C_4 \cosh \beta x \quad (7.46)$$

Using boundary conditions (7.24) and (7.45), we obtain the characteristic equation

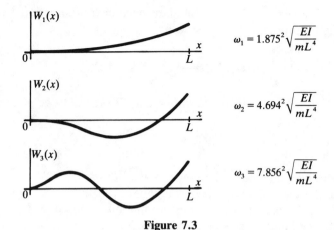

Figure 7.3

$$\cos \beta L \cosh \beta L = -1 \quad (7.47)$$

which yields a denumerably infinite set of solutions $\beta_r L$ $(r = 1, 2, \ldots)$ related to the natural frequencies by $\omega_r = \beta_r^2 (EI/mL^4)^{1/2}$. In addition, we obtain the denumerably infinite set of eigenfunctions

$$W_r(x) = A_r [(\sin \beta_r L - \sinh \beta_r L)(\sin \beta_r x - \sinh \beta_r x) \\ + (\cos \beta_r L + \cosh \beta_r L)(\cos \beta_r x - \cosh \beta_r x)], \quad r = 1, 2, \ldots \quad (7.48)$$

and we note that the eigenfunctions have not been normalized. The first three modes are shown in Fig. 7.3.

Closed-form eigensolutions are very rare indeed, and almost invariably they involve problems in which the operator $\mathscr{L}$ has constant coefficients, such as for uniform systems. The fact that the system is uniform does not guarantee closed-form eigensolutions, however. Indeed, if the cantilever beam considered above rotates about a vertical axis through the base with the uniform velocity Ω, such as in the case of the helicopter blade of Section 7.6, the centrifugal effects introduce space-dependent coefficients and no exact solution of the eigenvalue problem is possible [M26].

7.3 RAYLEIGH'S QUOTIENT

Let us consider once again the eigenvalue problem described by Eqs. (7.26) and (7.27), in which the operator $\mathscr{L}$ is self-adjoint, and define the Rayleigh quotient

$$R(u) = \frac{[u, u]}{(\sqrt{m}u, \sqrt{m}u)} \quad (7.49)$$

where u is a trial function, $[u, u]$ is the energy inner product and $(\sqrt{m}u, \sqrt{m}u)$ is the inner product of $\sqrt{m}u$ with itself. Using the expansion theorem, Eq. (7.42), it is possible to show that *Rayleigh's quotient has stationary values at the system eigenfunctions* and that *the stationary values are precisely the system eigenvalues* [M26]. Hence, as an alternative to solving the differential eigenvalue problem, Eqs. (7.26) and (7.27), one can consider rendering Rayleigh's quotient stationary. This approach is particularly attractive when the object is to obtain an approximate solution to the differential eigenvalue problem.

If the eigenvalues are ordered so that $\lambda_1 \le \lambda_2 \le \cdots$, then *Rayleigh's quotient provides an upper bound for the lowest eigenvalue*, or

$$R(u) \ge \lambda_1 \qquad (7.50)$$

In addition, we can write

$$\lambda_{s+1} = \max \min R(u), \quad (u, v_i) = 0, \quad i = 1, 2, \ldots, s \qquad (7.51)$$

where v_i are arbitrary functions. Equation (7.51) represents the *maximum-minimum characterization of the eigenvalues due to Courant and Fischer* and can be stated as follows: *The eigenvalue λ_{s+1} of the system described by Eqs. (7.26) and (7.27) is the maximum value which can be given to* min $R(u)$ *by the imposition of the s constraints* $(u, v_i) = 0$ $(i = 1, 2, \ldots, s)$ [M26].

It should be pointed out that the trial function u to be used in Eq. (7.49) can be from the space of admissible functions.

Example 7.2 Derive Rayleigh's quotient for the system of Example 7.1.

The boundary-value problem of Example 7.1 is defined by the differential equation

$$\frac{\partial}{\partial x}\left(EA\frac{\partial w}{\partial x}\right) = m\frac{\partial^2 w}{\partial t^2}, \quad 0 < x < L \qquad (a)$$

and the boundary conditions

$$w = 0 \quad \text{at } x = 0, \quad EA\frac{\partial w}{\partial x} + Kw = 0 \quad \text{at } x = L \qquad (b)$$

To derive the eigenvalue problem, let

$$w(x, t) = W(x)F(t) \qquad (c)$$

where $\ddot{F}(t) = -\lambda F(t)$, and reduce Eqs. (a) and (b) to

$$-\frac{d}{dx}\left(EA\frac{dW}{dx}\right) = \lambda mW \qquad (d)$$

$$W(0) = 0, \quad EA\frac{dW}{dx} + KW = 0 \quad \text{at } x = L \qquad (e)$$

Hence, the operator $\mathscr{L}$ is

$$\mathscr{L} = -\frac{d}{dx}\left(EA\frac{d}{dx}\right) \qquad (f)$$

Next, let us form

$$\int_0^L u\mathscr{L}u\, dx = -\int_0^L u\frac{d}{dx}\left(EA\frac{du}{dx}\right) dx \qquad (g)$$

where u is a comparison function. Integrating Eq. (g) by parts and considering boundary conditions (e), we obtain

$$[u, u] = -\int_0^L u\frac{d}{dx}\left(EA\frac{du}{dx}\right) dx = -uEA\frac{du}{dx}\bigg|_0^L + \int_0^L EA\left(\frac{du}{dx}\right)^2 dx$$

$$= \int_0^L EA\left(\frac{du}{dx}\right)^2 dx + Ku^2(L) \qquad (h)$$

Introducing Eq. (h) into Eq. (7.49), we obtain the Rayleigh quotient

$$R(u) = \frac{\int_0^L EA\left(\frac{du}{dx}\right)^2 dx + Ku^2(L)}{\int_0^L mu^2\, dx} \qquad (i)$$

7.4 THE RAYLEIGH–RITZ METHOD

More often than not, solving the eigenvalue problem, Eqs. (7.26) and (7.27), in closed-form is not possible (or not feasible) and the interest lies in an approximate solution. Such approximate solutions can be constructed by using comparison functions, and it is very important that the comparison functions be from a complete set (Section 7.2). Of course, comparison functions satisfy all the boundary conditions but not the differential equation, so that they are more abundant than the eigenfunctions. The solutions are constructed so that the differential equation is satisfied approximately.

As indicated in Section 7.3, the eigensolution can be obtained by a variational approach instead of the direct approach described by Eqs. (7.26) and (7.27). The variational approach becomes particularly attractive when the interest lies in approximate solutions of the eigenvalue problem, because a solution can be constructed from the space of admissible functions instead of the space of comparison functions. The advantage lies in the fact that admissible functions need satisfy only the geometric boundary conditions and, as pointed out in Section 7.2, this implies that admissible functions are appreciably more plentiful than comparison functions. Hence, let us consider the Rayleigh quotient (Section 7.3)

$$R(w) = \frac{[w, w]}{(\sqrt{m}w, \sqrt{m}w)} \qquad (7.52)$$

where w is a trial function from the space of admissible functions. For practical reasons, we cannot consider the entire space but only a finite-dimensional subspace, where the subspace is spanned by the functions $\phi_1, \phi_2, \ldots, \phi_n, \ldots$. The functions are such that: (i) any n elements $\phi_1, \phi_2, \ldots, \phi_n$ are linearly independent and (ii) the sequence $\phi_1, \phi_2, \ldots, \phi_n, \ldots$ is complete in energy.

The Rayleigh–Ritz method consists of selecting n functions $\phi_1, \phi_2, \ldots, \phi_n$ from a complete set and regarding them as a basis for an n-dimensional subspace S^n of the space of admissible functions. This implies that any element w^n in S^n can be expanded in the finite series

$$w^n = \sum_{i=1}^{n} u_i \phi_i \qquad (7.53)$$

where $u_1, u_2, \ldots, u_n$ are coefficients to be determined. Of course, w^n represents only an approximation of the exact solution w, where w^n tends to w as $n \to \infty$. Later in this section we examine the nature of the approximation. Inserting Eq. (7.53) into Eq. (7.52), we obtain

$$R(w^n) = \frac{[w^n, w^n]}{(\sqrt{m}w^n, \sqrt{m}w^n)} = \frac{\sum_{i=1}^{n}\sum_{j=1}^{n} u_i u_j [\phi_i, \phi_j]}{\sum_{i=1}^{n}\sum_{j=1}^{n} u_i u_j (\sqrt{m}\phi_i, \sqrt{m}\phi_j)} \qquad (7.54)$$

Next, let us recall Eq. (7.35) and introduce the notation

$$[\phi_i, \phi_j] = \int_0^L \sum_{k=0}^{P} a_k \frac{d^k \phi_i}{dx^k} \frac{d^k \phi_j}{dx^k} dx + \sum_{l=0}^{p-1} b_l \frac{d^l \phi_i}{dx^l} \frac{d^l \phi_j}{dx^l} \bigg|_0^L$$

$$= k_{ij} = k_{ji}, \quad i, j = 1, 2, \ldots, n \qquad (7.55a)$$

$$(\sqrt{m}\phi_i, \sqrt{m}\phi_j) = \int_0^L m \phi_i \phi_j \, dx = m_{ij} = m_{ji}, \quad i, j = 1, 2, \ldots, n \qquad (7.55b)$$

where k_{ij} and m_{ij} are symmetric *stiffness* and *mass coefficients*. Note that the symmetry of the stiffness and mass coefficients is a direct result of the system being self-adjoint. Substituting Eqs. (7.55) into Eq. (7.54), we can write Rayleigh's quotient in the form

$$R(u_1, u_2, \ldots, u_n) = \frac{\sum_{i=1}^{n} \sum_{j=1}^{n} k_{ij} u_i u_j}{\sum_{i=1}^{n} \sum_{i=1}^{n} m_{ij} u_i u_j} \qquad (7.56)$$

where we observe that the quotient depends only on the undetermined coefficients $u_1, u_2, \ldots, u_n$, because k_{ij} and m_{ij} ($i, j = 1, 2, \ldots, n$) are known constants. Introducing the n-vector

$$\mathbf{u} = \begin{bmatrix} u_1 & u_2 & \cdots & u_n \end{bmatrix}^T \qquad (7.57)$$

and the $n \times n$ matrices

$$K = [k_{ij}], \quad M = [m_{ij}] \qquad (7.58)$$

where K is the *stiffness matrix* and M is the *mass matrix*, both matrices being symmetric for a self-adjoint system, Eq. (7.56) can be written in the matrix form

$$R(\mathbf{u}) = \frac{\mathbf{u}^T K \mathbf{u}}{\mathbf{u}^T M \mathbf{u}} \qquad (7.59)$$

which has the appearance of the Rayleigh quotient for an n-degree-of-freedom discrete system. Hence, the effect of approximating the displacement by a series of the form (7.53) is to represent a distributed-parameter system by a discrete system, a process known as *discretization* (in space). Because the discretized model has a finite number of degrees of freedom, the representation of the solution by the series (7.53) implies *truncation* of the model as well. Clearly, the discretized model is only an approximation of the actual distributed model and it approaches the distributed model as $n \to \infty$. For practical reasons, however, n must remain finite, so that the discretized model is never an exact representation of the distributed model.

The conditions for the stationarity of Rayleigh's quotient are

$$\frac{\partial R}{\partial u_r} = 0, \quad r = 1, 2, \ldots, n \qquad (7.60)$$

It is not difficult to show that the satisfaction of the stationarity conditions are tantamount to the solution of the algebraic eigenvalue problem

$$K\mathbf{u} = \lambda^n M \mathbf{u} \qquad (7.61)$$

where λ^n is the stationary value of Rayleigh's quotient. Hence, spatial discretization has the effect of replacing a differential eigenvalue problem by an algebraic one. As pointed out earlier in this section, for a self-adjoint system, the matrices K and M are real and symmetric. Moreover, the positive definiteness (semidefiniteness) of the distributed system carries over to the discretized model.

The solution of the eigenvalue problem (7.61) was discussed in Sections 4.3 and 4.4. It consists of the eigenvalues λ_r^n and the associated eigenvectors $\mathbf{u}_r$ ($r = 1, 2, \ldots, n$). *These are the eigenvalues of the discretized model and not of the actual distributed model.* For this reason, we refer to λ_r^n as *estimated eigenvalues*, or *computed eigenvalues*. Of course, they provide estimates of the actual eigenvalues λ_r of the distributed system, more specifically estimates of the n lowest eigenvalues. The eigenvectors $\mathbf{u}_r$ can be used to obtain estimates of the eigenfunctions corresponding to the n lowest eigenvalues. Introducing the n-vector of space-dependent admissible functions

$$\boldsymbol{\phi} = [\phi_1 \quad \phi_2 \quad \cdots \quad \phi_n]^T \quad (7.62)$$

and recalling Eq. (7.57), Eq. (7.53) can be rewritten as

$$w^n = \boldsymbol{\phi}^T \mathbf{u} \quad (7.63)$$

so that the *estimated eigenfunctions*, or *computed eigenfunctions*, can be obtained from the discretized system eigenvectors by writing

$$w_r^n = \boldsymbol{\phi}^T \mathbf{u}_r, \quad r = 1, 2, \ldots, n \quad (7.64)$$

The question to be addressed next is how the estimated eigenvalues and eigenfunctions relate to the actual ones.

Let us assume that the actual and the estimated eigenvalues are ordered so that $\lambda_1 \leq \lambda_2 \leq \cdots$ and $\lambda_1^n \leq \lambda_2^n \leq \cdots \leq \lambda_n^n$, respectively. Because the set of admissible functions $\phi_1, \phi_2, \ldots, \phi_n, \ldots$ is complete, one can obtain the exact solution of the eigenvalue problem given by Eqs. (7.26) and (7.27) by simply letting $n \to \infty$ in series (7.53), at least in principle. Because n is finite, the discretized system can be regarded as being obtained from the actual system by imposing the constraints

$$u_{n+1} = u_{n+2} = \cdots = 0 \quad (7.65)$$

Because the lowest actual eigenvalue λ_1 is the minimum value that Rayleigh's quotient can taken, we conclude that

$$\lambda_1^n \geq \lambda_1 \quad (7.66)$$

Moreover, invoking the Courant and Fischer maximum-minimum principle, it can be shown that [M26]

$$\lambda_r^n \geq \lambda_r, \quad r = 1, 2, \ldots, n \tag{7.67}$$

so that *the computed eigenvalues provide upper bounds for the actual eigenvalues*. Unfortunately, the computed eigenfunctions w_r^n ($r = 1, 2, \ldots, n$) do not lend themselves to such meaningful comparison. However, it can be stated in general that the computed eigenvalues are better approximations to the actual eigenvalues than the computed eigenfunctions are to the actual eigenfunctions, a fact arising from the stationarity of Rayleigh's quotient [M26].

The Rayleigh–Ritz method calls for the use of the sequence

$$\begin{aligned} w^1 &= u_1 \phi_1 \\ w^2 &= u_1 \phi_1 + u_2 \phi_2 \\ w^3 &= u_1 \phi_1 + u_2 \phi_2 + u_3 \phi_3 \\ &\vdots \end{aligned} \tag{7.68}$$

obtained by increasing the number of terms in series (7.53) steadily, solving the eigenvalue problem (7.61) and observing the improvement in the computed eigenvalues. It should be clear from the onset that errors must be expected in the computed eigenvalues and eigenfunctions. The accuracy of the computed eigenvalues and eigenfunctions depends on the number n of terms in the series (7.53) and on the type of admissible functions ϕ_i ($i = 1, 2, \ldots, n$) used, but this accuracy is not the same for all computed eigenvalues and eigenfunctions. Indeed, lower eigenvalues tend to be more accurate than higher eigenvalues, and higher eigenvalues can be significantly in error. Hence, if the interest lies in a given number of accurate lower eigenvalues and eigenfunctions, then the number n of terms in series (7.53) must be appreciably larger; as a rule of thumb, it should be twice as large [M26].

The above statements are quite qualitative in nature. More quantitative statements (although still qualitative) can be made by invoking the inclusion principle. To this end, let us denote the stiffness and mass matrices derived on the basis of n terms in the series (7.53) by K^n and M^n, respectively. Similarly, we denote the stiffness and mass matrices derived on the basis of $n + 1$ terms by K^{n+1} and M^{n+1}, respectively. Because adding one term to series (7.53) does not affect the matrices K^n and M^n, we have

$$K^{n+1} = \begin{bmatrix} K^n & \times \\ & \times \\ \times & \times & \times \end{bmatrix}, \quad M^{n+1} = \begin{bmatrix} M^n & \times \\ & \times \\ \times & \times & \times \end{bmatrix} \tag{7.69}$$

where × denotes elements in one additional row and column. Denoting the eigenvalues associated with the eigenvalue problem defined by K^{n+1} and M^{n+1} by λ_r^{n+1} ($r = 1, 2, \ldots, n+1$), where $\lambda_1^{n+1} \leq \lambda_2^{n+1} \leq \cdots \leq \lambda_{n+1}^{n+1}$, the *inclusion principle* states that [M26]

$$\lambda_1^{n+1} \leq \lambda_1^n \leq \lambda_2^{n+1} \leq \lambda_2^n \leq \cdots \leq \lambda_n^{n+1} \leq \lambda_n^n \leq \lambda_{n+1}^{n+1} \qquad (7.70)$$

Hence, in addition to yielding lower estimates for the n lowest eigenvalues, the addition of an extra term in the series (7.53) produces an estimate of λ_{n+1}. Of course, the actual eigenvalues $\lambda_1, \lambda_2, \ldots$ provide lower bounds for the corresponding computed eigenvalues. Because the admissible functions $\phi_1, \phi_2, \ldots, \phi_n, \ldots$ are from a complete set, we can write

$$\lim_{n \to \infty} \lambda_r^n = \lambda_r, \quad r = 1, 2, \ldots, n \qquad (7.71)$$

Moreover, *as n increases, the computed eigenvalues approach the actual eigenvalues monotonically from above.*

Although convergence is guaranteed if the admissible functions $\phi_1, \phi_2, \ldots$ are from a complete set, the rate of convergence depends on the nature of the admissible functions. Hence, the choice of admissible functions can be important. As a first step in choosing admissible functions, we can consider existing complete sets of functions, such as power series, trigonometric functions, Bessel functions, Legendre polynomials, etc. In addition, we can generate admissible functions for a given system by solving a simpler eigenvalue problem characterized by simpler parameter distributions than those of the actual system and satisfying the same geometric boundary conditions as those of the given system. For example, as admissible functions for a rotating nonuniform cantilever beam, we can choose the eigenfunctions of a nonrotating uniform cantilever beam. Quite often, however, superior results can be obtained by using *quasi-comparison functions*. These are functions that act like admissible functions individually, but linear combinations thereof act more like comparison functions, as such linear combinations are capable of satisfying the natural boundary conditions as closely as desired [M56]. Otherwise, convergence is likely to be slow.

The admissible functions used in conjunction with the Rayleigh-Ritz method are global functions, in the sense that they are defined over the entire length of the domain $0 < x < L$. Moreover, they tend to be complicated functions, not easy to work with from a computational point of view. A Rayleigh-Ritz method not suffering from these drawbacks is the finite element method, discussed in Section 7.5.

Example 7.3 Consider the rod in axial vibration of Examples 7.1 and 7.2, in which the end $x = 0$ is clamped and the end $x = L$ is attached to a spring of stiffness K, and use the Rayleigh-Ritz method to obtain approximate eigensolutions for $n = 1, 2, \ldots, 10$. The system parameters are as follows:

$$EA(x) = \frac{6EA}{5}\left[1 - \frac{1}{2}\left(\frac{x}{L}\right)^2\right], \quad m(x) = \frac{6m}{5}\left[1 - \frac{1}{2}\left(\frac{x}{L}\right)^2\right], \quad K = \frac{EA}{L} \tag{a}$$

In Example 7.2, we obtained the Rayleigh quotient

$$R(w) = \frac{[w, w]}{(\sqrt{m}w, \sqrt{m}w)} = \frac{\int_0^L EA\left(\frac{dw}{dx}\right)^2 dx + Kw^2(L)}{\int_0^L mw^2 \, dx} \tag{b}$$

We use a Rayleigh–Ritz solution in the form

$$w = \sum_{i=1}^n u_i \phi_i, \quad n = 1, 2, \ldots, 10 \tag{c}$$

where ϕ_i are admissible functions. As admissible functions, we choose

$$\phi_i = \sin \frac{i\pi x}{2L}, \quad 1, 2, \ldots, 10 \tag{d}$$

Note that ϕ_i are actually quasi-comparison functions. Indeed, they satisfy the geometric boundary condition at $x = 0$, but individually they do not satisfy the natural boundary condition at $x = L$. However, a linear combination of ϕ_i is capable of satisfying the natural boundary condition at $x = L$.

Using Eqs. (7.55), we obtain the stiffness and mass coefficients

$$\begin{aligned} k_{ij} &= [\phi_i, \phi_j] = \int_0^L EA\phi_i'\phi_j' \, dx + K\phi_i(L)\phi_j(L) \\ m_{ij} &= (\sqrt{m}\phi_i, \sqrt{m}\phi_j) = \int_0^L m\phi_i\phi_j \, dx \end{aligned} \tag{e}$$

Computation of the coefficients is carried out by inserting Eqs. (d) into Eqs. (e).

The eigenvalue problem (7.61) was solved ten times, for $n = 1, 2, \ldots, 10$. The first four computed natural frequencies are listed in Table 7.1 below.

From Table 7.1, we conclude that convergence is very rapid. This can be attributed to the fact that the quasi-comparison functions $\phi_i = \sin i\pi x/2L$ represent a very good set of admissible functions. The above results were actually obtained by Meirovitch and Kwak [M56], who also present a comparison with results using other sets of admissible functions.

We also observe from Table 7.1 that, to obtain a given number of accurate computed natural frequencies, it is necessary to solve an eigenvalue problem of an order exceeding that number. For example, for $n = 4$ the first

TABLE 7.1

n	$\omega_1^n\sqrt{mL^2/EA}$	$\omega_2^n\sqrt{mL^2/EA}$	$\omega_3^n\sqrt{mL^2/EA})$	$\omega_4^n\sqrt{mL^2/EA}$
1	2.329652			
2	2.223595	5.984846		
3	2.216154	5.100072	11.092645	
4	2.215568	5.099571	8.153646	18.001143
5	2.215527	5.099528	8.116319	11.475165
6	2.215525	5.099525	8.116317	11.194091
7	2.215524	5.099525	8.116317	11.191050
8	2.215524	5.099525	8.116316	11.191047
9	2.215524	5.099525	8.116316	11.190147
10	2.215524	5.099525	8.116316	11.191047

two natural frequencies can be considered as accurate, the error in the third is about 0.5%, but the fourth is in error by more than 60%. To obtain four accurate computed natural frequencies, it is necessary to solve an eigenvalue problem of order $n = 7$.

7.5 THE FINITE ELEMENT METHOD

The finite element method can be regarded as a Rayleigh–Ritz method, although the finite element method has found uses in many areas of engineering and is not restricted to structural applications traditionally associated with the Rayleigh–Ritz method. To distinguish between the two, we refer to the method discussed in Section 7.4 as the "classical" Rayleigh–Ritz method. The basic difference between the classical Rayleigh–Ritz method and the finite element method lies in the type of admissible functions used. In the classical Rayleigh–Ritz method the admissible functions are global functions, i.e., they extend over the entire domain $0 \le x \le L$. Moreover, they tend to be complicated functions, often difficult to work with computationally. By contrast, in the finite element method, the admissible functions are local functions, defined over small subdomains and equal to zero identically everywhere else. In addition, they tend to be simple functions, such as low-degree polynomials. These two factors combine to give the finite element method the advantage of being easier to program on a digital computer. On the other hand, the discretized models obtained by the finite element method tend to be of higher order, so that a choice between the classical Rayleigh–Ritz method and the finite element method is not one-sided and depends on the nature of the system under consideration.

Let us consider a distributed-parameter system extending over the spatial domain $0 \le x \le L$ and divide the domain into n subdomains of length h each, so that $nh = L$. For simplicity, the subdomains are taken to be equal

290 DISTRIBUTED-PARAMETER STRUCTURES

in length, although this is not really necessary and in problems with large parameter variations it may not even be recommended. The subdomains are called *finite elements*, which explains the name of the method. In the finite element method, the admissible functions tend to be low-degree polynomials and they extend over a limited number of adjacent elements. The degree of the polynomials is dictated by the differentiability requirement, and hence it depends on the order $2p$ of the system. For example, for a second-order (in space) system, such as a rod in axial vibration, a shaft in torsional vibration or a string in transverse vibration, the admissible functions need be only once differentiable. This excludes sectionally constant functions but includes sectionally linear functions, such as those shown in Fig. 7.4. The functions $\phi_j(x)$ depicted in Fig. 7.4 are called *linear elements*, or more pictorially *roof functions*. The function $\phi_j(x)$ extends over two elements, $(j-1)h \leq x \leq jh$ and $jh \leq x \leq (j+1)h$, except for cases in which an element has a free end. The points $x = jh$ are called *nodes*, which is an unfortunate terminology in view of the fact that in vibrations nodes are points of zero displacement. The roof functions just described are *nearly orthogonal*, as ϕ_j is orthogonal to every other function ϕ_i except for $i = j \pm 1$. This near orthogonality holds regardless of any weighting function. All the functions ϕ_j can be regarded as being normalized in the sense that they have unit amplitudes, which has significant physical implications. Indeed, introducing the normalized roof functions into the equation

$$w^n = \sum_{j=1}^{n} u_j \phi_j \tag{7.72}$$

we can plot the function w^n as shown in Fig. 7.5. The implication is that: *the coefficients u_j are simply the displacements at the nodes $x_j = jh$*. This is in direct contrast with the classical Rayleigh–Ritz method, in which the coefficients u_j tend to be abstract quantities, having more mathematical than physical meaning.

The calculation of the stiffness coefficients k_{ij} and mass coefficients m_{ij} can be carried out by using Eqs. (7.55). The process can be simplified, however, by considering the finite elements separately. To this end, we refer to Fig. 7.6, which shows a typical element j together with the associated *nodal displacements*. The object is to evaluate the Rayleigh quotient by first

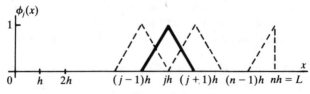

Figure 7.4

THE FINITE ELEMENT METHOD   291

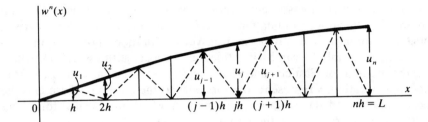

Figure 7.5

calculating the contribution of every finite element separately and then combining the results, where the latter operation is known as "assembling" the elements. Hence, let us write Rayleigh's quotient in the form

$$R = \frac{\sum_{j=1}^{n} N_j}{\sum_{j=1}^{n} D_j} \tag{7.73}$$

where N_j and D_j are the contributions of the jth element to the numerator and denominator of Rayleigh's quotient. For a second-order system, the energy inner product can be written in the general form

$$[w, w] = \int_0^L [s(w')^2 + kw^2]\, dx + Kw^2(L) \tag{7.74a}$$

and the weighted inner product in the form

$$(\sqrt{m}w, \sqrt{m}w) = \int_0^L mw^2\, dx \tag{7.74b}$$

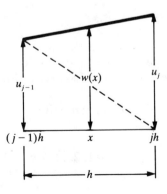

Figure 7.6

where s, k and m are space-dependent parameters and K is a constant parameter. The above formulation holds for rods in axial vibration, shafts in torsional vibration and strings in transverse vibration. For rods in axial vibration, the symbol s can be identified as the axial stiffness $EA(x)$, where E is the modulus of elasticity and $A(x)$ the cross-sectional area. For shafts in torsional vibration, s represents the torsional stiffness $GJ(x)$, where G is the shear modulus and $J(x)$ the area polar moment of inertia. Finally, for strings in transverse vibration, s is the string tension $T(x)$. Moreover, k and K are distributed spring constant and boundary spring constant, respectively, and m is either mass per unit length or mass polar moment of inertia per unit length. Then, it is easy to verify that

$$N_j = \int_{(j-1)h}^{jh} [s(w')^2 + kw^2]\, dx + \delta_{jn} K w^2(L), \quad j = 1, 2, \ldots, n \tag{7.75a}$$

$$D_j = \int_{(j-1)h}^{jh} mw^2\, dx, \quad j = 1, 2, \ldots, n \tag{7.75b}$$

Equations (7.75) are in terms of the *global coordinate* x. For computational purposes, it is more advantageous to work with *local coordinates*. To this end, we introduce the notation

$$j - \frac{x}{h} = \xi, \quad \frac{x}{h} - (j-1) = 1 - \xi \tag{7.76}$$

which can be regarded as nondimensional local coordinates and are known in the finite element terminology as *natural coordinates*. Then, introducing the notation

$$L_1 = \xi, \quad L_2 = 1 - \xi \tag{7.77}$$

we observe from Fig. 7.6 that the displacement can be written in terms of the nodal displacements u_{j-1} and u_j as follows:

$$w = w(\xi) = L_1 u_{j-1} + L_2 u_j \tag{7.78}$$

so that L_1 and L_2 play the role of admissible functions for the element j. The functions L_1 and L_2 can be identified as *linear interpolation functions*. They are also known as *shape functions*. For linear elements, they coincide with the natural coordinates.

Introducing Eqs. (7.77) and (7.78) into Eqs. (7.75), we can write

$$N_j = \begin{bmatrix} u_{j-1} \\ u_j \end{bmatrix}^T K_j \begin{bmatrix} u_{j-1} \\ u_j \end{bmatrix}, \quad D_j = \begin{bmatrix} u_{j-1} \\ u_j \end{bmatrix}^T M_j \begin{bmatrix} u_{j-1} \\ u_j \end{bmatrix}, \quad j = 1, 2, \ldots, n \tag{7.79}$$

where K_j and M_j ($j = 1, 2, \ldots, n$) are 2×2 *element stiffness* and *mass matrices* with the entries

$$k_{j11} = \frac{1}{h} \int_0^1 [s(\xi) + h^2 \xi^2 k(\xi)] \, d\xi \, ,$$

$$k_{j12} = k_{j21} = \frac{1}{h} \int_0^1 [-s(\xi) + h^2 \xi (1 - \xi) k(\xi)] \, d\xi \, ,$$

$$k_{j22} = \frac{1}{h} \int_0^1 [s(\xi) + h^2 (1 - \xi)^2 k(\xi)] \, d\xi + \delta_{jn} K \, ,$$

$$\qquad\qquad\qquad\qquad\qquad\qquad j = 1, 2, \ldots, n \quad (7.80)$$

$$m_{j11} = h \int_0^1 m(\xi) \xi^2 \, d\xi \, ,$$

$$m_{j12} = m_{j21} = h \int_0^1 m(\xi) \xi (1 - \xi) \, d\xi \, ,$$

$$m_{j22} = h \int_0^1 m(\xi) (1 - \xi)^2 \, d\xi \, ,$$

Noting from Fig. 7.5 that $u_0 = 0$, we must strike out the first row and column from the matrices K_1 and M_1, so that K_1 and M_1 are really scalars. If h is sufficiently small, then $s(\xi)$, $k(\xi)$ and $m(\xi)$ can be regarded as constant over each element.

Having the element stiffness and mass matrices, the next task is the *assembling process*. Introducing Eqs. (7.79) into Eq. (7.73), the Rayleigh quotient can be written in the form

$$R = \frac{\mathbf{u}^T K \mathbf{u}}{\mathbf{u}^T M \mathbf{u}} \qquad (7.81)$$

where $\mathbf{u} = [u_1 \ u_2 \ \cdots \ u_n]^T$ is the *nodal vector* and K and M are real and symmetric *global stiffness* and *mass matrices*, respectively, and are obtained from the element matrices K_j and M_j ($j = 1, 2, \ldots, n$) in the schematic form

$$K = \begin{bmatrix} K_1 & & & & \\ & K_2 & & & \\ & & K_3 & & \\ & & & \ddots & \\ & & & & K_{n-1} \\ & & & & & K_n \end{bmatrix} \quad M = \begin{bmatrix} M_1 & & & & \\ & M_2 & & & \\ & & M_3 & & \\ & & & \ddots & \\ & & & & M_{n-1} \\ & & & & & M_n \end{bmatrix} \quad (7.82)$$

The shaded areas denote entries that are the sum of the entries corresponding to nodal displacements shared by two adjacent elements. Note that the use of a local basis, in which the admissible functions extend over two adjacent elements, results in banded matrices K and M, where the bandwidth is equal to three.

The requirement that Rayleigh's quotient, Eq. (7.81), be stationary yields the real symmetric eigenvalue problem

$$K\mathbf{u} = \lambda^n M\mathbf{u} \tag{7.83}$$

which was discussed in Sections 4.3 and 4.4.

In the case of bending vibration, the denominator of Rayleigh's quotient retains the same form as that given by Eq. (7.74b) but the numerator, which represents the energy inner product, has a different form. We consider

$$[w, w] = \int_0^L [EI(w'')^2 + P(w')^2]\, dx \tag{7.84}$$

where EI is the flexural rigidity, in which I is the area moment of inertia, and P is a tensile axial force. The presence of w'' in $[w, w]$ suggests that linear elements must be ruled out. In fact, quadratic elements must also be ruled out, because both the displacement and the slope must be continuous at both nodes.

As nodal coordinates, we use displacements and rotations, as shown in Fig. 7.7. The vector of nodal displacements for the jth element can be written in the form

$$\mathbf{w}_j = [w_{j-1} \quad h\theta_{j-1} \quad w_j \quad h\theta_j]^T \tag{7.85}$$

where the rotations have been multiplied by h to preserve the units of displacement for all the components of $\mathbf{w}$. As a nondimensional natural

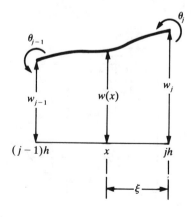

Figure 7.7

coordinate, we retain $\xi = j - x/h$. Interpolation functions satisfying the requirements of the problem are the *Hermite cubics*

$$L_1 = 3\xi^2 - 2\xi^3, \quad L_2 = \xi^2 - \xi^3$$
$$L_3 = 1 - 3\xi^2 + 3\xi^3, \quad L_4 = -\xi + 2\xi^2 - \xi^3 \quad (7.86)$$

Introducing the vector of interpolation functions

$$\mathbf{L} = [L_1 \ L_2 \ L_3 \ L_4]^T \quad (7.87)$$

The displacement in the jth element can be written as

$$w = \mathbf{L}^T \mathbf{w}_j \quad (7.88)$$

Following a procedure similar to that used earlier in this section, Eq. (7.84) can be written in the form

$$[w, w] = \sum_{j=1}^{n} \int_{(j-1)h}^{jh} [EI(w'')^2 + P(w')^2] \, dx = \sum_{j=1}^{n} \mathbf{w}_j^T K_j \mathbf{w}_j \quad (7.89)$$

where K_j is the element stiffness matrix given by

$$K_j = h \int_0^1 \left(\frac{EI_j}{h^4} \mathbf{L}'' \mathbf{L}''^T + \frac{P_j}{h^4} \mathbf{L}' \mathbf{L}'^T \right) d\xi, \quad j = 1, 2, \ldots, n \quad (7.90)$$

in which I_j and P_j are, respectively, the area moment of inertia and the tensile force over the jth element. Moreover, the weighted inner product becomes

$$(\sqrt{m}w, \sqrt{m}w) = \sum_{j=1}^{n} \int_{(j-1)h}^{jh} mw^2 \, dx = \sum_{j=1}^{n} \mathbf{w}_j^T M_j \mathbf{w}_j \quad (7.91)$$

where M_j is the element mass matrix having the form

$$M_j = h \int_0^1 m_j \mathbf{L} \mathbf{L}^T \, d\xi, \quad j = 1, 2, \ldots, n \quad (7.92)$$

in which m_j is the mass density over the jth element.

As an illustration, consider the case of a uniform cantilever beam rotating with the constant angular velocity Ω about a vertical axis through the fixed end. The beam can be regarded as representing a rotating helicopter blade. It is assumed that $EI = \text{constant}$, $m = \text{constant}$, in which case the axial force has the expression

$$P(x) = \int_x^L m\Omega^2 \zeta \, d\zeta = \frac{1}{2} m\Omega^2 L^2 \left(1 - \frac{x^2}{L^2}\right) \tag{7.93}$$

so that the force over the jth element is

$$P_j(\xi) = \frac{1}{2} m\Omega^2 L^2 \left[1 - \left(\frac{h}{L}\right)^2 (j-\xi)^2\right], \quad 0 \le \xi \le 1 \tag{7.94}$$

Introducing Eqs. (7.86) and (7.94) into Eq. (7.90), we obtain the element stiffness matrices

$$K_j = \frac{EI}{h^3} \begin{bmatrix} 12 & 6 & -12 & 6 \\ & 4 & -6 & 2 \\ \text{symm} & & 12 & -6 \\ & & & 4 \end{bmatrix} + \frac{1}{2} m\Omega^2 Ln \left(\frac{1}{30} \left[1 - \left(\frac{j}{h}\right)^2\right] \right.$$

$$\times \begin{bmatrix} 36 & -3 & -36 & 3 \\ & 4 & -3 & 1 \\ \text{symm} & & 36 & -3 \\ & & & 4 \end{bmatrix} - \frac{j}{30n} \begin{bmatrix} 36 & 0 & -36 & 6 \\ & 6 & 0 & -1 \\ \text{symm} & & 36 & -36 \\ & & & 2 \end{bmatrix}$$

$$+ \frac{1}{210 n^2} \begin{bmatrix} 72 & -6 & -72 & -15 \\ & 18 & 6 & -3 \\ \text{symm} & & 72 & 15 \\ & & & 4 \end{bmatrix} \right), \quad j = 1, 2, \ldots, n \tag{7.95}$$

Moreover, introducing Eqs. (7.86) into Eq. (7.92), we obtain the element mass matrices

$$M_j = \frac{mL}{420n} \begin{bmatrix} 156 & 22 & 54 & -13 \\ & 4 & 13 & -3 \\ \text{symm} & & 156 & -22 \\ & & & 4 \end{bmatrix}, \quad j = 1, 2, \ldots, n \tag{7.96}$$

The assembling process is similar to that described earlier.

The finite element method can be used to model trusses and frames. In such cases, it is common practice to regard each member in the truss (or frame) as a single element with the nodes coinciding with the joints [M44]. For improved accuracy the members can be divided into smaller elements. The finite element method can be used for two- and three-dimensional members, but the details are more involved. Interested readers are urged to consult the book by Huebner and Thornton [H22].

In recent years, a new version of the finite element method has been developed. It is known as the *hierarchical finite element method* and it combines the advantages of both the classical Rayleigh–Ritz method and the finite element method. As in the finite element method, the admissible functions are local functions consisting of low-degree polynomials defined over finite elements. In improving the accuracy, however, the finite element mesh is kept the same and the number of polynomials over the elements is increased, which is similar to the way in which accuracy is increased in the classical Rayleigh–Ritz method. Details of the hierarchical finite element method are given by Meirovitch [M44].

Example 7.4 Solve the eigenvalue problem of Example 7.3 by the finite element method using $n = 1, 2, \ldots, 10$ linear elements. Compare the first four computed natural frequencies with those obtained in Example 7.3 by the classical Rayleigh–Ritz method and discuss the results.

Using the first of Eqs. (7.76), we can express the stiffness and mass distributions in terms of local coordinates. Hence, from Eqs. (a) of Example 7.3, we can write

$$s(\xi) = \frac{6EA}{5}\left[1 - \frac{1}{2}\left(\frac{h}{L}\right)^2 (j-\xi)^2\right],$$

$$m(\xi) = \frac{6m}{5}\left[1 - \frac{1}{2}\left(\frac{h}{L}\right)^2 (j-\xi)\right], \quad K = \frac{EA}{L} \quad (a)$$

Introducing Eqs. (a) into Eqs. (7.80), we obtain the element stiffness and mass matrices

$$K_j = \frac{6EA}{5L}\left[n - \frac{1}{6n}(1 - 3j + 3j^2)\right]\begin{bmatrix} 1 & -1 \\ -1 & 1 \end{bmatrix} + \delta_{jn}\frac{EA}{L}\begin{bmatrix} 0 & 0 \\ 0 & 1 \end{bmatrix},$$

$$j = 1, 2, \ldots, n \quad (b)$$

$$M_j = \frac{mL}{5n}\begin{bmatrix} 2 & 1 \\ 1 & 2 \end{bmatrix} - \frac{mL}{100n^3}\begin{bmatrix} 2(6 - 15j + 10j^2) & 3 - 10j + 10j^2 \\ 3 - 10j + 10j^2 & 2(1 - 5j + 10j^2) \end{bmatrix},$$

$$j = 1, 2, \ldots, n \quad (c)$$

in which $n = L/h$. Using the scheme (7.82), in which we recall that K_1 and M_1 contain one entry each, we obtain

$$K = \frac{6EAn}{5L}\begin{bmatrix} 2 & -1 & & & & \\ & 2 & -1 & & 0 & \\ & & 2 & \ddots & & \\ & & & \ddots & 2 & -1 \\ \text{symm} & & & & & 1 + (5/6n) \end{bmatrix}$$

$$-\frac{EA}{5Ln}\begin{bmatrix} 8 & -7 & & & & \\ & 26 & -19 & & 0 & \\ & & 56 & \ddots & & \\ & & & \ddots & 2(4 - 6n + 3n^2) & -(1 - 3n + 3n^2) \\ \text{symm} & & & & & 1 - 3n + 3n^2 \end{bmatrix} \quad (d)$$

TABLE 7.2

n	$\omega_1^n \sqrt{mL^2/EA}$	$\omega_2^n \sqrt{mL^2/EA}$	$\omega_3^n \sqrt{mL^2/EA}$	$\omega_4^n \sqrt{mL^2/EA}$
1	2.672612			
2	2.325511	6.271634		
3	2.264690	5.683449	9.884047	
4	2.243258	5.431282	9.414906	13.451681
5	2.233301	5.311806	8.983983	13.262952
6	2.227880	5.246761	8.724168	12.750707
7	2.224607	5.207580	8.563522	12.363970
8	2.222481	5.182184	8.458553	12.095809
9	2.221023	5.164795	8.386491	11.907522
10	2.219979	5.152367	8.334964	11.771584

$$M = \frac{mL}{5n}\begin{bmatrix} 4 & 1 & & & & \\ & 4 & 1 & & 0 & \\ & & 4 & & & \\ & \text{symm} & & \ddots & & \\ & & & & 4 & 1 \\ & & & & & 2 \end{bmatrix}$$

$$-\frac{mL}{100n^3}\begin{bmatrix} 44 & 23 & & & & \\ & 164 & 63 & & 0 & \\ & & 364 & \ddots & & \\ & \text{symm} & & 2(37-50n+20n^2) & 3-10n+10n^2 \\ & & & & 2(1-5n+10n^2) \end{bmatrix} \quad \text{(e)}$$

The eigenvaue problem is obtained by inserting Eqs. (d) and (e) into Eq. (7.83). The eigenvalue problem must be solved 10 times, for $n = 1, 2, \ldots, 10$. It was actually solved by Meirovitch and Kwak [M56], and the first four computed natural frequencies are given in Table 7.2 above.

Comparing the results in Tables 7.1 and 7.2, we conclude that the classical Rayleigh–Ritz method in conjunction with admissible functions in the form of the quasi-comparison functions $\phi_i = \sin i\pi x/2L$ yields considerably better accuracy than the finite element method in conjunction with linear interpolation functions. In fact, the first two natural frequencies computed by the classical Rayleigh–Ritz method with $n = 3$ are more accurate than those computed by the finite element method with $n = 10$. Better accuracy can be obtained by the finite element method using higher-degree interpolation functions [M56].

7.6 THE METHOD OF WEIGHTED RESIDUALS

The Rayleigh–Ritz method is based on the stationarity of Rayleigh's quotient, which is guaranteed for self-adjoint systems. A method applicable to non-self-adjoint systems, as well as to self-adjoint systems, is the *method of weighted residuals*. The term does not imply a single method but a family of methods, and is based on the idea of reducing to zero average errors incurred in approximate solutions.

Let us consider a differential eigenvalue problem of the form given by Eq. (7.26), except that here the operator $\mathscr{L}$ can be self-adjoint or non-self-adjoint. Then, if $w^n(x)$ is a given trial function not satisfying Eq. (7.26), we can define the error incurred by using $w^n(x)$ instead of an actual solution w as

$$R(w^n, x) = \mathscr{L}w^n - \Lambda^n m w^n, \quad 0 < x < L, \quad (7.97)$$

where R is known as the *residual*. The trial function $w^n(x)$ is normally taken from the space of comparison functions (Section 7.2), so that it satisfies all the boundary conditions of the problem. The object is to drive the residual to zero. To this end, we consider a weighting function $v(x)$ and insist that the integral of the weighted residual be zero, or

$$\int_0^L v(\mathscr{L}w^n - \Lambda^n m w^n)\, dx = 0 \quad (7.98)$$

which implies that v is orthogonal to R.

The method of weighted residuals consists of assuming an approximate solution w^n of Eq. (7.26) in the form of the series

$$w^n = \sum_{j=1}^n a_j \phi_j(x) \quad (7.99)$$

where $a_1, a_2, \ldots, a_n$ are undetermined coefficients and $\phi_1, \phi_2, \ldots, \phi_n$ are given members of a complete set of comparison functions. Moreover, the procedure consists of choosing n weighting functions $\psi_1, \psi_2, \ldots, \psi_n$ from another complete set. Then the coefficients $a_1, a_2, \ldots, a_n$ and the parameters Λ^n are determined by imposing the conditions that the weighting functions be orthogonal to R, or

$$\int_0^L \psi_i R\, dx = \int_0^L \psi_i(\mathscr{L}w^n - \Lambda^n m w^n)\, dx = 0, \quad i = 1, 2, \ldots, n \quad (7.100)$$

As the integer n increases without bounds, Eqs. (7.100) can be satisfied for all ψ_i if and only if the residual is zero, or

$$\lim_{n \to \infty} (\mathscr{L}w^n - \Lambda^n m w^n) = \mathscr{L}w - \lambda m W = 0 \quad (7.101)$$

so that in the limit Λ^n and w^n converge to the actual solution λ and W.

Introducing Eq. (7.99) into Eqs. (7.100), we obtain the algebraic eigenvalue problem

$$\int_0^L \psi_i \left(\sum_{j=1}^n a_j \mathcal{L} \phi_j - \Lambda^n \sum_{j=1}^n a_j m \phi_j \right) dx = \sum_{j=1}^n (k_{ij} - \Lambda^n m_{ij}) a_j = 0,$$

$$i = 1, 2, \ldots, n \quad (7.102)$$

where

$$k_{ij} = \int_0^L \psi_i \mathcal{L} \phi_j \, dx, \quad m_{ij} = \int_0^L m \psi_i \phi_j \, dx, \quad i, j = 1, 2, \ldots, n$$

$$(7.103\text{a, b})$$

are constant coefficients, generally nonsymmetric. Equations (7.102) can be written in the matrix form

$$K\mathbf{a} = \Lambda^n M \mathbf{a} \quad (7.104)$$

where $\mathbf{a} = [a_1 \; a_2 \; \cdots \; a_n]^T$ is an n-dimensional vector of coefficients and $K = [k_{ij}]$ and $M = [m_{ij}]$ are $n \times n$ are coefficient matrices, generally nonsymmetric. The solution of Eq. (7.104) yields n estimated eigenvalues $\Lambda_1^n, \Lambda_2^n, \ldots, \Lambda_n^n$ and n associated eigenvectors $\mathbf{a}_1, \mathbf{a}_2, \ldots, \mathbf{a}_n$. The eigenvectors can be inserted into Eq. (7.99) to produce the estimated eigenfunctions

$$w_r^n = \mathbf{a}_r^T \boldsymbol{\phi}(x), \quad r = 1, 2, \ldots, n \quad (7.105)$$

where $\boldsymbol{\phi}(x) = [\phi_1(x) \; \phi_2(x) \; \cdots \; \phi_n(x)]^T$ is the n-dimensional vector of comparison functions.

The Galerkin method and the collocation method belong to the family of weighted residuals methods. In the first the weighting functions are the same as the comparison functions used in expansion (7.99), $\psi_i = \phi_i$, and in the second they are spatial Dirac delta functions, $\psi_i = \delta(x - x_i)$ [M26].

7.7 SUBSTRUCTURE SYNTHESIS

Substructure synthesis is a method whereby a structure is regarded as an assemblage of substructures, each of which modeled separately and made to act as a single structure by imposing certain geometric compatibility at boundaries between two adjacent substructures. In its inception, the method was generally referred to as component-modes synthesis [H27], because the motion of each of the substructures (referred to as components) was represented by a finite number of modes, which implies that one must solve some kind of eigenvalue problem for the substructure. However, the method can be regarded as a Rayleigh–Ritz method, so that, in the spirit of Rayleigh–Ritz, one need represent the motion of each substructure by admissible functions and substructure modes are not really necessary [M29]. To be sure, substructure modes are likely to belong to the class of admissible functions, but for the most part they are difficult to obtain and they may not be easy to work with computationally.

Let us consider the structure of Fig. 7.8 and concentrate our attention on substructure s. The kinetic energy associated with substructure s has the general expression

$$T_s = \frac{1}{2} \int_{D_s} m_s \dot{\mathbf{u}}_s^T \dot{\mathbf{u}}_s \, dD_s \qquad (7.106)$$

where $\mathbf{u}_s = \mathbf{u}_s(x, y, z, t)$ is the displacement vector, m_s is the mass density and D_s is the domain of extension of the substructure. Similarly, the potential energy can be written in the symbolic form

$$V_s = \frac{1}{2} [\mathbf{u}_s, \mathbf{u}_s] \qquad (7.107)$$

where $[\mathbf{u}_s, \mathbf{u}_s]$ is an energy integral of the type (7.36). Moreover, the Rayleigh dissipation function for substructure s can be written as [M26]

$$\mathscr{F}_s = \frac{1}{2} \int_{D_s} c_s \dot{\mathbf{u}}_s^T \dot{\mathbf{u}}_s \, dD_s \qquad (7.108)$$

where c_s is a distributed damping coefficient. Finally, the virtual work for substructure s has the form

$$\delta W_s = \int_{D_s} \mathbf{f}_s^T \delta \mathbf{u}_s \, dD_s \qquad (7.109)$$

where $\mathbf{f}_s$ is the distributed force vector and $\delta \mathbf{u}_s$ is the virtual displacement vector.

Next, we represent the displacement vector $\mathbf{u}_s$ by

$$\mathbf{u}_s = \Phi_s(x, y, z)\boldsymbol{\zeta}_s(t) \qquad (7.110)$$

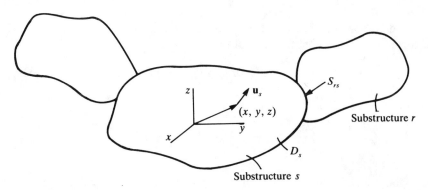

Figure 7.8

where $\Phi_s(x, y, z)$ is a matrix of space-dependent admissible functions and $\zeta_s(t)$ is a time-dependent generalized displacement vector. The dimension of the vector ζ_s is equal to the number of admissible functions used to represent $\mathbf{u}_s$. Introducing Eq. (7.110) into Eq. (7.106), we can rewrite the kinetic energy in the discretized form

$$T_s = \frac{1}{2} \dot{\zeta}_s^T M_s \dot{\zeta}_s \tag{7.111}$$

where

$$M_s = \int_{D_s} m_s \Phi_s^T \Phi_s \, dD_s \tag{7.112}$$

is the mass matrix for substructure s. Similarly, introducing Eq. (7.110) into Eq. (7.107), we obtain the discretized form of the potential energy

$$V_s = \frac{1}{2} \zeta_s^T K_s \zeta_s \tag{7.113}$$

where

$$K_s = [\Phi_s, \Phi_s] \tag{7.114}$$

is the stiffness matrix for substructure s. Moreover, inserting Eq. (7.110) into Eq. (7.108), we obtain the discretized Rayleigh dissipation function

$$\mathscr{F}_s = \frac{1}{2} \dot{\zeta}_s^T C_s \dot{\zeta}_s \tag{7.115}$$

where

$$C_s = \int_{D_s} c_s \Phi_s^T \Phi_s \, dD_s \tag{7.116}$$

is the damping matrix for substructure s. Finally, introducing Eq. (7.110) into Eq. (7.109), we obtain the discretized virtual work

$$\delta W_s = \mathbf{Z}_s^T \delta \zeta_s \tag{7.117}$$

where

$$\mathbf{Z}_s = \int_{D_s} \Phi_s^T \mathbf{f}_s \, dD_s \tag{7.118}$$

is a generalized force vector for substructure s, which excludes damping forces.

The equations of motion for substructure s can be obtained by means of Lagrange's equations, which can be written in the symbolic form

$$\frac{d}{dt}\left(\frac{\partial T_s}{\partial \dot{\zeta}_s}\right) - \frac{\partial T_s}{\partial \zeta_s} + \frac{\partial V_s}{\partial \zeta_s} + \frac{\partial \mathcal{F}_s}{\partial \dot{\zeta}_s} = \mathbf{Z}_s, \quad s = 1, 2, \ldots \quad (7.119)$$

Introducing Eqs. (7.111), (7.113) and (7.115) into Eq. (7.119), we obtain the equations of motion for substructure s

$$M_s \ddot{\zeta}_s(t) + C_s \dot{\zeta}_s(t) + K_s \zeta_s(t) = \mathbf{Z}_s(t), \quad s = 1, 2, \ldots \quad (7.120)$$

At this point, we begin the assembling process. Introducing the matrices

$$M^d = \text{block-diag } M_s, \quad C^d = \text{block-diag } C_s, \quad K^d = \text{block-diag } K_s,$$
$$s = 1, 2, \ldots \quad (7.121)$$

as well as the vectors

$$\zeta(t) = [\zeta_1^T(t) \quad \zeta_2^T(t) \quad \cdots \quad \zeta_s^T(t) \quad \cdots]^T \quad (7.122a)$$

$$\mathbf{Z}(t) = [\mathbf{Z}_1^T(t) \quad \mathbf{Z}_2^T(t) \quad \cdots \quad \mathbf{Z}_s^T(t) \quad \cdots]^T \quad (7.122b)$$

we can write the equations of motion for the collection of substructures in the form

$$M^d \ddot{\zeta}(t) + C^d \dot{\zeta}(t) + K^d \zeta(t) = \mathbf{Z}(t) \quad (7.123)$$

Equation (7.123) represents a set of disconnected substructure equations, which explains the use of the superscript d. Hence, there is nothing in Eq. (7.123) that makes the substructures act together as a single structure. But, in reality, the structure does act as a whole and the division into substructures is merely for modeling purposes. To force the individual substructures to act as a single structure, we must invoke the geometric compatibility conditions at the boundaries between any two substructures.

Let us consider two adjacent substructures s and r and denote the boundary between them by S_{rs}. It should be clear from the onset that it is not possible to satisfy geometric compatibility at every point of S_{rs} exactly. Indeed, to satisfy the geometric compatibility exactly, one must satisfy an infinity of such conditions. These conditions take the form of constraint equations relating the components of $\zeta(t)$, and the vector $\zeta(t)$ is finite-dimensional. Hence, we can only impose a finite number of geometric compatibility conditions, and their number must be smaller than the dimension of $\zeta(t)$. Assuming that the dimension of $\zeta(t)$ is m and that the number of constraint equations is c, the number of independent generalized coordinates is $n = m - c$, where n is the number of degrees of freedom of the model. There are various ways of ensuring geometric compatibility at the boundaries S_{rs}. The most common one is to require that the displacements and rotations at given points of S_{rs} are the same, or

$$\mathbf{u}_r = \mathbf{u}_s, \quad S_{rs}^* \subset S_{rs} \tag{7.124a}$$

$$\frac{\partial \mathbf{u}_r}{\partial \mathbf{n}_r} = \frac{\partial \mathbf{u}_s}{\partial \mathbf{n}_s}, \quad S_{rs}^* \subset S_{rs} \tag{7.124b}$$

where S_{rs}^* is a subset of S_{rs}, in which $\mathbf{n}_r$ is the outward normal to substructure r at a boundary point. Another method [M29] is to regard $\mathbf{u}_r - \mathbf{u}_s$ and $\partial \mathbf{u}_r / \partial \mathbf{n}_r - \partial \mathbf{u}_s / \partial \mathbf{n}_s$ at every point of S_{rs} as errors. Then, introducing certain weighting functions g_{rsi} and h_{rsj}, we ensure geometric compatiblity by setting the weighted average errors equal to zero on S_{rs}, or

$$\int_{S_{rs}} g_{rsi}(\mathbf{u}_r - \mathbf{u}_s) \, dS_{rs} = 0, \quad i = 1, 2, \ldots, M_{rs} \tag{7.125a}$$

$$\int_{S_{rs}} h_{rsj}\left(\frac{\partial \mathbf{u}_r}{\partial \mathbf{n}_r} - \frac{\partial \mathbf{u}_s}{\partial \mathbf{n}_s}\right) dS_{rs} = 0, \quad j = 1, 2, \ldots, N_{rs} \tag{7.125b}$$

where M_{rs} and N_{rs} are integers corresponding to the number of functions g_{rsi} and h_{rsj}, respectively. Note that by letting g_{rsi} and h_{rsj} be spatial Dirac delta functions, geometric compatibility is ensured at given points of S_{rs}.

The question remains as to the nature of the approximation, and in particular the relation between the eigensolution of the actual structure and that of the model. To answer this question, it is convenient to introduce the concept of an intermediate structure. The model can be regarded as representing an intermediate structure, in the sense that it lies somewhere between the totally disconnected structure and the actual structure. Because constraints tend to raise the natural frequencies of a structure and because Eqs. (7.125) represent constraints on the disconnected structure, the computed eigenvalues of the (undamped) intermediate structure are upper bounds for the (undamped) disconnected structure. Hence, the possibility exists that the computed eigenvalues of the intermediate structure are lower than the corresponding ones of the actual structure. The computed eigenvalues can still be made to converge to those of the actual structure, but it is necessary to consider two limiting processes, one in which the number of admissible functions for each substructure is increased and the other in which the number of constraint equations for each boundary S_{rs} is increased [M29].

7.8 RESPONSE OF UNDAMPED STRUCTURES

The approach to the determination of the response of an undamped structure depends on whether or not the system admits a closed-form solution to the eigenvalue problem, Eqs. (7.26) and (7.27). We consider first the case in which it is possible to obtain a closed-form solution to the eigenvalue problem.

The boundary-value problem derived in Section 7.1 can be generalized by considering the partial differential equation

$$\mathcal{L}w(x,t) + m(x)\ddot{w}(x,t) = f(x,t), \quad 0 < x < L \tag{7.126}$$

where $\mathcal{L}$ is a self-adjoint spatial differential operator of order $2p$. The displacement $w(x,t)$ must satisfy the boundary conditions

$$B_i w(x,t) = 0, \quad i = 1, 2, \ldots, p \quad \text{at } x = 0, L \tag{7.127}$$

as well as the initial conditions

$$w(x,0) = w_0(x), \quad \dot{w}(x,0) = v_0(x) \tag{7.128}$$

where $w_0(x)$ and $v_0(x)$ are given functions.

We propose to obtain the solution of Eqs. (7.126)–(7.128) by modal analysis. To this end, we must first solve the eigenvalue problem, Eqs. (7.26) and (7.27), for the eigenvalues $\lambda_j = \omega_j^2$ and eigenfunctions $W_j(x)$ ($j = 1, 2, \ldots$). Then, we invoke the expansion theorem and assume the solution in the form

$$w(x,t) = \sum_{j=1}^{\infty} W_j(x)\eta_j(t) \tag{7.129}$$

where $\eta_j(t)$ are *natural coordinates*. The eigenfunctions have been normalized so as to satisfy

$$\int_0^L mW_j W_k \, dx = \delta_{jk}, \quad \int_0^L W_j \mathcal{L} W_k \, dx = \omega_j^2 \delta_{jk}, \quad j, k = 1, 2, \ldots \tag{7.130}$$

so that $\eta_j(t)$ actually represent *normal coordinates*. It is assumed that the first r eigenfunctions represent rigid-body modes with zero natural frequencies. Introducing Eq. (7.129) into Eq. (7.126), multiplying the result by W_k, integrating over the domain $0 < x < L$ and considering the orthonormality relations, Eqs. (7.130), we obtain the *modal equations*

$$\ddot{\eta}_k(t) = N_k(t), \quad k = 1, 2, \ldots, r \tag{7.131a}$$

$$\ddot{\eta}_k(t) + \omega_k^2 \eta_k(t) = N_k(t), \quad k = r+1, r+2, \ldots \tag{7.131b}$$

where

$$N_k(t) = \int_0^L W_k(x) f(x,t) \, dx, \quad k = 1, 2, \ldots \tag{7.132}$$

are *modal forces*. Equations (7.131) represent an infinite set of independent second-order ordinary differential equations. They are subject to the initial conditions

$$\eta_k(0) = \int_0^L mW_k w_0 \, dx, \quad \dot{\eta}_k(0) = \int_0^L mW_k v_0 \, dx, \quad k = 1, 2, \ldots \tag{7.133}$$

Equations similar to Eqs. (7.131) were encountered in Chapter 4 in connection with discrete systems. The solution of the equations is

$$\eta_k(t) = \int_0^t \left[\int_0^\tau N_k(\sigma) \, d\sigma \right] d\tau + \eta_k(0) + t\dot{\eta}_k(0), \quad k = 1, 2, \ldots, r \tag{7.134a}$$

$$\eta_k(t) = \frac{1}{\omega_k} \int_0^t N_k(\tau) \sin \omega_k(t - \tau) \, d\tau + \eta_k(0) \cos \omega_k t$$

$$+ \frac{\dot{\eta}_k(0)}{\omega_k} \sin \omega_k t, \quad k = r+1, r+2, \ldots \tag{7.134b}$$

The formal solution is completed by introducing Eqs. (7.134) into Eq. (7.129).

Next, we turn our attention to the case in which the differential eigenvalue problem, Eqs. (7.26) and (7.27), does not admit a closed-form solution, so that we must be content with an approximate solution of the boundary-value problem, Eqs. (7.126)–(7.128). To this end, we choose a set of comparison functions ϕ_k ($k = 1, 2, \ldots, n$) and consider an approximate solution of Eq. (7.126) in the form

$$w(x, t) = \sum_{k=1}^n \phi_k(x) q_k(t) \tag{7.135}$$

where $q_k(t)$ are generalized coordinates. Inserting Eq. (7.135) into Eq. (7.126), multiplying the result by $\phi_j(x)$ ($j = 1, 2, \ldots, n$) and integrating over the domain, we obtain

$$\sum_{k=1}^n m_{jk} \ddot{q}_k(t) + \sum_{k=1}^n k_{jk} q_k(t) = Q_j(t), \quad j = 1, 2, \ldots, n \tag{7.136}$$

where

$$m_{jk} = \int_0^L m\phi_j \phi_k \, dx = m_{kj}, \quad j, k = 1, 2, \ldots, n \tag{7.137a}$$

$$k_{jk} = \int_0^L \phi_j \mathcal{L} \phi_k \, dx = k_{kj}, \quad j, k = 1, 2, \ldots, n \tag{7.137b}$$

are symmetric mass and stiffness coefficients and

$$Q_j(t) = \int_0^L \phi_j(x) f(x, t)\, dx, \quad j = 1, 2, \ldots, n \tag{7.138}$$

are generalized forces.

Equations (7.136) represent a set of n simultaneous second-order ordinary differential equations, replacing the partial differential equation (7.126) and the boundary conditions (7.127). Hence, the above procedure represents a discretization in space of the boundary-value problem. Moreover, because in the process the infinitely-many-degrees-of-freedom distributed system is approximated by an n-degree-of-freedom system, the approach also represents a truncation procedure.

Equations (7.136) can be written in the matrix form

$$M\ddot{\mathbf{q}}(t) + K\mathbf{q}(t) = \mathbf{Q}(t) \tag{7.139}$$

where $M = [m_{jk}] = M^T$ and $K = [k_{jk}] = K^T$ are symmetric mass and stiffness matrices, $\mathbf{q}(t) = [q_1(t) \quad q_2(t) \quad \cdots \quad q_n(t)]^T$ is a generalized displacement vector and $\mathbf{Q}(t) = [Q_1(t) \quad Q_2(t) \quad \cdots \quad Q_n(t)]^T$ an associate generalized force vector. Equation (7.139) resembles entirely Eq. (4.88), so that the solution of the equation can be obtained by the modal approach of Section 4.6. This involves the solution of the eigenvalue problem in terms of the two real symmetric matrices M and K, as discussed in Section 4.5, which results in the matrix $\Lambda = \text{diag}[\lambda_r] = \text{diag}[\omega_r^2]$ of natural frequencies squared and the modal matrix Q. The modal matrix can be normalized so as to satisfy the orthonormality relations

$$Q^T M Q = I, \quad Q^T K Q = \Lambda \tag{7.140a, b}$$

Then, using the linear transformation

$$\mathbf{q}(t) = Q\boldsymbol{\eta}(t) \tag{7.141}$$

where $\boldsymbol{\eta}(t)$ is an n-vector of normal coordinates, the solution of Eq. (7.139) can be obtained by following the steps outlined in Section 4.6.

It should be pointed out that the procedure presented above represents in essence an application of the classical Rayleigh–Ritz method, or the finite element method, toward obtaining an approximate solution to a boundary-value problem. In view of this, it is appropriate to discuss a closely related procedure having all the advantages of the variational approach to the Rayleigh–Ritz method (Section 7.4), but which is perhaps easier to grasp. This approach is known as the *assumed-modes method* and in some texts it is actually referred to as the Rayleigh–Ritz method.

The assumed-modes method is based on the use of the Lagrange's equations to derive the equations of motion. To this end, we consider the kinetic energy

$$T(t) = \frac{1}{2}(\sqrt{m(x)}\dot{w}(x,t), \sqrt{m(x)}\dot{w}(x,t)) = \frac{1}{2}\int_0^L m(x)\dot{w}^2(x,t)\,dx \quad (7.142)$$

the potential energy

$$V(t) = \frac{1}{2}[w(x,t), w(x,t)] = \frac{1}{2}\int_0^L \sum_{k=0}^p a_k(x)\left[\frac{\partial^k w(x,t)}{\partial x^k}\right]^2 dx$$

$$+ \sum_{l=0}^{p-1} b_l(x)\left[\frac{\partial^l w(x,t)}{\partial x^l}\right]^2 \Big|_0^L \quad (7.143)$$

and the virtual work

$$\overline{\delta W}(t) = \int_0^L f(x,t)\,\delta w(x,t)\,dx \quad (7.144)$$

where $f(x,t)$ is the distributed force. Then, an approximate solution for $w(x,t)$ is assumed in the form of the series given by Eq. (7.135). Inserting Eq. (7.135) into Eq. (7.142), we can write the kinetic energy in the discretized form

$$T(t) = \frac{1}{2}\sum_{i=1}^n \sum_{j=1}^n m_{ij}\dot{q}_1(t)\dot{q}_j(t) \quad (7.145)$$

where

$$m_{ij} = (\sqrt{m}\phi_i, \sqrt{m}\phi_j) = \int_0^L m\phi_i\phi_j\,dx, \quad i,j = 1,2,\ldots,n \quad (7.146)$$

are symmetric mass coefficients. Similarly, introducing Eq. (7.135) into Eq. (7.143), we obtain the discretized potential energy

$$V(t) = \frac{1}{2}\sum_{i=1}^n \sum_{j=1}^n k_{ij}q_i(t)q_j(t) \quad (1.147)$$

in which

$$k_{ij} = [\phi_i, \phi_j] = \int_0^L \sum_{k=0}^p a_k \frac{d^k\phi_i}{dx^k}\frac{d^k\phi_j}{dx^k}\,dx + \sum_{l=0}^{p-1} b_l \frac{d^l\phi_i}{dx^l}\frac{d^l\phi_j}{dx^l}\Big|_0^L,$$

$$i,j = 1,2,\ldots,n \quad (7.148)$$

are symmetric stiffness coefficients. Finally, inserting Eq. (7.135) into Eq. (7.144), we obtain the discretized virtual work

$$\overline{\delta W}(t) = \sum_{i=1}^n Q_i(t)\,\delta q_i(t) \quad (7.149)$$

where

$$Q_i(t) = \int_0^L f(x,t)\phi_i(x)\,dx, \quad i = 1, 2, \ldots, n \tag{7.150}$$

are generalized forces. Then, recalling that $L = T - V$ and inserting Eqs. (7.145) and (7.147) into Lagrange's equations, Eqs. (2.37), we obtain the equations of motion in the form of Eq. (7.139), in which the mass matrix M has the entries given by Eqs. (7.146), the stiffness matrix K has the entries given by Eqs. (7.148) and the generalized force vector $\mathbf{Q}(t)$ has the components given by Eqs. (7.150).

Comparing Eqs. (7.146) and (7.148) with Eqs. (7.55), we observe that the expressions for the mass and stiffness coefficients obtained by the assumed-modes method are identical to those obtained by the Rayleigh–Ritz method based on the variational approach. Hence, although not rigorously correct, referring to the assumed-modes method as the Rayleigh–Ritz method can be justified to some extent. It should be pointed out that the assumed-modes method in conjunction with Lagrange's equations can also be regarded as being based on a variational approach, by virtue of the fact that Lagrange's equations were derived from Hamilton's principle.

The question remains as to the nature of the trial functions $\phi_k(x)$ used in Eq. (7.135). Clearly, these functions are not modes, so that the title "assumed-modes method" is misleading. In fact, in view of Eqs. (7.148), it is sufficient that $\phi_k(t)$ be merely admissible functions. Moreover, as in the Rayleigh–Ritz method, if the problem involves natural boundary conditions, then the possibility of using quasi-comparison functions (Section 7.4) to improve convergence should be considered.

At this point, a word of caution is in order. As demonstrated in Sections 7.4 and 7.5, in using the Rayleigh–Ritz method to approximate a distributed system by a finite-dimensional one, the eigensolution associated with the finite-dimensional system experiences significant errors in the higher eigenvalues. In turn, this implies that the solution (7.135) is likely to contain errors. To reduce these errors, it is advisable to use a relatively large number of terms in series (7.135). Then, the solution $\mathbf{q}(t)$ of Eq. (7.139) can be truncated by rewriting Eq. (7.141) in the form

$$\mathbf{q}(t) = Q_l \boldsymbol{\eta}(t) \tag{7.151}$$

where Q_l is an $n \times l$ matrix of normal modes belonging to accurate eigenvalues and $\boldsymbol{\eta}(t)$ is now an l-vector of normal coordinates, in which $l < n$. Of course, l must be sufficiently large that the solution contains all the modes likely to participate in the response.

7.9 DAMPED STRUCTURES

In the foregoing discussion of distributed structures, there was an implicit assumption that damping was sufficiently small that it could be ignored. In this section, we propose to examine the effect of damping on the response of structures. To this end, we reconsider the boundary-value problem and rewrite the partial differential equation, Eq. (7.126), in the form

$$\mathscr{L}w(x,t) + \mathscr{C}\dot{w}(x,t) + m(x)\ddot{w}(x,t) = f(x,t), \quad 0 < x < L \quad (7.152)$$

where $\mathscr{C}$ is a damping operator. The boundary conditions and initial conditions remain in the form of Eqs. (7.127) and (7.128), respectively.

In general, damping causes difficulties in the vibration analysis of structures, because in this case the system does not lend itself to the decoupling procedure described in Section 7.8. There is one case, however, in which the modal analysis of Section 7.8 does work. This is the special case of viscous damping known as *proportional damping*, in which the damping operator $\mathscr{C}$ is a linear combination of the stiffness operator $\mathscr{L}$ and the mass distribution m of the form

$$\mathscr{C} = \alpha_1 \mathscr{L} + \alpha_2 m \quad (7.153)$$

where α_1 and α_2 are constant coefficients. Following the procedure of Section 7.8, we assume a solution of Eq. (7.152) in the form of the series (7.129). Then, the orthonormality conditions (7.130) are supplemented by

$$\int_0^L W_j \mathscr{C} W_k \, dx = 2\zeta_j \omega_j \delta_{jk}, \quad j, k = 1, 2, \ldots \quad (7.154)$$

where

$$2\zeta_j \omega_j = \alpha_1 \omega_j^2 + \alpha_2, \quad j = 1, 2, \ldots \quad (7.155)$$

Damping of this type does not affect the rigid-body modes, so that Eqs. (7.131a) retain their form. On the other hand, Eqs. (7.131b) must be replaced by

$$\ddot{\eta}_k(t) + 2\zeta_k \omega_k \dot{\eta}_k(t) + \omega_k^2 \eta_k(t) = N_k(t), \quad k = r+1, r+2, \ldots$$
(7.156)

where r is the number of rigid-body modes. At this point, we observe that the notation on the right side of Eqs. (7.154) was chosen so that Eqs. (7.156) have the same structure as the equation of motion of a single-degree-of-freedom damped system [M44]. From [M44], we can write the solution of Eqs. (7.156) as follows:

$$\eta_k(t) = \frac{1}{\omega_{dk}} \int_0^t N_k(t-\tau) e^{-\zeta_k \omega_k \tau} \sin \omega_{dk} \tau \, d\tau$$

$$+ \frac{\eta_k(0)}{(1-\zeta_k^2)^{1/2}} e^{-\zeta_k \omega_k t} \cos(\omega_{dk} t - \psi_k)$$

$$+ \frac{\dot{\eta}_k(0)}{\omega_{dk}} e^{-\zeta_k \omega_k t} \sin \omega_{dk} t, \quad k = r+1, r+2, \ldots \quad (7.157)$$

in which

$$\omega_{dk} = (1-\zeta_k^2)^{1/2} \omega_k, \quad \psi_k = \tan^{-1} \frac{\zeta_k}{(1-\zeta_k^2)^{1/2}}, \quad k = r+1, r+2, \ldots \quad (7.158)$$

are frequencies of damped oscillation and phase angles, respectively. Moreover, $N_k(t)$ are given by Eqs. (7.132) and $\eta_k(0)$ and $\dot{\eta}_k(0)$ are given by Eqs. (7.133). Equations (7.157) replace Eqs. (7.134b).

The case of proportional damping includes a very important type of damping in structural dynamics, namely, structural damping. In the case in which the external excitation $f(x, t)$ is harmonic, structural damping can be treated as if it were viscous. We propose to examine this case and, to this end, we consider the harmonic external excitation

$$f(x, t) = F(x) e^{i\Omega t} \quad (7.159)$$

where $F(x)$ is a complex amplitude and Ω is the excitation frequency. Concentrating on the steady-state solution, we can write

$$\dot{w}(x, t) = i\Omega w(x, t) \quad (7.160)$$

so that Eq. (7.152) becomes

$$\mathcal{L} w(x, t) + i\Omega \mathcal{C} w(x, t) + m(x) \ddot{w}(x, t) = F(x) e^{i\Omega t} \quad (7.161)$$

It is customary to assume that the operator $\mathcal{C}$ is proportional to the operator $\mathcal{L}$, or

$$\mathcal{C} = \frac{\gamma}{\Omega} \mathcal{L} \quad (7.162)$$

where γ is a structural damping factor. Introducing Eq. (7.162) into Eq. (7.161), we obtain

$$(1 + i\gamma) \mathcal{L} w(x, t) + m(x) \ddot{w}(x, t) = F(x) e^{i\Omega t} \quad (7.163)$$

The solution of Eqs. (7.163) can be derived by the modal analysis of Section 7.8. Following the usual steps, we conclude that the modal equations (7.156) reduce to

$$\ddot{\eta}_k(t) + (1 + i\gamma)\omega_k^2 \eta_k(t) = N_k e^{i\Omega t}, \quad k = r+1, r+2, \ldots \quad (7.164)$$

where

$$N_k = \int_0^L W_k(x) F(x)\, dx, \quad k = 1, 2, \ldots \quad (7.165)$$

Equations (7.164) have the steady-state solution

$$\eta_k(t) = \frac{N_k e^{i\Omega t}}{(1 + i\gamma)\omega_k^2 - \Omega^2}, \quad k = r+1, r+2, \ldots \quad (7.166)$$

It should be stressed that the analogy between structural and viscous damping is valid only for harmonic excitation.

In the general case of viscous damping, i.e., in the case in which the operator $\mathscr{C}$ does not satisfy an orthogonality relation of the type (7.154), no closed-form solution of Eq. (7.152) is possible. In this case, we seek an approximate solution of Eq. (7.152) in the form of Eq. (7.135). Following the steps used in Section 7.8, we obtain the equations of motion in the matrix form

$$M\ddot{\mathbf{q}}(t) + C\dot{\mathbf{q}}(t) + K\mathbf{q}(t) = \mathbf{Q}(t) \quad (7.167)$$

where the damping matrix C has the entries

$$c_{jk} = \int_0^L \phi_j \mathscr{C} \phi_k\, dx, \quad j, k = 1, 2, \ldots, n \quad (7.168)$$

The remaining quantities in Eq. (7.167) were defined in Section 7.8.

Equation (7.167) is of the type discussed in Section 4.1 and its solution was presented in Section 4.6.

CHAPTER 8

CONTROL OF DISTRIBUTED STRUCTURES

The motion of distributed structures is described by variables depending not only on time but also on space. As a result, the motion is governed by partial differential equations to be satisfied inside a given domain defining the structure and by boundary conditions to be satisfied at points bounding this domain. In essence, distributed structures are infinite-dimensional systems, so that control of distributed structures presents problems not encountered in lumped-parameter systems. Indeed, for the most part, the control theory was developed for lumped-parameter systems, and many of the concepts are not applicable to distributed systems. The situation is considerably better in using modal control, which amounts to controlling a structure by controlling its modes. In this case, many of the concepts developed for lumped-parameter structures do carry over to distributed-parameter structures, as both types of structures can be described in terms of modal coordinates. The main difficulty arises in computing the control gains, as this implies infinite-dimensional gain matrices. This question can be obviated by using the independent modal-space control method, but this requires a distributed control force, which can be difficult to implement. Implementation by point actuators is possible, but this implies control of a reduced number of modes, i.e., modal truncation. Another approach to the control of distributed structures is direct output feedback control, whereby the sensors are collocated with the actuators, and the actuator force at a given point depends only on the sensor signal at the same point. Here the difficulty is in deciding on suitable control gains. When the eigenvalue problem does not admit a closed-form solution, such as in the case of arbitrary damping, the problem of model truncation must be approached with care.

8.1 CLOSED-LOOP PARTIAL DIFFERENTIAL EQUATION OF MOTION

We are concerned with the problem of controlling a distributed structure whose behavior is governed by the partial differential equation (Section 7.9)

$$\mathcal{L}w(x, t) + \mathcal{C}\dot{w}(x, t) + m(x)\ddot{w}(x, t) = f(x, t), \quad 0 < x < L \quad (8.1)$$

where $w(x, t)$ is the displacement of a typical point x inside the domain of the structure and at time t, $\mathcal{L}$ is a homogeneous differential stiffness operator of order $2p$, $\mathcal{C}$ is a homogeneous differential damping operator, $m(x)$ is the mass density and $f(x, t)$ is a distributed control force. The solution $w(x, t)$ of Eq. (8.1) is subject to the boundary conditions

$$B_i w(x, t) = 0, \quad x = 0, L, \quad i = 1, 2, \ldots, p \quad (8.2)$$

where B_i are differential operators of maximum order $2p - 1$.

If $f(x, t)$ depends explicitly on x and t, as indicated in Eq. (8.1), then the control is said to be *open-loop*. In this case, Eqs. (8.1) and (8.2) describe an ordinary response problem, which can be solved by the approach discussed in Section 7.9. However, if the object is to drive the motion of the structure to zero, it is unlikely that an open-loop function $f(x, t)$ capable of achieving this task can be found, so that open-loop control must be ruled out as an effective approach to the control of structures.

In view of the above, we wish to consider feedback control. By analogy with lumped-parameter systems, we can consider distributed control of the form

$$f(x, t) = f(w(x, t), \dot{w}(x, t)), \quad 0 < x < L \quad (8.3)$$

where f is in general a nonlinear function of the displacement $w(x, t)$ and velocity $\dot{w}(x, t)$. Inserting Eq. (8.3) into Eq. (8.1), we obtain the *closed-loop* partial differential equation for nonlinear control

$$\mathcal{L}w(x, t) + \mathcal{C}\dot{w}(x, t) + m(x)\ddot{w}(x, t) - f(w(x, t), \dot{w}(x, t)) = 0, \quad 0 < x < L \quad (8.4)$$

It turns out that nonlinear distributed control is not feasible, so that we consider linear control in the form of proportional and rate feedback control, or

$$f(x, t) = -\mathcal{G}(x)w(x, t) - \mathcal{H}(x)\dot{w}(x, t) \quad (8.5)$$

where $\mathcal{G}(x)$ and $\mathcal{H}(x)$ are control gain operators. Inserting Eq. (8.5) into

Eq. (8.1), we obtain the closed-loop partial differential equation for linear control

$$\mathscr{L}^*w(x,t) + \mathscr{C}^*\dot{w}(x,t) + m\ddot{w}(x,t) = 0, \quad 0 < x < L \qquad (8.6)$$

where

$$\mathscr{L}^* = \mathscr{L} + \mathscr{G}, \quad \mathscr{C}^* = \mathscr{C} + \mathscr{H} \qquad (8.7\text{a,b})$$

are closed-loop stiffness and damping operators, respectively. Hence, the problem of linear control of a distributed structure reduces to the determination of operators $\mathscr{G}$ and $\mathscr{H}$ such that the motion of the structure approaches zero asymptotically. Unfortunately, there are no algorithms capable of producing operators $\mathscr{G}$ and $\mathscr{H}$. Later in this chapter, we consider an approach achieving the same result without requiring explicit expressions for $\mathscr{G}$ and $\mathscr{H}$.

8.2 MODAL EQUATIONS FOR UNDAMPED STRUCTURES

Control of a structure using the formulation based on a partial differential equation is not feasible, so that a different approach is desirable. In particular, we propose to replace the partial differential equation by a set of ordinary differential equations. We examine first the case in which the open-loop differential eigenvalue problem admits a closed-form solution (Section 7.2). To this end, we consider an undamped structure, in which case Eq. (8.1) reduces to

$$\mathscr{L}w(x,t) + m(x)\ddot{w}(x,t) = f(x,t), \quad 0 < x < L \qquad (8.8)$$

The boundary conditions remain in the form of Eqs. (8.2).

The open-loop differential eigenvalue problem for the undamped system has the form (Section 7.2)

$$\mathscr{L}W(x) = \lambda m W(x), \quad 0 < x < L \qquad (8.9)$$

where $W(x)$ is subject to the boundary conditions

$$B_i W(x) = 0, \quad x = 0, L, \quad i = 1, 2, \ldots, p \qquad (8.10)$$

The solution of the eigenvalue problem consists of a denumerably infinite set of eigenvalues λ_r and associated eigenfunctions $W_r(x)$ ($r = 1, 2, \ldots$). We assume that the operator $\mathscr{L}$ is self-adjoint (Section 7.2) and nonnegative-definite, so that the eigenvalues are real and nonnegative. We denote them by $\lambda_r = \omega_r^2$ ($r = 1, 2, \ldots$), where ω_r are the natural frequencies. Moreover, the eigenfunctions are orthogonal and can be normalized so as to satisfy the orthonormality relations (Section 7.2)

$$(W_s, mW_r) = \delta_{rs}, \quad (W_s, \mathscr{L} W_r) = \omega_r^2 \delta_{rs}, \quad r, s = 1, 2, \ldots \quad (8.11\text{a,b})$$

where, according to Eq. (7.29), the symbol (,) denotes an inner product and δ_{rs} is the Kronecker delta.

Using the expansion theorem (Section 7.2), the displacement $w(x, t)$ can be expressed as the linear combination

$$w(x, t) = \sum_{r=1}^{\infty} W_r(x) \eta_r(t) \quad (8.12)$$

where $\eta_r(t)$ ($r = 1, 2, \ldots$) are the normal coordinates, also known as *modal coordinates*. Similarly, we can expand the distributed force $f(x, t)$ in the series

$$f(x, t) = \sum_{r=1}^{\infty} m(x) W_r(x) f_r(t) \quad (8.13)$$

where

$$f_r(t) = (W_r(x), f(x, t)), \quad r = 1, 2, \ldots \quad (8.14)$$

are referred to as *modal forces*, or *modal controls*. Then, inserting Eqs. (8.12) and (8.13) into Eq. (8.8), multiplying through by $W_s(x)$, integrating over the domain of the structure and considering Eqs. (8.11), we obtain

$$\ddot{\eta}_r(t) + \omega_r^2 \eta_r(t) = f_r(t), \quad r = 1, 2, \ldots \quad (8.15)$$

which are the desired *modal equations*.

In control, it is customary to work with state equations instead of configuration equations such as Eqs. (8.15). To this end, we adjoin the identities $\dot{\eta}_r(t) \equiv \dot{\eta}_r(t)$ to Eqs. (8.15), so that, introducing the rth modal state vector $\mathbf{w}_r(t) = [\eta_r(t) \quad \dot{\eta}_r(t)]^T$, Eqs. (8.15) can be replaced by

$$\dot{\mathbf{w}}_r(t) = \Lambda_r \mathbf{w}_r(t) + \mathbf{B}_r f_r(t), \quad r = 1, 2, \ldots \quad (8.16)$$

where

$$\Lambda_r = \begin{bmatrix} 0 & 1 \\ -\omega_r^2 & 0 \end{bmatrix}, \quad \mathbf{B}_r = \begin{bmatrix} 0 \\ 1 \end{bmatrix}, \quad r = 1, 2, \ldots \quad (8.17)$$

are coefficient matrices. Equations (8.16) are known as *modal state equations*.

In state feedback control, it is necessary to measure the system output and to estimate the state from the output. Consistent with the modal state equations, we assume that the modal states are related to the modal measurements $y_r(t)$ by

$$y_r(t) = \mathbf{C}_r^T \mathbf{w}_r(t), \quad r = 1, 2, \ldots \tag{8.18}$$

where in the case of displacement measurements

$$\mathbf{C}_r = [1 \quad 0]^T, \quad r = 1, 2, \ldots \tag{8.19a}$$

and in the case of velocity measurements

$$\mathbf{C}_r = [0 \quad 1]^T, \quad r = 1, 2, \ldots \tag{8.19b}$$

The connection between modal measurements and actual measurements is discussed in Section 8.3.

8.3 MODE CONTROLLABILITY AND OBSERVABILITY

The concept of state controllability was introduced in Section 3.12 in conjunction with lumped-parameter systems. In this section, we wish to examine the concept in the context of distributed systems. To this end, we refer to developments in Section 3.12 and introduce the *modal controllability matrix*

$$\mathbb{C}_r = [\mathbf{B}_r \quad \Lambda_r \mathbf{B}_r] = \begin{bmatrix} 0 & 1 \\ 1 & 0 \end{bmatrix}, \quad r = 1, 2, \ldots \tag{8.20}$$

in which we considered Eqs. (8.17). Equations (8.20) permit us to state that *the distributed system is modal-state controllable if and only if each and every controllability matrix $\mathbb{C}_r$ is of full rank 2*; this is clearly the case. The preceding statement implies, of course, that each and every modal control $f_r(t)$ is nonzero, in which case the application of the controllability criterion is a trivial formality. Note that an infinity of modal controls $f_r(t)$ is guaranteed by an actual distributed control function $f(x, t)$, as indicated by Eqs. (8.14), but a distributed control function is not always necessary.

Next, we consider the concept of observability first introduced in Section 3.13. Referring to developments in Section 3.13 and considering the modal output equations, Eqs. (8.18), we can introduce the modal observability matrix, defined as

$$\mathbb{O}_r = [\mathbf{C}_r \quad \Lambda_r^T \mathbf{C}_r], \quad r = 1, 2, \ldots \tag{8.21}$$

and state that *the distributed system is modal-state observable if and only if each and every observability matrix $\mathbb{O}_r$ is of full rank* 2. For displacement measurements

$$\mathbb{O}_r = \begin{bmatrix} 1 & 0 \\ 0 & 1 \end{bmatrix}, \quad r = 1, 2, \ldots \tag{8.22a}$$

and for velocity measurements

$$\mathbb{O}_r = \begin{bmatrix} 0 & -\omega_r^2 \\ 1 & 0 \end{bmatrix}, \quad r = 1, 2, \ldots \tag{8.22b}$$

so that the system is in general observable with either displacement measurements or velocity measurements. Notable exceptions are semidefinite systems, which admit rigid-body modes with zero eigenvalues. Indeed, semidefinite systems are not observable with velocity measurements alone, as in this case the modal observability matrices corresponding to the rigid-body modes do not have rank 2.

At this point, let us return to the relation between the modal outputs and the actual outputs. Multiplying Eq. (8.12) by $m(x)W_s(x)$, integrating over the domain of the structure and considering the orthonormality conditions (8.11a), we can write for displacement measurements

$$y_r(t) = \eta_r(t) = \int_0^L m(x)W_r(x)w(x, t)\, dx, \quad r = 1, 2, \ldots \tag{8.23a}$$

and for velocity measurements

$$y_r(t) = \dot{\eta}_r(t) = \int_0^L m(x)W_r(x)\dot{w}(x, t)\, dx, \quad r = 1, 2, \ldots \tag{8.23b}$$

From Eqs. (8.23), we conclude that an infinity of modal displacement observations is guaranteed by a distributed displacement measurement $w(x, t)$ and an infinity of modal velocity observations is guaranteed by a distributed velocity measurement $\dot{w}(x, t)$. Because Eqs. (8.23) permit the extraction of modal displacements and velocities from measurements of actual displacements and velocities, they are referred to as *modal filters* [M30].

8.4 CLOSED-LOOP MODAL EQUATIONS

Let us return now to the partial differential equation for undamped structures, Eq. (8.8). Considering the distributed feedback control force $f(x, t)$ given by Eq. (8.5) and retracing the steps leading from Eq. (8.8) to the modal equations, Eqs. (8.15), we obtain

$$\ddot{\eta}_s(t) + \sum_{r=1}^{\infty} h_{sr}\dot{\eta}_r(t) + \sum_{r=1}^{\infty} (\omega_s^2 \delta_{sr} + g_{sr})\eta_r(t) = 0, \quad s = 1, 2, \ldots \tag{8.24}$$

where

$$g_{sr} = (W_s, \mathcal{G}W_r), \quad h_{sr} = (W_s, \mathcal{H}W_r), \quad r, s = 1, 2, \ldots \tag{8.25a,b}$$

Equations (8.24) represent the *closed-loop modal equations*. In contrast, Eqs. (8.15) in which the modal forces $f_r(t)$ depend explicitly on the time t and not on $\eta_r(t)$ and/or $\dot{\eta}_r(t)$ are referred to as *open-loop modal equations*. The open-loop modal equations, Eqs. (8.15), are independent. In general, $g_{sr} \neq 0$ and $h_{sr} \neq 0$ for $s \neq r$, so that Eqs. (8.24) represent an infinite set of simultaneous second-order ordinary differential equations. Hence, the effect of feedback control is to couple the open-loop modal equations. Physically, the term $g_{sr}\eta_r(t)$ implies a generalized spring force and the term $h_{sr}\dot{\eta}_r(t)$ a generalized viscous damping force. The fact that the cross-product terms are not zero implies that the feedback control provides nonproportional augmenting stiffness and nonproportional impressed viscous damping [M26], respectively. Because feedback control characterized by $g_{sr} \neq 0$ and $h_{sr} \neq 0$ for $s \neq r$ destroys the independence of the open-loop modal equations, we refer to this case as *coupled control*.

The closed-loop modal equations can be written in the compact form

$$\ddot{\boldsymbol{\eta}}(t) + H\dot{\boldsymbol{\eta}}(t) + (\Omega^2 + G)\boldsymbol{\eta}(t) = \mathbf{0} \tag{8.26}$$

where $\boldsymbol{\eta}(t) = [\eta_1(t) \quad \eta_2(t) \quad \cdots]^T$ is the infinite-dimensional configuration vector, $\Omega = \text{diag}[\omega_r]$ is the infinite-order diagonal matrix of natural frequencies and $G = [g_{sr}]$ and $H = [h_{sr}]$ are square control gain matrices of infinite order. Note that in the general case the matrices G and H are not only nondiagonal but they are unlikely to be even symmetric.

Before the behavior of the closed-loop system can be established, it is necessary to determine the control gain operators $\mathcal{G}$ and $\mathcal{H}$ or the associated gain matrices G and H. However, there are no algorithms capable of producing the operators $\mathcal{G}$ and $\mathcal{H}$ or the infinite-order matrices G and H. Hence, distributed feedback control realized through coupled control is not possible.

For future reference, we would like to cast Eq. (8.26) in state form. To this end, we introduce the 2∞-dimensional modal state vector $\mathbf{w}(t) = [\boldsymbol{\eta}^T(t) \mid \dot{\boldsymbol{\eta}}^T(t)]^T$, so that, adjoining the identity $\dot{\boldsymbol{\eta}}(t) \equiv \dot{\boldsymbol{\eta}}(t)$, Eq. (8.26) can be rewritten in the state form

$$\dot{\mathbf{w}}(t) = A\mathbf{w}(t) \tag{8.27}$$

where

$$A = \left[\begin{array}{c|c} 0 & I \\ \hline -(\Omega^2 + G) & -H \end{array}\right] \tag{8.28}$$

is the coefficient matrix of order 2∞, in which I is the identity matrix of infinite order. The problem of determining the control gain matrices G and H remains. In this regard, we could consider pole allocation or optimal control. As shown in Chapter 6, in the pole allocation method the problem reduces to the solution of a set of nonlinear algebraic equations, which is not feasible for infinite-dimensional systems. Similarly, for optimal control using a quadratic performance index, one is faced with the solution of a matrix Riccati equation of order 2∞, which is not possible.

8.5 INDEPENDENT MODAL-SPACE CONTROL

There is one case in which distributed feedback control is possible, namely the one in which the operators $\mathcal{G}$ and $\mathcal{H}$ satisfy the eigenvalue problems

$$\mathcal{G}W_r(x) = g_r m(x) W_r(x), \quad \mathcal{H}W_r(x) = h_r m(x) W_r(x), \quad r = 1, 2, \ldots \tag{8.29a,b}$$

which imply that $\mathcal{G}$ and $\mathcal{H}$ are such that

$$(W_s, \mathcal{G}W_r) = g_r \delta_{rs}, \quad (W_s, \mathcal{H}W_r) = h_r \delta_{rs}, \quad r, s = 1, 2, \ldots \tag{8.30a,b}$$

In this case the closed-loop modal equations, Eqs. (8.24), reduce to the *independent* set

$$\ddot{\eta}_s(t) + h_s \dot{\eta}_s(t) + (\omega_s^2 + g_s) \eta_s(t) = 0, \quad s = 1, 2, \ldots \tag{8.31}$$

Because of the independence of the closed-loop modal equations, this type of control is called *independent modal-space control* (IMSC), first encountered in Section 6.14 in conjunction with lumped-parameter systems. It is characterized by modal control forces of the form

$$f_s(t) = -g_s \eta_s(t) - h_s \dot{\eta}_s(t), \quad s = 1, 2, \ldots \tag{8.32}$$

In open-loop response problems, the coordinates $\eta_s(t)$ corresponding to independent equations of motion are called *natural* [M26]. Because IMSC guarantees the independence of the closed-loop equations, we refer to IMSC as *natural control*.

The fact that both the open-loop and closed-loop modal equations are independent has very important implications. Indeed, this implies that *the open-loop eigenfunctions W_s are closed-loop eigenfunctions as well.* Hence, in natural control, the control effort is directed entirely to altering the eigenvalues, leaving the eigenfunctions unaltered. In this regard, it should be recalled that the stability of a linear system is determined by the system eigenvalues, with the eigenfunctions playing no role, so that in natural control, no control effort is used unnecessarily.

One question remaining is how to determine the modal gains g_s and h_s ($s = 1, 2, \ldots$). Two of the most widely used techniques are pole allocation and optimal control:

8.5.1 Pole Allocation

In the pole allocation method, the closed-loop poles are selected in advance and the gains are determined so as to produce these poles. In the IMSC, the procedure is exceedingly simple. Denoting the closed-loop eigenvalue associated with the sth mode by $-\alpha_s + i\beta_s$, the solution of Eqs. (8.31) can be written as

$$\eta_s(t) = c_s e^{(-\alpha_s + i\beta_s)t}, \quad s = 1, 2, \ldots \quad (8.33)$$

Inserting Eqs. (8.33) into Eqs. (8.31) and separating the real and imaginary parts, we obtain the modal gains

$$g_s = \alpha_s^2 + \beta_s^2 - \omega_s^2, \quad h_s = 2\alpha_s, \quad s = 1, 2, \ldots \quad (8.34)$$

To guarantee asymptotic stability, however, it is only necessary to impart the open-loop eigenvalues some negative real part and it is not necessary to alter the frequencies. This can be achieved by letting $\beta_s = \sqrt{\lambda_s} = \omega_s$ ($s = 1, 2, \ldots$), where ω_s is the sth natural frequency of the open-loop system. Hence, the *frequency-preserving control gains* are

$$g_s = \alpha_s^2, \quad h_s = 2\alpha_s, \quad s = 1, 2, \ldots \quad (8.35)$$

8.5.2 Optimal Control

In optimal control, the closed-loop poles are determined by minimizing a given performance index. Consistent with previous developments, we are interested in constant gains and, to this end, we consider the performance functional

$$J = \int_0^\infty [(\dot{w}, m\dot{w}) + (w, \mathcal{L}w) + (f, rf)] \, dt \quad (8.36)$$

where the various quantities are as defined in Eq. (8.1), except for $r = r(x)$ which is a weighting function assumed to satisfy [M35]

$$(f, rf) = \sum_{r=1}^{\infty} R_r f_r^2 \qquad (8.37)$$

where R_r are modal weights. Inserting Eqs. (8.12) and (8.37) into Eq. (8.36) and recalling Eqs. (8.11), we obtain

$$J = \sum_{r=1}^{\infty} J_r \qquad (8.38)$$

where

$$J_r = \int_0^{\infty} (\dot{\eta}_r^2 + \omega_r^2 \eta_r^2 + R_r f_r^2)\, dt, \quad r = 1, 2, \ldots \qquad (8.39)$$

are modal performance indices. Because in IMSC the modal control f_r is independent of any other modal control, it follows that

$$\min J = \min \sum_{r=1}^{\infty} J_r = \sum_{r=1}^{\infty} \min J_r \qquad (8.40)$$

so that the minimization can be carried out independently for each mode.

The minimization of J_r leads to a 2×2 matrix Riccati equation that in the steady-state case can be solved in closed form (Section 6.15*), yielding the modal control gains

$$\begin{aligned}&g_r = -\omega_r^2 + \omega_r(\omega_r^2 + R_r^{-1})^{1/2}, \\ &h_r = [R_r^{-1} - 2\omega_r^2 + 2\omega_r(\omega_r^2 + R_r^{-1})^{1/2}]^{1/2},\end{aligned} \qquad r = 1, 2, \ldots \qquad (8.41)$$

Because no constraint has been imposed on the control function $f = f(x, t)$, the solution defined by Eqs. (8.13), (8.32) and (8.41) is *globally optimal*, and is unique because the solution to the linear optimal control problem is unique [M35].

It should be pointed out that the solution presented above requires distributed sensors and actuators. Indeed, inserting Eqs. (8.32) into Eq. (8.13), we obtain the distributed feedback control force

$$f(x, t) = -\sum_{r=1}^{\infty} m(x) W_r(x)[g_r \eta_r(t) + h_r \dot{\eta}_r(t)] \qquad (8.42)$$

---

*Note that, because of a difference in the definition of the modal coordinates, the gain g_r is obtained by multiplying the corresponding gain in Eqs. (6.279) by ω_r.

Equation (8.42) indicates that control implementation requires the entire infinity of modal displacements $\eta_r(t)$ and modal velocities $\dot{\eta}_r(t)$ ($r = 1, 2, \ldots$) for feedback. This, in turn, implies a distributed sensor, as can be concluded from Eqs. (8.23). Note that, inserting Eq. (8.12) into Eq. (8.5) and comparing the results with Eq. (8.42), we can verify Eqs. (8.29). At this point, we observe that the gain operators $\mathscr{G}$ and $\mathscr{H}$ are never determined explicitly, nor is it necessary to do so, as the determination of the modal gains g_r and h_r ($r = 1, 2, \ldots$) is sufficient to produce the feedback control density function $f(x, t)$.

As indicated above, globally optimal control of a distributed structure requires a distributed actuator, which represents an infinite-dimensional controller. If distributed actuation is not feasible, then the distributed actuator force given by Eq. (8.42) can be approximated by means of a finite-dimensional actuator force. Meirovitch and Silverberg [M35] show how the distributed actuator force $f(x, t)$ can be approximated by a finite number of point forces or a finite number of sectionally uniform forces.

It was also indicated above that distributed control implementation requires a distributed sensor, i.e., an infinite-dimensional sensor. If distributed sensing is not feasible, then modal state estimation can be carried out by using a discrete number of measurements in conjunction with independent modal-space Luenberger observers or Kalman filters [O5].

8.6 COUPLED CONTROL

As can be concluded from Section 8.5, independent control of the entire infinity of modes characterizing a distributed structure requires a distributed actuator and a distributed sensor. Approximate implementation of independent modal control can be carried out by means of a finite number of sensors and actuators. The question can be raised, however, whether it is possible to design directly a control system using a finite number of actuators and sensors to control distributed structures, instead of using the finite number of actuators and sensors to approximate distributed actuators and sensors. In this section, we propose to address this question.

We consider the problem of controlling a distributed structure by means of a finite number m of discrete actuators acting at the points $x = x_i$ ($i = 1, 2, \ldots, m$) of the structure. Discrete actuators can be treated as distributed by writing

$$f(x, t) = \sum_{i=1}^{m} F_i(t) \delta(x - x_i), \quad 0 < x < L \tag{8.43}$$

where $F_i(t)$ are force amplitudes and $\delta(x - x_i)$ are spatial Dirac delta functions. Introducing Eq. (8.43) into Eq. (8.8), we obtain

324 CONTROL OF DISTRIBUTED STRUCTURES

$$\mathscr{L}w(x, t) + m(x)\ddot{w}(x, t) = \sum_{i=1}^{m} F_i(t)\delta(x - x_i), \quad 0 < x < L \quad (8.44)$$

Control of the structure in terms of a partial differential equation is not feasible, however, so that once again we transform the partial differential equation, Eq. (8.44) in this case, into a set of modal equations. Inserting Eq. (8.12) into Eq. (8.44), multiplying by $W_s(x)$, integrating over the domain of the structure and considering the orthonormality relations, Eqs. (8.11), we have

$$\ddot{\eta}_r(t) + \omega_r^2 \eta_r(t) = \sum_{i=1}^{m} W_r(x_i)F_i(t), \quad r = 1, 2, \ldots \quad (8.45)$$

Equations (8.45) can be written in the matrix form

$$\ddot{\boldsymbol{\eta}}(t) + \Omega^2 \boldsymbol{\eta}(t) = B'\mathbf{F}(t) \quad (8.46)$$

where $\Omega = \text{diag}(\omega_1 \ \omega_2 \ \cdots)$,

$$B' = [b'_{ri}] = [W_r(x_i)] \quad (8.47)$$

is an $\infty \times m$ matrix known as the *modal participation matrix* and $\mathbf{F}(t) = [F_1(t) \ F_2(t) \ \cdots \ F_m(t)]^T$ is the control vector. As in Section 8.4, Eq. (8.46) can be written in the modal state form

$$\dot{\mathbf{w}}(t) = A\mathbf{w}(t) + B\mathbf{F}(t) \quad (8.48)$$

where $\mathbf{w}(t) = [\boldsymbol{\eta}^T(t) \mid \dot{\boldsymbol{\eta}}^T(t)]^T$ is the modal state vector and

$$A = \begin{bmatrix} 0 & I \\ \hline -\Omega^2 & 0 \end{bmatrix}, \quad B = \begin{bmatrix} 0 \\ \hline B' \end{bmatrix} \quad (8.49a,b)$$

are $2\infty \times 2\infty$ and $2\infty \times m$ coefficient matrices. For linear feedback control, the control vector is related to the modal state vector according to

$$\mathbf{F}(t) = -G\mathbf{w}(t) \quad (8.50)$$

where G is an $m \times 2\infty$ control gain matrix. No confusion should arise from using the same notation here as for the $\infty \times \infty$ gain matrix defined in Section 8.4. Determination of infinite-dimensional gain matrices is not possible, so that control of the entire infinity of modes is not feasible, nor is it necessary. Indeed, higher modes have only minimal participation in the motion, as they are difficult to excite. Moreover, various assumptions made in deriving the equation of motion of a structure limit the validity of the theory to lower modes only. Hence, we propose to control only a limited number of lower modes, where the number is sufficiently large that the accuracy will not suffer.

In view of the above, we propose to control n modes only. Ordinarily, the lower modes are the ones in need of control. Hence, we assume that the displacement can be approximated by

$$w(x, t) \cong \sum_{r=1}^{n} W_r(x)\eta_r(t) \qquad (8.51)$$

so that, retracing the steps leading to Eq. (8.48), we obtain

$$\dot{\mathbf{w}}_C(t) = A_C \mathbf{w}_C(t) + B_C \mathbf{F}(t) \qquad (8.52)$$

where $\mathbf{w}_C$ is a $2n$-dimensional modal state vector and

$$A_C = \left[\begin{array}{c|c} 0 & I_C \\ \hline -\Omega_C^2 & 0 \end{array}\right], \quad B_C = \left[\begin{array}{c} 0 \\ \hline B_C' \end{array}\right] \qquad (8.53\text{a,b})$$

are $2n \times 2n$ and $2n \times m$ matrices, respectively, in which the notation is obvious. Equation (8.52) represents a $2n$-order discrete system, and control of the system can be carried out by one of the methods presented in Chapter 6. Of course, in this case Eq. (8.50) must be replaced by

$$\mathbf{F}(t) = -G_C \mathbf{w}_C(t) \qquad (8.54)$$

where G_C is an $m \times 2n$ control gain matrix. Note that, in using the control law given by Eq. (8.54), the closed-loop modal equations are not independent, so that this procedure represents *coupled control*.

There remains the question of modal state estimation. To this end, we consider point sensors and let $y_i(t)$ be the outputs from s sensors at points $x = x_i$ ($i = 1, 2, \ldots, s$). Then, if displacement sensors are used

$$y_i(t) = w(x_i, t) = \sum_{r=1}^{n} W_r(x_i)\eta_r(t), \quad i = 1, 2, \ldots, s \qquad (8.55\text{a})$$

and if velocity sensors are used

$$y_i(t) = \dot{w}(x_i, t) = \sum_{r=1}^{n} W_r(x_i)\dot{\eta}_r(t), \quad i = 1, 2, \ldots, s \qquad (8.55\text{b})$$

Introducing the notation

$$C_C' = [c_{ir}'] = [W_r(x_i)], \quad i = 1, 2, \ldots, s; \quad r = 1, 2, \ldots, n \qquad (8.56)$$

where we note that the entries in C_C' are transposed relative to the entries in B_C' (disregarding the dimensions of the matrices, which are in general different), we can write Eqs. (8.55) in the compact form

$$y(t) = C_c w_c(t) \tag{8.57}$$

where in the case of displacement sensors

$$C_C = [C_C' \mid 0] \tag{8.58a}$$

and in the case of velocity sensors

$$C_C = [0 \mid C_C'] \tag{8.58b}$$

To estimate the full controlled modal state $w_C(t)$ from the output $y(t)$, we can consider a Luenberger observer (Section 6.12) or a Kalman–Bucy filter (Section 6.13). As can be concluded from Sections 6.12 and 6.13, Luenberger observers are to be used for low noise-to-signal ratios and Kalman–Bucy filters for high noise-to-signal ratios. In either case, we can write the modal observer in the form

$$\dot{\hat{w}}_C(t) = A_C \hat{w}_C(t) + B_C F(t) + K(t)[y(t) - C_C \hat{w}_C(t)] \tag{8.59}$$

where $\hat{w}_C(t)$ is the estimated controlled modal state. In the case of the Luenberger observer, the observer gain matrix $K(t)$ can be obtained by a pole allocation technique. In the case of the Kalman–Bucy filter, $K(t)$ can be determined optimally by solving a matrix Riccati equation, which assumes that the noise intensities associated with the actuators and sensors are known (Section 6.13). Upon obtaining the estimated controlled modal state from Eq. (8.59), we compute the feedback control forces by writing

$$F(t) = -G_C \hat{w}_C(t) \tag{8.60}$$

In the above, we treated the distributed system as if it were discrete. In the process, we ignored the uncontrolled modes. At this point, we wish to examine the effect of the control forces on the uncontrolled modes. To this end, we refer to the uncontrolled modes as *residual* and denote them by the subscript R. Then, we define

$$w(t) = [w_C^T(t) \mid w_R^T(t)]^T, \quad A = \begin{bmatrix} A_C & 0 \\ \hline 0 & A_R \end{bmatrix}, \quad B = \begin{bmatrix} B_C \\ \hline B_R \end{bmatrix} \tag{8.61a,b,c}$$

so that, inserting Eqs. (8.61) into Eq. (8.48) and considering Eq. (8.60), we obtain

$$\dot{w}_C(t) = A_C w_C(t) - B_C G_C \hat{w}_C(t) \tag{8.62a}$$

$$\dot{w}_R(t) = A_R w_R(t) - B_R G_C \hat{w}_C(t) \tag{8.62b}$$

Moreover, inserting Eqs. (8.57) and (8.60) into Eq. (8.59), we can write the observer equation in the form

$$\dot{\hat{\mathbf{w}}}_C(t) = (A_C - B_C G_C)\hat{\mathbf{w}}_C(t) + K(t)C_C[\mathbf{w}_C(t) - \hat{\mathbf{w}}_C(t)] \qquad (8.63)$$

Next, we introduce the error vector

$$\mathbf{e}_C(t) = \hat{\mathbf{w}}_C(t) - \mathbf{w}_C(t) \qquad (8.64)$$

so that Eqs. (8.62) and (8.63) can be rearranged as

$$\dot{\mathbf{w}}_C(t) = (A_C - B_C G_C)\mathbf{w}_C(t) - B_C G_C \mathbf{e}_C(t) \qquad (8.65a)$$

$$\dot{\mathbf{w}}_R(t) = -B_R G_C \mathbf{w}_C(t) + A_R \mathbf{w}_R(t) - B_R G_C \mathbf{e}_C(t) \qquad (8.65b)$$

$$\dot{\mathbf{e}}_C(t) = (A_C - KC_C)\mathbf{e}_C(t) \qquad (8.65c)$$

where $K(t)$ was assumed to be constant. Equations (8.65) can be written in the matrix form

$$\begin{bmatrix} \dot{\mathbf{w}}_C(t) \\ \dot{\mathbf{w}}_R(t) \\ \dot{\mathbf{e}}_C(t) \end{bmatrix} = \begin{bmatrix} A_C - B_C G_C & 0 & -B_C G_C \\ -B_R G_C & A_R & -B_R G_C \\ 0 & 0 & A_C - KC_C \end{bmatrix} \begin{bmatrix} \mathbf{w}_C(t) \\ \mathbf{w}_R(t) \\ \mathbf{e}_C(t) \end{bmatrix} \qquad (8.66)$$

Because of the block-triangular nature of the coefficient matrix, the eigenvalues of the closed-loop system are determined by the submatrices $A_C - B_C G_C$, A_R and $A_C - KC_C$. The term $-B_R G_C$ is responsible for the excitation of the residual modes by the control forces and is known as *control spillover* [B4]. Because the term has no effect on the eigenvalues of the closed-loop system, we conclude that *control spillover cannot destabilize the system*, although it can cause some degradation in the system performance. As in the case of lumped systems, we observe that the *separation principle* (Section 6.12) is valid here as well.

Equation (8.57) implies that the sensors measure only the contribution of the controlled modes to the motion of the structure. In reality, however, the sensor signals will include contributions from all the modes, so that the proper expression for the output vector is

$$\mathbf{y}(t) = C\mathbf{w}(t) = C_C \mathbf{w}_C(t) + C_R \mathbf{w}_R(t) \qquad (8.67)$$

where the notation is obvious. In this case, the observer equation, Eq. (8.63), must be replaced by

$$\dot{\hat{\mathbf{w}}}_C(t) = (A_C - B_C G_C)\hat{\mathbf{w}}_C(t) + KC_C[\mathbf{w}_C(t) - \hat{\mathbf{w}}_C(t)] + KC_R \mathbf{w}_R(t) \tag{8.68}$$

so that the error equation, Eq. (8.65c), becomes

$$\dot{\mathbf{e}}_C(t) = (A_C - KC_C)\mathbf{e}_C(t) + KC_R \mathbf{w}_R(t) \tag{8.69}$$

Combining Eq. (8.69) with Eqs. (8.65a) and (8.65b), we can write

$$\begin{bmatrix} \dot{\mathbf{w}}_C(t) \\ \hdashline \dot{\mathbf{w}}_R(t) \\ \hdashline \dot{\mathbf{e}}_C(t) \end{bmatrix} = \begin{bmatrix} A_C - B_C G_C & 0 & -B_C G_C \\ \hdashline -B_R G_C & A_R & -B_R G_C \\ \hdashline 0 & KC_R & A_C - KC_C \end{bmatrix} \begin{bmatrix} \mathbf{w}_C(t) \\ \hdashline \mathbf{w}_R(t) \\ \hdashline \mathbf{e}_C(t) \end{bmatrix} \tag{8.70}$$

This time, however, the separation principle is no longer valid, as the term KC_R causes the closed-loop system eigenvalues to be affected by the observer. This effect is known as *observation spillover* and can produce instability in the residual modes [B4]. This is particularly true if the actuators and sensors are not collocated. The residual modes are particularly vulnerable because, in the absence of observation spillover, the eigenvalues associated with these modes lie on the imaginary axis and have no stability margin. Note, however, that a small amount of damping inherent in the structure is often sufficient to overcome the observation spillover effect [M32]. At any rate, observation spillover can be eliminated if the sensor signals are prefiltered so as to screen out the contribution of the uncontrolled modes.

The effect of the observation spillover can be greatly reduced by using a large number of sensors. Indeed, if the displacement $w(x_i, t)$ is measured at a large number s of points x_i, it is possible to reconstruct an approximate displacement profile $\hat{w}(x, t)$ through spatial interpolation by writing

$$\hat{w}(x, t) = \sum_{i=1}^{s} L_i(x) w(x_i, t) \tag{8.71}$$

where $L_i(x)$ are interpolation functions. Then, using Eqs. (8.23a), we can estimate the controlled modal coordinates as follows:

$$\hat{\eta}_r(t) = \int_0^L m(P) W_r(x) \hat{w}(x, t) \, dx, \quad r = 1, 2, \ldots, n \tag{8.72}$$

Equations (8.72) represent modal filters based on discrete measurements. Modal velocity estimates can be produced by measuring velocities $\dot{w}(x_i, t)$ and generating the velocity profile $\hat{\dot{w}}(x, t)$ through interpolation and inserting this velocity profile into the modal filters, Eqs. (8.72). Alternatively, it is possible to estimate modal velocities from modal displacements by using modal Luenberger observers or modal Kalman–Bucy filters [O5].

In the above discussion, we treated the vector $\mathbf{w}_R(t)$ of residual modal coordinates as if it had finite dimension when in fact it is infinite-dimensional. In practice, however, only a finite number of modes can be excited, so that $\mathbf{w}_R(t)$ can be treated as if it were finite-dimensional without incurring much error, provided the dimension of $\mathbf{w}_R(t)$ is sufficiently large.

Example 8.1 Use three point actuators to control the lowest eight modes of bending vibration of a uniform beam pinned at both ends. Plot the response of the controlled and uncontrolled modes to a unit impulse applied at $x = a = 0.43$ and examine the control spillover effect. Assume that there are sufficient sensors to permit perfect estimation of the modal states associated with the controlled modes. For simplicity, let the value of the mass density, bending stiffness and beam length be equal unity, $m = 1$, $EI = 1$, $L = 1$.

The natural frequencies and the normalized modes of vibrations are

$$\omega_r = (r\pi)^2, \quad r = 1, 2, \ldots \tag{a}$$

and

$$W_r(x) = \sqrt{2} \sin r\pi x, \quad r = 1, 2, \ldots \tag{b}$$

respectively. The modal state equations are given by Eq. (8.52), in which the coefficient matrices are given by Eqs. (8.53). For the example at hand, the submatrices in Eqs. (8.53) have the expressions

$$\Omega_C^2 = \mathrm{diag}\,[\pi^4 \quad (2\pi)^4 \quad \cdots \quad (8\pi)^4] \tag{c}$$

and

$$B_C' = [W_r(x_i)] = [\sqrt{2} \sin r\pi x_i], \quad r = 1, 2, \ldots, 8; \quad i = 1, 2, 3 \tag{d}$$

To ensure controllability, we choose the actuator locations

$$x_1 = 0.15, \quad x_2 = 0.55, \quad x_3 = 0.73 \tag{e}$$

which yields

$$B_C' = \begin{bmatrix} 0.6420 & 1.3968 & 1.0608 \\ 1.1441 & -0.4370 & -1.4031 \\ \vdots & \vdots & \vdots \\ -0.8313 & 1.3450 & -0.6813 \end{bmatrix} \tag{f}$$

The three-dimensional control vector is given by Eq. (8.54), where G_C is a 3×16 control gain matrix. We determine the gain matrix by considering linear optimal control (Section 6.4). To this end, we assume that the final time is infinitely large, so that the performance measure, Eq. (6.47), can be written as

$$J = \frac{1}{2} \int_0^\infty (\mathbf{w}_C^T Q \mathbf{w}_C + \mathbf{F}^T R \mathbf{F}) \, dt \tag{g}$$

in which Q and R are coefficient matrices yet to be selected. Using Eq. (6.59), the control gain matrix has the expression

$$G_C = R^{-1} B_C^T K \tag{h}$$

where the 16×16 matrix K is obtained by solving the steady-state matrix Ricatti equation, Eq. (6.74),

$$-Q - A_C^T K - K A_C + K B_C R^{-1} B_C^T K = 0 \tag{i}$$

As coefficient matrices in Eq. (g), we choose

$$Q = \left[\begin{array}{c|c} \Omega_C^2 & 0 \\ \hline 0 & I_C \end{array}\right], \quad R = I_R \tag{j}$$

where I_C is 8×8 and I_R is 3×3. The choice of Q renders the first term in the performance measure equal to the total energy. Moreover, by letting R be the identity matrix, we give equal weight to all three actuators.

Inserting Eqs. (c) and (f) into Eqs. (8.53), considering Eqs. (j) and solving Eq. (i), we obtain the matrix K as follows:

$$K = \begin{bmatrix} 74.9199 & -0.2172 & \cdots & -0.0001 \\ -0.2172 & 1185.8751 & \cdots & 0.0014 \\ \vdots & \vdots & & \vdots \\ -0.0001 & 0.0014 & \cdots & 0.8214 \end{bmatrix} \tag{k}$$

Introducing Eq. (f) in conjunction with Eq. (8.53b), the second of Eqs. (j) and Eq. (k) into Eq. (h), we obtain the control gain matrix

$$G_C = \begin{bmatrix} 0.3699 & -0.2224 & \cdots & -0.6828 \\ 0.6629 & -1.2696 & \cdots & 1.1047 \\ 0.4504 & -1.6382 & \cdots & -0.5596 \end{bmatrix} \tag{l}$$

Then, solving the eigenvalue problem associated with the closed-loop coefficient matrix $A_C - B_C G_C$, we obtain the closed-loop poles listed in Table 8.1.

To determine the response, we note that the effect of the impulse is to generate an initial velocity at every point of the beam. To obtain the corresponding initial modal velocities, we refer to Eq. (8.8), recall that $\hat{f}_0 = 1$, $EI = 1$ and $m = 1$ and write the uncontrolled beam equation in the form

COUPLED CONTROL   331

Table 8.1. Closed-Loop Poles

r	Closed-Loop Poles ρ_r
1,2	$-1.3175 \pm i9.8692$
3,4	$-1.3164 \pm i39.4784$
5,6	$-1.4439 \pm i88.8264$
7,8	$-1.1452 \pm i157.9137$
9,10	$-1.3393 \pm i246.7402$
11,12	$-1.2705 \pm i355.3054$
13,14	$-0.5876 \pm i483.6106$
15,16	$-1.2174 \pm i631.6543$

$$\frac{\partial^4 w(x,t)}{\partial x^4} + \frac{\partial^2 w(x,t)}{\partial t^2} = f(x,t) = \hat{f}_0 \delta(x-a)\delta(t) = \delta(x-0.43)\delta(t),$$

$$0 < x < 1 \quad (m)$$

Letting the response have the expression

$$w(x,t) = \sum_{r=1}^{\infty} W_r(x)\eta_r(t) = \sum_{r=1}^{\infty} \sqrt{2} \sin r\pi x \, \eta_r(t) \quad (n)$$

inserting Eq. (n) into Eq. (m), multiplying by $W_s(x) = \sqrt{2} \sin s\pi x$, integrating over the domain $0 < x < 1$ and considering the orthogonality of the eigenfunctions, we obtain the modal equations

$$\ddot{\eta}_r(t) + \omega_r^2 \eta_r(t) = \int_0^1 f(x,t) W_r(x) \, dx = \sqrt{2} \sin 0.43 r\pi \, \delta(t), \quad r = 1, 2, \ldots \quad (o)$$

which yield the equivalent initial modal velocities

$$\dot{\eta}_r(0+) = \frac{\sqrt{2} \sin 0.43 r\pi}{\omega_r} = \frac{\sqrt{2} \sin 0.43 r\pi}{(r\pi)^2}, \quad r = 1, 2, \ldots \quad (p)$$

and we observe that the initial modal velocities go down rapidly with r.

The response of the controlled modes is obtained by solving the closed-loop state equation

$$\dot{\mathbf{w}}_C(t) = (A_C - B_C G_C)\mathbf{w}_C(t) \quad (q)$$

which is subject to the initial conditions $\eta_r(0) = 0$ and $\dot{\eta}_r(0)$ given by Eqs. (p) with $r = 1, 2, \ldots, 8$. Then, the response of the uncontrolled modes is obtained by solving

$$\dot{\mathbf{w}}_R(t) = A_R \mathbf{w}_R(t) - B_R G_C \mathbf{w}_C(t) \tag{r}$$

where $\mathbf{w}_R = [\eta_9 \quad \eta_{10} \quad \cdots \quad \dot{\eta}_9 \quad \dot{\eta}_{10} \quad \cdots]^T$ and

$$A_R = \begin{bmatrix} 0 & I_R \\ \hline -\Omega_R^2 & 0 \end{bmatrix}, \quad B_R = \begin{bmatrix} 0 \\ \hline B_R' \end{bmatrix} \tag{s}$$

in which

$$\Omega_R = \operatorname{diag}[(9\pi)^2 \quad (10\pi)^2 \quad \cdots] \tag{t}$$

and

$$B_R' = [\sqrt{2} \sin r\pi x_i], \quad r = 9, 10, \ldots; \quad i = 1, 2, 3 \tag{u}$$

Of course, for practical reasons the uncontrolled modal state vector $\mathbf{w}_R(t)$ must be truncated. This presents no problem, as the higher modes are not excited anyway.

The response of the controlled modes is shown in Fig. 8.1 and that of the uncontrolled modes is shown in Fig. 8.2. Both were obtained by the method based on the transition matrix (Section 3.9). It is clear that the largest contribution to the response $w(x, t)$ of the beam comes from the lowest mode. Control spillover does exist, as can be seen from Fig. 8.2, but it is

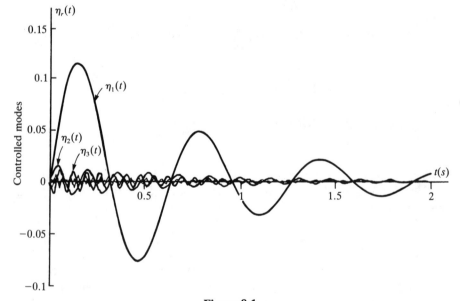

Figure 8.1

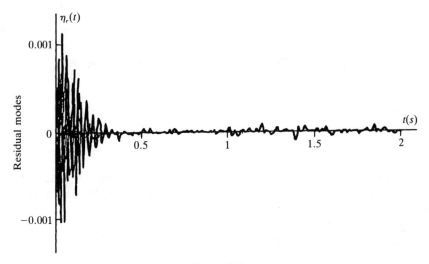

Figure 8.2

reasonably small and tends to disappear as the controlled modes decay. In this regard, note that Figs. 8.1 and 8.2 use different scales for the modal amplitudes.

8.7 DIRECT OUTPUT FEEDBACK CONTROL

The problem of estimating the modal states for feedback in modal control of distributed structures can prove troublesome. As pointed out in Section 8.6, the problem of observation spillover is potentially more serious than the problem of control spillover, as it can lead to instability. Hence, a procedure not requiring modal state estimation appears desirable.

One approach not requiring modal state estimation is direct output feedback control, whereby the sensors are collocated with the actuators and a given actuator force is a function of the sensor output at the same point. We introduced direct output feedback control in Section 6.14 in conjunction with lumped-parameter structures, but it is for distributed-parameter structures that the method is particularly useful [B7].

For simplicity, we assume that the control law is linear, although nonlinear control laws can be used. We consider m discrete actuators acting at the points $x = x_i$ $(i = 1, 2, \ldots, m)$, where the force amplitudes are given by

$$F_i(t) = -g_i w(x_i, t) - h_i \dot{w}(x_i, t), \quad i = 1, 2, \ldots, m \qquad (8.73)$$

in which g_i and h_i $(i = 1, 2, \ldots, m)$ are actual control gains. Clearly, the gains must be positive. As in Section 8.6, we can treat the actuators as distributed by writing

334 CONTROL OF DISTRIBUTED STRUCTURES

$$f(x, t) = -\sum_{i=1}^{m} [g_i w(x, t) + h_i \dot{w}(x, t)] \delta(x - x_i), \quad 0 < x < L \quad (8.74)$$

To make the connection with the approach of Section 8.1, we can regard the operators $\mathcal{G}$ and $\mathcal{H}$ defined in Eq. (8.5) as having the expressions

$$\mathcal{G}(x) = \sum_{i=1}^{m} g_i \delta(x - x_i), \quad \mathcal{H}(x) = \sum_{i=1}^{m} h_i \delta(x - x_i), \quad 0 < x < L \quad (8.75a,b)$$

The closed-loop partial differential equation can be obtained by inserting Eqs. (8.75) into Eqs. (8.6) and (8.7). We shall not pursue this approach, but turn our attention to the modal equations.

We showed in Section 8.4 that the closed-loop modal equations for an undamped structure have the form

$$\ddot{\eta}_s(t) + \sum_{r=1}^{\infty} h_{sr} \dot{\eta}_r(t) + \sum_{r=1}^{\infty} (\omega_s^2 \delta_{sr} + g_{sr}) \eta_r(t) = 0, \quad s = 1, 2, \ldots \quad (8.76)$$

where

$$g_{sr} = (W_s, \mathcal{G} W_r), \quad h_{sr} = (W_s, \mathcal{H} W_r), \quad r, s = 1, 2, \ldots \quad (8.77a,b)$$

Introducing Eqs. (8.75) into Eqs. (8.77), we obtain

$$g_{sr} = \sum_{i=1}^{m} g_i W_r(x_i) W_s(x_i), \quad h_{sr} = \sum_{i=1}^{m} h_i W_r(x_i) W_s(x_i), \quad r, s = 1, 2, \ldots \quad (8.78a,b)$$

The modal equations can be written in the compact form

$$\ddot{\boldsymbol{\eta}}(t) + H\dot{\boldsymbol{\eta}}(t) + (\Omega^2 + G)\boldsymbol{\eta}(t) = \mathbf{0} \quad (8.79)$$

where $\boldsymbol{\eta}(t)$ is the infinite-dimensional configuration vector, $\Omega = \text{diag}[\omega_r]$ is the infinite-order diagonal matrix of natural frequencies and

$$G = \sum_{i=1}^{m} g_i \mathbf{W}(x_i) \mathbf{W}^T(x_i), \quad H = \sum_{i=1}^{m} h_i \mathbf{W}(x_i) \mathbf{W}^T(x_i) \quad (8.80a,b)$$

are symmetric control gain matrices of infinite order, where $\mathbf{W}(x_i) = [W_1(x_i) \; W_2(x_i) \cdots]^T$ is an infinite-dimensional vector of eigenfunctions evaluated at $x = x_i$. The matrices $\mathbf{W}(x_i)\mathbf{W}^T(x_i)$ represent outer products of given vectors, so that they have rank one. The matrices G and H are linear combinations of the matrices $\mathbf{W}(x_i)\mathbf{W}^T(x_i)$ with coefficients g_i and h_i, respectively. Because g_i and h_i are positive, it follows that the matrices G and H are positive semidefinite with maximum rank equal to m.

From Eq. (8.79), we conclude that the effect of displacement feedback is to increase the structure stiffness, thus increasing the natural frequencies. On the other hand, the effect of velocity feedback is to provide viscous damping. Considering the approach of Section 6.14, Eq. (8.79) can be shown to yield

$$\frac{d}{dt}(T + V_{CL}) = -\dot{\boldsymbol{\eta}}^T H \dot{\boldsymbol{\eta}} \qquad (8.81)$$

where

$$T = \frac{1}{2}\dot{\boldsymbol{\eta}}^T\dot{\boldsymbol{\eta}}, \quad V_{CL} = \frac{1}{2}\boldsymbol{\eta}^T(\Omega^2 + G)\boldsymbol{\eta} \qquad (8.82\text{a,b})$$

in which T is the kinetic energy expressed in terms of modal velocities and V_{CL} is a closed-loop potential energy expressed in terms of modal displacements, where "closed-loop" is to be interpreted in the sense that it includes the effect of displacement feedback. If damping is pervasive, i.e., if it couples all the modal equations of motion, then the term on the right side of Eq. (8.81) is negative at all times, so that the structure dissipates energy continuously. Hence, in this case the structure is guaranteed to be asymptotically stable. This is true regardless of whether the structure is subjected to displacement feedback. It follows that, if the object is to ensure asymptotic stability, then velocity feedback is sufficient. Notable exceptions are unrestrained structures, for which the operator $\mathcal{L}$ is only positive semidefinite. Such structures are referred to as semidefinite [M26] and admit rigid-body modes characterized by zero natural frequencies. Semidefinite structures cannot be stabilized with velocity feedback alone, and displacement feedback is essential. To verify this statement, let us consider the case of velocity feedback alone, in which case $G = 0$. Then, if the structure admits a rigid-body mode, say W_r, the associated natural frequency ω_r is zero. It follows that Eq. (8.81) can be satisfied if $\dot{\boldsymbol{\eta}} = \mathbf{0}$, $\eta_i = 0$ for $i \neq r$ and $\eta_r = \text{constant} \neq 0$. The implication is that the system can come to rest in a position other than the origin of state space, which is due to the fact that rigid-body modes cannot be controlled with velocity feedback alone.

It should be pointed out that, because there is only a finite number m of point actuators, the possibility exists that all actuators can be placed at the nodes of one of the higher modes. In this case, this particular higher mode will not be controlled.

Once again the problem is that of determining the control gains. The problem is different here because there is only a finite number of gains g_i and h_i ($i = 1, 2, \ldots, m$) and the system is infinite-dimensional. There is no computational algorithm permitting the computation of the control gains in conjunction with either pole allocation or optimal control, so that one must consider model truncation. Even for the truncated model, the situation remains questionable. The reason for this is that pole allocation and optimal control most likely will require gain matrices with entries independent of each other while direct outback feedback control implies that the entries of G and H are not independent, as can be seen from Eqs. (8.80). In fact, there is some question whether arbitrary pole placement is possible for direct output feedback control. This statement can be explained by the fact that, for given preselected closed-loop poles, it is not possible to guarantee that all the control gains h_i $(i = 1, 2, \ldots, m)$ are positive, thus causing some of the uncontrolled modes to become unstable due to control spillover [M46]. Moreover, because the entries of G and H are not independent, there is some question as to the nature of optimal control in the presence of constraints on the control gains.

8.8 SYSTEMS WITH PROPORTIONAL DAMPING

In Sections 8.2–8.7, we made the asssumption that the system possessed no inherent damping. At this point we wish to relax this assumption and include the damping effect. To this end, we return to Eq. (8.1) and assume that the motion of the distributed structure is described by the partial differential equations

$$\mathscr{L}w(x, t) + \mathscr{C}\dot{w}(x, t) + m(x)\ddot{w}(x, t) = f(x, t), \quad 0 < x < L \quad (8.83)$$

where the various quantities are as defined in Section 8.1. Then, following the approach of Section 8.1, we conclude that the closed-loop partial differential equation for damped systems has the form

$$\mathscr{L}^*w(x, t) + \mathscr{C}^*\dot{w}(x, t) + m\ddot{w}(x, t) = 0, \quad 0 < x < L \quad (8.84)$$

where $\mathscr{L}^*$ and $\mathscr{C}^*$ are closed-loop stiffness and damping operators, as defined by Eqs. (8.7a) and (8.7b), respectively.

As pointed out in Section 8.2, designing controls for a structure based on a partial differential equation is not feasible, so that once again we propose to investigate the use of modal equations. An attempt to use the pattern of Section 8.2 will reveal difficulties from the onset. Indeed, the open-loop eigenvalue problem no longer has the simple form given by Eqs. (8.9) and (8.10), so that in general the solution is no longer real. In fact, a closed-form solution of the differential eigenvalue problem for damped systems is generally not feasible. A notable exception is the case in which damping is of the proportional type. This implies that the damping operator $\mathscr{C}$ is a linear combination of the stiffness operator $\mathscr{L}$ and the mass density m, or

SYSTEMS WITH PROPORTIONAL DAMPING

$$\mathscr{C} = \alpha_1 \mathscr{L} + \alpha_2 m \tag{8.85}$$

where α_1 and α_2 are proportionality constants of appropriate dimensions. In this case, the eigenfunctions of the associated undamped system can still be used to decouple the open-loop equations. Indeed, using the expansion theorem, Eq. (8.12), and retracing the steps of Section 8.2, we obtain the modal equations

$$\ddot{\eta}_r(t) + 2\zeta_r \omega_r \dot{\eta}_r(t) + \omega_r^2 \eta_r(t) = f_r(t), \quad r = 1, 2, \ldots \tag{8.86}$$

where

$$\zeta_r = \frac{1}{2\omega_r}(\alpha_1 \omega_r^2 + \alpha_2), \quad r = 1, 2, \ldots \tag{8.87}$$

are modal damping factors. The remaining quantities in Eqs. (8.86) are as defined in Section 8.2. The modal state equations retain the general form

$$\dot{\mathbf{w}}_r(t) = \Lambda_r \mathbf{w}_r(t) + \mathbf{B}_r f_r(t), \quad r = 1, 2, \ldots \tag{8.88}$$

but here the coefficient matrix Λ_r has the form

$$\Lambda_r = \begin{bmatrix} 0 & 1 \\ -\omega_r^2 & -2\zeta_r \omega_r \end{bmatrix}, \quad r = 1, 2, \ldots \tag{8.89}$$

From Section 8.3, we conclude that the controllability matrix becomes

$$\mathbb{C}_r = [\mathbf{B}_r \quad \Lambda_r \mathbf{B}_r] = \begin{bmatrix} 0 & 1 \\ 1 & -2\zeta_r \omega_r \end{bmatrix}, \quad r = 1, 2, \ldots \tag{8.90}$$

Moreover, the observability matrix for displacement measurements retains the form (8.22a), but for velocity measurements it becomes

$$\mathbb{O}_r = [\mathbf{C}_r \quad \Lambda_r^T \mathbf{C}_r] = \begin{bmatrix} 0 & \omega_r^2 \\ 1 & -2\zeta_r \omega_r \end{bmatrix}, \quad r = 1, 2, \ldots \tag{8.91}$$

Hence, all the conclusions reached in Section 8.3 concerning modal-state controllability and observability remain valid here.

Following the pattern of Section 8.4, the closed-loop modal state equations can be shown to retain the form (8.27), but the coefficient matrix now is

$$A = \left[\begin{array}{c|c} & I \\ \hline -(\Omega^2 + G) & -(2\zeta\Omega + H) \end{array} \right] \tag{8.92}$$

where $\zeta\Omega = \operatorname{diag}(\zeta_r \omega_r)$.

If the control gain operators $\mathscr{G}$ and $\mathscr{H}$ satisfy Eqs. (8.29a) and (8.29b), respectively, then the closed-loop modal equations are still independent, but this time they have the form

$$\ddot{\eta}_s(t) + (2\zeta_s\omega_s + h_s)\dot{\eta}_s(t) + (\omega_s^2 + g_s)\eta_s(t) = 0, \quad s = 1, 2, \ldots \tag{8.93}$$

In using IMSC, the poles can be placed with the same ease as in Section 8.5 and the modal control gains corresponding to the closed-loop pole $-\alpha_s + i\beta_s$ can be shown to have the expressions

$$g_s = \alpha_s^2 + \beta_s^2 - \omega_s^2, \quad h_s = 2\alpha_s - 2\zeta_s\omega_s, \quad s = 1, 2, \ldots \tag{8.94a,b}$$

from which we conclude that damping affects only the rate feedback. Consistent with physical intuition, Eq. (8.94b) shows that internal damping reduces the need for rate feedback. Similarly, using the same performance index as in Section 8.5, it is not difficult to show that optimal IMSC is characterized by the modal control gains

$$g_r = -\omega_r^2 + \omega_r(\omega_r^2 + R_r^{-1})^{1/2},$$

$$h_r = -2\zeta_r\omega_r + [4\zeta_r^2\omega_r^2 + R_r^{-1} - 2\omega_r^2 + 2\omega_r(\omega_r^2 + R_r^{-1})^{1/2}]^{1/2},$$

$$r = 1, 2, \ldots \tag{8.95}$$

Once again, we observe that damping affects the rate feedback only and, moreover, it reduces the need for rate feedback.

Damping has another beneficial effect, in addition to reducing the need for rate feedback control. Indeed, we recall from Section 8.6 that damping increases the stability margin to the extent that the possibility of observation spillover instability is virtually eliminated.

The question arises as to what happens if damping is not of the proportional type. In this case, if damping is relatively small, it is possible to treat it as if it were proportional by simply ignoring the coupling terms in the modal equations [M26] On the other hand, if damping is not small the situation becomes significantly more involved. This case is discussed in Section 8.10.

8.9 CONTROL OF DISCRETIZED STRUCTURES

In our preceding discussion of control of distributed structures, we assumed that the eigenvalue problem, Eqs. (8.9) and (8.10), possesses a closed-form solution. Then, using the expansion theorem, Eq. (8.12), we replaced the partial differential equation governing the motion of the distributed system,

CONTROL OF DISCRETIZED STRUCTURES   339

Eq. (8.8), by an infinite set of ordinary differential equations known as modal equations.

More often than not, structures tend to be sufficiently complex that a closed-form solution of the eigenvalue problem is not possible. In such cases, the approach to the control of distributed structures must be modified. It is clear from Chapter 7 that the only alternative is spatial discretization of the structure, which amounts to developing a discrete model approximating the distributed structure. In essence, this implies truncation of the system. As pointed out in Section 7.4, however, this truncation will introduce errors, as no discrete model can serve as an entirely accurate representation of a distributed structure. Hence, care must be exercised in the manner in which the discrete model is used.

Let us consider once again an undamped distributed structure described by the partial differential equation

$$\mathscr{L}w(x, t) + m(x)\ddot{w}(x, t) = f(x, t), \quad 0 < x < L \qquad (8.96)$$

where $\mathscr{L}$ is a self-adjoint differential operator of order $2p$, $m(x)$ the distributed mass and $f(x, t)$ the distributed control force. The displacement $w(x, t)$ is subject to the boundary conditions

$$B_i w(x, t) = 0, \quad x = 0, L, \quad i = 1, 2, \ldots, p \qquad (8.97)$$

where B_i are differential operators of maximum order $2p - 1$. Equations (8.96) and (8.97) have the same form as in Section 8.2. Unlike the system considered in Section 8.2, however, here no closed-form solution to the corresponding eigenvalue problem is known to exist, so that the interest lies in an approximate solution. To this end, we consider either the classical Rayleigh–Ritz method (Section 7.4) or the finite element method (Section 7.5). In either case, we assume an approximate solution of Eq. (8.96) in the form of the finite series

$$w(x, t) = \sum_{k=1}^{n} \phi_k(x) q_k(t) \qquad (8.98)$$

where $\phi_k(x)$ are comparison functions (Section 7.2) and $q_k(t)$ are generalized coordinates. We recall that in the classical Rayleigh–Ritz method $\phi_k(t)$ are global functions and $q_k(t)$ are abstract quantities, whereas in the finite element method $\phi_k(x)$ are local interpolation functions and $q_k(t)$ are actual displacements of the structure at the nodal points. Introducing Eq. (8.98) into Eq. (8.96), multiplying both sides of the resulting equation by $\phi_j(x)$ and integrating over the domain, we obtain

$$\sum_{k=1}^{n} m_{jk}\ddot{q}_k(t) + \sum_{k=1}^{n} k_{jk}q_k(t) = Q_k(t), \quad j = 1, 2, \ldots, n \qquad (8.99)$$

where

$$m_{jk} = \int_0^L m\phi_j\phi_k \, dx = m_{kj}, \quad j, k = 1, 2, \ldots, n \qquad (8.100a)$$

$$k_{jk} = \int_0^L \phi_j \mathscr{L}\phi_k \, dx = k_{kj}, \quad j, k = 1, 2, \ldots, n \qquad (8.100b)$$

are symmetric mass and stiffness coefficients, respectively, and

$$Q_j(t) = \int_0^L \phi_j(x) f(x, t) \, dx, \quad j = 1, 2, \ldots, n \qquad (8.101)$$

are generalized forces. Equations (8.99) constitute a simultaneous set of n second-order ordinary differential equations representing our discretized structure.

Equations (8.99) can be written in the matrix form

$$M\ddot{\mathbf{q}}(t) + K\mathbf{q}(t) = \mathbf{Q}(t) \qquad (8.102)$$

where $M = [m_{jk}] = M^T$ and $K = [k_{jk}] = K^T$ are symmetric mass and stiffness matrices, $\mathbf{q}(t) = [q_1(t) \quad q_2(t) \quad \cdots \quad q_n(t)]^T$ is a generalized displacement vector and $\mathbf{Q}(t) = [Q_1(t) \quad Q_2(t) \quad \cdots \quad Q_n(t)]^T$ is an associated generalized control vector. Equation (8.102) represents a $2n$-order discrete system, so that one can proceed with the design of controls by one of the methods discussed in Chapter 6. As pointed out above, however, any discrete model of distributed structure contains errors. Hence, a procedure capable of reducing modeling errors is highly desirable. To this end, we turn our attention to the eigenvalue problem.

To derive the eigenvalue problem for the discrete model described by Eq. (8.102), we consider the associated homogeneous equation and assume a solution in the form

$$\mathbf{q}(t) = e^{i\omega t}\mathbf{u} \qquad (8.103)$$

where $\mathbf{u}$ is a constant vector. Then, inserting Eq. (8.103) into Eq. (8.102), letting $\mathbf{Q}(t) = \mathbf{0}$ and dividing through by $e^{i\omega t}$, we obtain the eigenvalue problem

$$K\mathbf{u} = \omega^2 M\mathbf{u} \qquad (8.104)$$

Equation (8.104) possesses n independent solutions consisting of the computed eigenvalues ω_r^2 and associated eigenvectors $\mathbf{u}_r$ $(r = 1, 2, \ldots, n)$. The quantities ω_r can be identified as the computed natural frequencies and $\mathbf{u}_r$ as the corresponding natural modes. The eigenvectors are orthogonal with respect to the matrices M and K and can be normalized so as to satisfy the orthonormality relations

$$\mathbf{u}_s^T M \mathbf{u}_r = \delta_{rs}, \quad \mathbf{u}_s^T K \mathbf{u}_r = \omega_r^2 \delta_{rs}, \quad r, s = 1, 2, \ldots, n \qquad (8.105)$$

The computed eigenvectors can be used to derive computed eigenfunctions. Indeed, using Eqs. (8.98) and (8.103), we can write the computed eigenfunctions in the form

$$W_r(x) = \sum_{k=1}^{n} \phi_k(x)u_{kr} = \boldsymbol{\phi}^T(x)\mathbf{u}_r, \quad r = 1, 2, \ldots, n \quad (8.106)$$

where u_{kr} is the kth component of the eigenvector $\mathbf{u}_r$ and $\boldsymbol{\phi} = [\phi_1 \ \phi_2 \ \cdots \ \phi_n]^T$ is an n-vector of comparison functions. The computed eigenfunctions satisfy the same orthonormality conditions as the actual eigenfunctions. To show this, we rewrite Eqs. (8.100) in the matrix form

$$M = \int_0^L m\boldsymbol{\phi}\boldsymbol{\phi}^T \, dx, \quad K = \int_0^L \boldsymbol{\phi}\mathscr{L}\boldsymbol{\phi}^T \, dx \quad (8.107a,b)$$

Then, inserting Eqs. (8.107) into Eqs. (8.105) and recalling Eqs. (8.106), we obtain the orthonormality conditions for the computed eigenfunctions

$$\mathbf{u}_s^T M \mathbf{u}_r = \mathbf{u}_s^T \left(\int_0^L m\boldsymbol{\phi}\boldsymbol{\phi}^T \, dx \right) \mathbf{u}_r = \int_0^L m W_s W_r \, dx = \delta_{rs}, \quad r, s = 1, 2, \ldots, n$$
(8.108a)

$$\mathbf{u}_s^T K \mathbf{u}_r = \mathbf{u}_s^T \left(\int_0^L \boldsymbol{\phi}\mathscr{L}\boldsymbol{\phi}^T \, dx \right) \mathbf{u}_r = \int_0^L W_s \mathscr{L} W_r \, dx = \omega_r^2 \delta_{rs},$$

$$r, s = 1, 2, \ldots, n \quad (8.108b)$$

which have the same form as the orthonormality conditions for the actual eigenfunctions, Eqs. (8.11), with the notable exception that the set of computed eigenfunctions is finite.

The computed eigenfunctions can be treated as if they were actual. Indeed, by analogy with Eq. (8.12), we can approximate the solution of Eq. (8.96) by the linear combination

$$w(x, t) = \sum_{r=1}^{n} W_r(x)\eta_r(t) \quad (8.109)$$

where $\eta_r(t)$ are the modal coordinates. Then, following the pattern established in Section 8.2, the distributed control force can be approximated by

$$f(x, t) = \sum_{r=1}^{n} m(x)W_r(x)f_r(t) \quad (8.110)$$

where

342 CONTROL OF DISTRIBUTED STRUCTURES

$$f_r(t) = \int_0^L W_r(x) f(x, t)\, dx, \quad r = 1, 2, \ldots, n \tag{8.111}$$

are the modal controls. Similarly, the modal equations have the form

$$\ddot{\eta}_r(t) + \omega_r^2 \eta_r(t) = f_r(t), \quad r = 1, 2, \ldots, n \tag{8.112}$$

Again, Eqs. (8.112) describe the behavior of the actual distributed system only approximately.

At this point, we wish to examine the problem of modeling errors and to suggest ways of reducing these errors. It is typical of discretized models that the higher computed eigenvalues and eigenvectors are inaccurate. As a rule of thumb, less than one half of such eigensolutions are accurate [M26], so that one can question the wisdom of basing controls on inaccurate model equations. As it turns out, this is not necessary. Indeed, we recall that even in the case of structures admitting a closed-form solution of the eigenvalue problem, i.e., when all the modal equations are accurate, it is not always feasible to control the entire infinity of modes. This is certainly the case when controls are to be implemented by only a finite number of point actuators and sensors. In view of this, we propose to divide the n modeled modes into n_C controlled modes and n_R residual modes, $n_C + n_R = n$. The residual modes are modeled uncontrolled modes, which join an infinity of unmodeled uncontrolled modes.

Next, we turn our attention to the control design. First, we consider a distributed control force in conjunction with the IMSC method. Following the pattern established in Section 8.5, we write the modal controls for the controlled modes in the form.

$$f_r(t) = -g_r \eta_r(t) - h_r \dot{\eta}_r(t), \quad r = 1, 2, \ldots, n_C \tag{8.113}$$

where g_r and h_r are modal control gains. In the case of pole allocation, they are given by Eqs. (8.34) or (8.35), and in the case of optimal control in conjunction with modal quadratic performance measures they are given by Eqs. (8.41). Then, the distributed control force can be written as

$$f(x, t) = -\sum_{r=1}^{n_C} m(x) W_r(x) [g_r \eta_r(t) + h_r \dot{\eta}_r(t)], \quad 0 < x < L \tag{8.114}$$

The above distributed force will drive the controlled modes to zero. The question remains as to what happens to the residual modes. To answer this question, we insert Eq. (8.114) into Eq. (8.111) for the residual modes, consider Eqs. (8.108a) and obtain the modal forces

$$f_s(t) = \int_0^L W_s(x) f(x,t)\, dx$$
$$= -\sum_{r=1}^{n_C} [g_r \eta_r(t) + h_r \dot{\eta}_r(t)] \int_0^L m(x) W_s(x) W_r(x)\, dx = 0,$$
$$s = n_C + 1, n_C + 2, \ldots, n \quad (8.115)$$

Hence, when a distributed control force is used in conjunction with IMSC, the modal forces for the residual modes are zero, so that there is no control spillover into the residual modes. This result is based on the orthogonality of the computed eigenfunctions, which holds true even though the computed eigenfunctions may not be accurate representations of the actual eigenfunctions.

The distributed control force can be approximated by a finite number of point forces or a finite number of sectionally uniform forces, as discussed by Meirovitch and Silverberg [M35]. Another approach is to design modal controls for point actuators by IMSC directly. As in Section 8.6, we consider control by m distrete actuators acting at the points $x = x_i$ $(i = 1, 2, \ldots, m)$ and treat them as distributed by writing

$$f(x,t) = \sum_{i=1}^m F_i(t) \delta(x - x_i) \quad (8.116)$$

where $F_i(t)$ are the control force amplitudes and $\delta(x - x_i)$ are spatial Dirac delta functions. Inserting Eq. (8.116) into Eqs. (8.111), we obtain the relation between the modal forces and the actuator forces

$$f_r(t) = \int_0^L W_r(x) f(x,t)\, dx = \sum_{i=1}^m F_i(t) \int_0^L W_r(x) \delta(x - x_i)\, dx$$
$$= \sum_{i=1}^m W_r(x_i) F_i(t), \quad r = 1, 2, \ldots, n_C \quad (8.117)$$

Inserting Eqs. (8.117) into the first n_C of Eqs. (8.112), we can write the modal equations in the vector form

$$\ddot{\boldsymbol{\eta}}_C(t) + \Omega_C^2 \boldsymbol{\eta}_C(t) = B_C' \mathbf{F}(t) \quad (8.118)$$

where $\boldsymbol{\eta}_C = [\eta_1 \quad \eta_2 \quad \cdots \quad \eta_{n_C}]^T$ is the controlled modal vector, $\Omega_C = \text{diag}(\omega_1 \quad \omega_2 \quad \cdots \quad \omega_{n_C})$ is the matrix of natural frequencies for the controlled modes, $B_C' = [W_r(x_i)]$ is an $n_C \times m$ modal participation matrix and $\mathbf{F}(t) = [F_1(t) \quad F_2(t) \quad \cdots \quad F_m(t)]^T$ is the actuator force vector.

In IMSC, the modal controls are determined first and the actuator forces are then computed as linear combinations of the modal controls. In essence, we need the inverse relation to that given by Eqs. (8.117). To this end, we write Eqs. (8.117) in the vector form

$$\mathbf{f}_C(t) = B'_C \mathbf{F}(t) \tag{8.119}$$

where $\mathbf{f}_C(t) = [f_1(t) \quad f_2(t) \quad \cdots \quad f_{n_C}(t)]^T$ is the modal control force vector. Hence, the desired inverse relation is

$$\mathbf{F}(t) = (B'_C)^\dagger \mathbf{f}_C(t) \tag{8.120}$$

where, assuming that $m < n_C$

$$(B'_C)^\dagger = [(B'_C)^T B'_C]^{-1} (B'_C)^T \tag{8.121}$$

is the pseudo-inverse of B'_C. We note, however, that a pseudo-inverse is not an actual inverse, so that $B'_C (B'_C)^\dagger \neq I$. This implies that if the number of actuators is smaller than the number of controlled modes, then the above process does not yield genuine IMSC. To obtain genuine IMSC, i.e., to be able to design a control force for each controlled mode independently, we must have as many actuators as the number of controlled modes, $m = n_C$, in which case $(B'_C)^\dagger = (B'_C)^{-1}$, so that the pseudo-inverse represents an actual inverse. In this case, using Eq. (8.120) and recalling Eqs. (8.113), the actual control vector can be written in the form

$$\mathbf{F}(t) = -(B'_C)^{-1} [G'_C \boldsymbol{\eta}_C(t) + H'_C \dot{\boldsymbol{\eta}}_C(t)] \tag{8.122}$$

where

$$G'_C = \mathrm{diag}(g_1 \quad g_2 \quad \cdots \quad g_{n_C}), \quad H'_G = \mathrm{diag}(h_1 \quad h_2 \quad \cdots \quad h_{n_C}) \tag{8.123}$$

are diagonal modal gain matrices. As indicated earlier in this section, the modal control gains can be determined by the pole allocation method or by optimal control in conjunction with modal quadratic performance measures.

Implementation of the control law given by Eq. (8.122) requires estimates of the controlled modal state vector $\mathbf{w}_C(t) = [\boldsymbol{\eta}_C^T(t) \mid \dot{\boldsymbol{\eta}}_C^T(t)]^T$. As pointed out earlier in this chapter, such estimates can be obtained by means of a Luenberger observer or a Kalman–Bucy filter, depending on the noise-to-signal ratio.

The question remains as to how the above control affects the residual modes. By analogy with Eq. (8.118), the equation for the residual modes has the vector form

$$\ddot{\boldsymbol{\eta}}_R(t) + \Omega_R^2 \boldsymbol{\eta}_R(t) = B'_R \mathbf{F}(t) \tag{8.124}$$

where, assuming that $\mathbf{F}(t)$ is an n_C-vector, B'_R is an $n_R \times n_C$ modal participation matrix. The remaining notation is obvious. Inserting Eq. (8.122) into Eq. (8.124), we obtain

$$\ddot{\boldsymbol{\eta}}_R(t) + \Omega_R^2 \boldsymbol{\eta}_R(t) = -B'_R(B'_C)^{-1}[G'_C \boldsymbol{\eta}_G(t) + H'_C \dot{\boldsymbol{\eta}}_C(t)] \qquad (8.125)$$

It is clear from Eq. (8.125) that there is control spillover into the residual modes, so that some performance degradation can be expected. The extent of the control spillover depends on the matrix product $B'_R(B'_C)^{-1}$. In this regard, it may be possible to place the actuators throughout the structure so as to minimize the entries of the matrix product. A word of caution is in order here. Indeed, Eq. (8.125) is unable to give a reliable picture of the extent of spillover because Ω_R is not very accurate and B'_R is computed on the basis of inaccurate modes. Still, one may be able to use the equation to obtain some information concerning the spillover into the lower residual modes. The higher residual modes are more difficult to excite and they may not matter. This is particularly true in view of the fact that any system has some measure of damping, albeit small, and damping tends to cause the higher residual modes to decay faster than the lower ones. Of course, the control spillover diminishes as the control modes are regulated.

When the number of actuators is smaller than the number of controlled modes, $m < n_C$, independent control is not possible. In this case, we can design controls by using the procedure for coupled control described in Section 8.6. In fact, the developments are exactly as those in Section 8.6, the main difference being that here the dimension of the controlled state vector is $2n_C$ and that of the residual state vector is $2n_R$, as opposed to $2n$ and infinite, respectively. As with independent control, any modal information relating to the residual modes is not accurate and must be treated with caution.

In direct output feedback, a given actuator force depends on the state at the same point, so that discretization by the finite element method appears to offer certain advantages. At the same time, because direct output feedback amounts to controlling a set of points throughout the distributed structure, the suitability of basing the control design on modal equations comes into question. In view of this, we propose to return to the discretized model described by Eq. (8.102) and examine its nature as it relates to the finite element method in more detail. To this end, we assume that the nodes are at the points $x = x_i$ ($i = 1, 2, \ldots, n$) and use Eq. (8.98) to obtain the nodal displacements (in the sense of the finite element method)

$$w(x_i, t) = \sum_{k=1}^{n} \phi_k(x_i) q_k(t), \quad i = 1, 2, \ldots, n \qquad (8.126)$$

But, the interpolation functions $\phi_k(x)$ are such that (Section 7.5)

$$\phi_k(x_i) = \delta_{ik}, \quad i, k = 1, 2, \ldots, n \tag{8.127}$$

so that Eqs. (8.126) yield

$$w(x_i, t) = q_i(t), \quad i = 1, 2, \ldots, n \tag{8.128}$$

Hence, in the finite element method the generalized displacements represent actual displacements of the nodal points, which is a significant advantage of the finite element method over the classical Rayleigh–Ritz method.

Next, we assume that control is implemented by means of m discrete actuators acting at the arbitrary points $x = x_l$ ($l = 1, 2, \ldots, m$), not necessarily coinciding with nodal points, and we note that in general $m \le n$. The discrete forces $F_l(t)$ generated by these actuators can be treated as distributed, as can be concluded from Eq. (8.116). Inserting Eq. (8.116) into Eqs. (8.101), we obtain the nodal forces

$$Q_j(t) = \sum_{l=1}^{m} F_l(t) \int_0^L \phi_j(x)\delta(x - x_l)\, dx = \sum_{l=1}^{m} \phi_j(x_l) F_l(t) \tag{8.129}$$

which can be written in the matrix form

$$\mathbf{Q}(t) = B'\mathbf{F}(t) \tag{8.130}$$

where

$$B' = [\phi_{jl}] = [\phi_j(x_l)], \quad j = 1, 2, \ldots, n; \quad l = 1, 2, \ldots, m \tag{8.131}$$

is an $n \times m$ matrix. Introducing the m-vector of displacements at the actuator locations

$$\mathbf{w}'(t) = [w(x_1, t) \quad w(x_2, t) \quad \cdots \quad w(x_m, t)]^T \tag{8.132a}$$

and the $m \times m$ diagonal control gain matrices

$$G'' = \text{diag}(g_1 \quad g_2 \quad \cdots \quad g_m), \quad H'' = \text{diag}(h_1 \quad h_2 \quad \cdots \quad h_m)$$
$$\tag{8.132b,c}$$

Eqs. (8.73) can be written in the matrix form

$$\mathbf{F}(t) = -G''\mathbf{w}' - H''\dot{\mathbf{w}}' \tag{8.133}$$

To obtain the closed-loop equation, we must express the above control law in terms of nodal forces and nodal displacements and velocities. To this end, we consider once again Eq. (8.98) and write

$$w(x_l, t) = \sum_{k=1}^{n} \phi_k(x_l) q_k(t) \qquad (8.134)$$

which can be written in the matrix form

$$\mathbf{w}'(t) = (B')^T \mathbf{q}(t) \qquad (8.135)$$

Inserting Eq. (8.135) into Eq. (8.133) and the result into Eq. (8.130), we obtain

$$\mathbf{Q}(t) = -B'G''(B')^T \mathbf{q}(t) - B'H''(B')^T \dot{\mathbf{q}}(t) \qquad (8.136)$$

Finally, introducing Eq. (8.136) into Eq. (8.102), we obtain the closed-loop equation

$$M\ddot{\mathbf{q}}(t) + B'H''(B')^T \dot{\mathbf{q}}(t) + [K + B'G''(B')^T]\mathbf{q}(t) = 0 \qquad (8.137)$$

An analysis paralleling that in Section 8.7 reveals that the system (8.137) has very good stability characteristics. Indeed, it can be verified that the matrices $B'H''(B')^T$ and $B'G''(B')^T$ are positive semidefinite. Because the matrices M and K guarantee coupling of all the equations in the set, the matrix $B'H''(B')^T$ provides pervasive damping, so that the closed-loop system is asymptotically stable. This is true regardless of whether the uncontrolled system is positive definite or only positive semidefinite, because both matrices M and $K + B'G''(B')^T$ are positive definite.

It should be pointed out that models derived by the finite element method tend to be characterized by a large number of degrees of freedom. Moreover, unlike the modal approach, model truncation can cause problems. In this regard, one can use mass condensation [M26], but this must be considered more of an art than a method, as intimate knowledge of the structure is needed before a rational decision can be made as to the nodal displacements to be eliminated. Of course, because of the simple nature of direct feedback, model truncation is not really necessary for control design. In computer simulation of the closed-loop response, however, a lower-order model is desirable. The closed-loop response can be computed by the transition matrix approach (Section 3.9), but difficulties can be encountered in computing the transition matrix for high-order systems. The alternative of using the transformation based on the right and left eigenvectors to simplify the computation of the series for the exponential term does not offer much comfort, as it requires the solution of two complex eigenvalue problems of order $2n$.

As pointed out on several occasions, a discretized model is only an approximation of a distributed structure and not the structure itself. Hence, one can expect the dynamic behavior of the structure itself to differ somewhat from that of the model, so that care should be exercised in using a discretized model. In using modal equations to design controls, we used a second truncation for control purposes, going from an n-degree-of-freedom modeled system to an n_C-degree-of-freedom controlled system. This second truncation amounts simply to retaining n_C modal equations for control, where n_C is sufficiently large to include all the modes in need of control and n is sufficiently large to guarantee the accuracy of the n_C modes and possibly that of some lower residual modes. This is true regardless of whether the discretization is carried out by the classical Rayleigh–Ritz method or the finite element method. Direct output feedback control is not a modal method and we based the control design on the coupled discretized model, Eq. (8.102). This model does not permit separation between accurate and inaccurate modal information. As the number of elements increases, however, the discretized model tends to respond like the actual structure. For this reason, one must make sure that the number of elements is not too small.

8.10 STRUCTURES WITH GENERAL VISCOUS DAMPING

In Section 8.1, we posed the problem of controlling a distributed structure described by the partial differential equation

$$\mathscr{L}w(x, t) + \mathscr{C}\dot{w}(x, t) + m(x)\ddot{w}(x, t) = f(x, t) \qquad (8.138)$$

where $w(x, t)$ is the displacement, $\mathscr{L}$ a stiffness operator, $\mathscr{C}$ a damping operator, $m(x)$ the mass density and $f(x, t)$ the distributed control force. The displacement is subject to given boundary conditions. Sections 8.2–8.7 were concerned primarily with undamped structures admitting a closed-form solution to the differential eigenvalue problem, and Section 8.8 extended the discussion to the special case in which the damping operator $\mathscr{C}$ is a linear combination of the stiffness operator $\mathscr{L}$ and the mass density m. We recall that this latter case can be treated by the same techniques as undamped structures.

At this point, we wish to consider the general case of damping, in which no closed-form solution to the eigenvalue problem can be expected, so that system discretization is necessary. As in Section 8.9, we assume an approximate solution of Eq. (8.138) in the form of the finite series

$$w(x, t) = \sum_{k=1}^{n} \phi_k(x) q_k(t) \qquad (8.139)$$

where $\phi_k(x)$ are comparison functions and $q_k(t)$ are generalized coordinates. Following the pattern established in Section 8.9, we approximate the partial differential equation, Eq. (8.138), by the set of n ordinary differential equations

STRUCTURES WITH GENERAL VISCOUS DAMPING

$$\sum_{k=1}^{n} m_{jk}\ddot{q}_k(t) + \sum_{k=1}^{n} c_{jk}\dot{q}_k(t) + \sum_{k=1}^{n} k_{jk}q_k(t) = Q_j(t), \quad j = 1, 2, \ldots, n \quad (8.140)$$

where m_{jk} and k_{jk} are mass and stiffness coefficients given in Eqs. (8.100) and $Q_j(t)$ are generalized forces given by Eqs. (8.101). In addition,

$$c_{jk} = \int_0^L \phi_j \mathscr{C} \phi_k \, dx, \quad j, k = 1, 2, \ldots, n \quad (8.141)$$

are damping coefficients.

Equations (8.140) can be written in the matrix form

$$M\ddot{\mathbf{q}}(t) + C\dot{\mathbf{q}}(t) + K\mathbf{q}(t) = \mathbf{Q}(t) \quad (8.142)$$

where the various quantities are as defined in Section 8.9. In addition, $C = [c_{jk}]$ is an $n \times n$ damping matrix. We consider a modal approach to the design of controls for the distributed structure on the basis of the discretized model, Eq. (8.142). To this end, we must first solve the eigenvalue problem associated with Eq. (8.142) and then derive the modal equations. Because the damping operator $\mathscr{C}$ is not a linear combination of the stiffness operator $\mathscr{L}$ and the mass density m, the damping matrix C is not a linear combination of the stiffness matrix K and the mass matrix M. It follows that the classical approach used in Section 8.9 for undamped systems will not work here, so that we must use a more general approach, namely, the approach based on state equations used in Section 6.15 for discrete systems. Hence, following the pattern established in Section 6.15, we introduce the state vector $\mathbf{x}(t) = [\mathbf{q}^T(t) | \dot{\mathbf{q}}^T(t)]^T$, adjoin the identity $\dot{\mathbf{q}}(t) \equiv \dot{\mathbf{q}}(t)$ and rewrite Eq. (8.142) in the state form

$$\dot{\mathbf{x}}(t) = A\mathbf{x}(t) + B\mathbf{Q}(t) \quad (8.143)$$

where

$$A = \left[\begin{array}{c|c} 0 & I \\ \hline -M^{-1}K & -M^{-1}C \end{array}\right], \quad B = \left[\begin{array}{c} 0 \\ \hline M^{-1} \end{array}\right] \quad (8.144\mathrm{a,b})$$

are coefficient matrices. To derive the modal state equations, we first solve the eigenvalue problem associated with A. In fact, because A is not symmetric, we must solve two eigenvalue problems, one associated with A and the other with A^T. The first yields the diagonal matrix of eigenvalues $\Lambda = \mathrm{diag}(\lambda_1 \ \lambda_2 \ \cdots \ \lambda_{2n})$ and the matrix of right eigenvectors U and the second yields once again the matrix Λ of eigenvalues and the matrix V of left eigenvectors (Section 3.11). The two matrices of eigenvectors are biorthogonal and we assume that they have been normalized so as to satisfy the biorthonormality relations

$$V^T U = I, \quad V^T A U = \Lambda \qquad (8.145a,b)$$

Then, introducing the linear transformation

$$\mathbf{x}(t) = U\boldsymbol{\zeta}(t) \qquad (8.146)$$

into Eq. (8.143), multiplying both sides of the resulting equation on the left by V^T and considering Eqs. (8.145), we obtain the desired modal state equations

$$\dot{\boldsymbol{\zeta}}(t) = \Lambda\boldsymbol{\zeta}(t) + \mathbf{Z}(t) \qquad (8.147)$$

where

$$\mathbf{Z}(t) = V^T B \mathbf{Q}(t) = V_L^T M^{-1} \mathbf{Q}(t) \qquad (8.148)$$

is the modal control vector, in which V_L is an $n \times 2n$ matrix representing the lower half of the matrix V.

Equations (8.147) can be used to design controls for the discretized system. As indicated in Section 8.9, however, the algebraic eigenvalue problem resulting from the discretization of a distributed system is not capable of yielding completely accurate modal information, and indeed the higher modes are progressively more inaccurate. This has been demonstrated for real symmetric matrices, admitting real eigenvalues and eigenvectors [M26]. The matrix A given by Eq. (8.144a) is real but not symmetric, and its eigensolutions are generally complex. Although many of the properties of the eigenvalue problem for real symmetric matrices discussed in Section 7.4 are not shared by the eigenvalue problem for real nonsymmetric matrices, it is inherent in the discretization process that the higher eigensolutions are in error. Hence, using the same arguments as in Section 8.9, we propose to control only $2n_C$ modes, where n_C is sufficiently large to include all the modes in need of control. At the same time, we choose n in Eq. (8.139) sufficiently large to ensure that A yields $2n_C$ accurate eigensolutions.

Consistent with the above discussion, we divide the modal equations, Eq. (8.147), into the two sets

$$\dot{\boldsymbol{\zeta}}_C(t) = \Lambda_C \boldsymbol{\zeta}_C(t) + \mathbf{Z}_C(t) \qquad (8.149a)$$

$$\dot{\boldsymbol{\zeta}}_R(t) = \Lambda_R \boldsymbol{\zeta}_R(t) + \mathbf{Z}_R(t) \qquad (8.149b)$$

where the subscripts C and R refer to controlled and uncontrolled, or residual modes, respectively. We note from Eq. (8.148) that

$$\mathbf{Z}_C(t) = V_{LC}^T M^{-1} \mathbf{Q}(t), \quad \mathbf{Z}_R(t) = V_{LR}^T M^{-1} \mathbf{Q}(t) \qquad (8.150)$$

where V_{LC} and V_{LR} are submatrices of V_L of dimensions $n \times 2n_C$ and $n \times 2n_R$, respectively, in which $n_C + n_R = n$. In the case of a distributed control force, we can use Eq. (8.101) to write

$$\mathbf{Q}(t) = \int_0^L \boldsymbol{\phi}(x) f(x, t) \, dx \qquad (8.151)$$

and in the case of n point actuators, Eqs. (8.101) and (8.116) yield

$$\mathbf{Q}(t) = \Phi^T \mathbf{F}(t) \qquad (8.152)$$

where

$$\Phi = [\phi_{ij}] = [\phi_i(x_j)] \qquad (8.153)$$

is an $m \times n$ participation matrix and $\mathbf{F}(t)$ is an m-vector of control force amplitudes.

The control law in the case of point actuators can be written in the general form

$$\mathbf{F}(t) = -G_C \boldsymbol{\zeta}_C(t) \qquad (8.154)$$

where G_C is a control gain matrix. Then, inserting Eq. (8.154) into Eq. (8.149a), the closed-loop equation for the controlled modes becomes

$$\dot{\boldsymbol{\zeta}}_C(t) = (\Lambda_C - V_{LC}^T M^{-1} \Phi^T G_C) \boldsymbol{\zeta}_C(t) \qquad (8.155)$$

and the equation for the residual modes takes the form

$$\dot{\boldsymbol{\zeta}}_R(t) = \Lambda_R \boldsymbol{\zeta}_R(t) - V_{LR}^T M^{-1} \Phi^T G_C \boldsymbol{\zeta}_C(t) \qquad (8.156)$$

which indicates that there is control spillover, as should be expected. Determination of the control gains was discussed in Section 6.15 and will not be repeated here.

CHAPTER 9

A REVIEW OF LITERATURE ON STRUCTURAL CONTROL

In the past several decades, there has been an increasing interest in dynamics and control of structures across a broad spectrum of engineering disciplines. Consistent with this, the literature on the subject has experienced a virtual explosion, as can be concluded from the bibliography on noise and vibration control cited below [G18]. The present review is more limited in scope and it concentrates on articles on structural control published in archive journals. To complement this review, the References section following this chapter includes a number of pertinent survey papers. These surveys deal with a variety of subjects and are reviewed briefly here, listed in chronological order of appearance. Robinson [R7] presents a survey of theoretical and computational methods in the field of optimal control of distributed-parameter systems described by partial differential equations and integral equations. Included in the survey there is a small number of papers on structural control, and in particular on attitude control of flexible space structures. Swaim [S53] discusses control systems for alleviating aircraft rigid-body and elastic mode responses to turbulence. In an early review paper, Thompson and Kass [T3] recognize active flutter supression in aircraft as an emerging technology offering practical solutions to the problems of unstable structural modes. Sandell et al. [S8] survey the control theory on decentralized and hierarchical control, and methods of control of large-scale systems. Decentralized control is beginning to attract attention as a method for structural control, although applications to date remain limited. Croopnick et al. [C16] present a literature survey in the areas of attitude control, vibration control and shape control, as they apply to space structures. Meirovitch and Öz [M22] assess various methods for the active control of space structures with a view to the problems of high dimensionality and modeling. Attention is called to the difference between structural modeling requirements and control algorithms capability. Balas [B13] presents a survey of the state of the art in the control of space structures, including such topics as control framework, generation of reduced-order models and controllers and spillover compensation. Newsom et al. [N2] review analytical methods and associated tools for active control, with particular emphasis on flutter suppression in aircraft. Both classical and modern control methods are reviewed. Nurre et al. [N8] present a comprehensive survey of problems in dynamic modeling and control of space structures, focusing on the areas of computation time, substructuring, efficient modeling of changeable configuration, nonlinear analysis and model verification. Bryson [B38] reviews new concepts in control theory, with emphasis on problems associated with flight vehicles. Miller et al. [M61] present a survey of vibration control methods for civil structures. After reviewing the application of control theory, the paper concentrates on pulse-control methods. A very extensive bibliography, containing in excess of 1700 titles and compiled by Guicking [G18], is concerned with noise and vibration control.

Additional information can be obtained from proceedings of two symposia on structural control held in Waterloo, Ontario, Canada (publication dates 1980 and 1987) and edited by H. H. E. Leipholz, as well as seven biannual symposia on dynamics and control of structures held in Blacksburg, VA (1977–1989) and edited by L. Meirovitch. Yet another source of information is a monograph edited by Atluri and Amos [A20] and including contributions in the areas of computations, modeling, reduced-order controllers, control of distributed structures, adaptive control and simultaneous structural and control optimization.

The references included in this review are divided into various classes according to the topics covered. This classification, however, must be regarded as being loosely defined, as a given paper can belong in several categories. Moreover, many papers do not lend themselves to easy classification. In each category, the papers are reviewed in chronological order. Care should be exercised in attempting to use this order to establish precedence of ideas, however, as it often takes one year or more for a paper presented at a conference to appear in an archive publication. This may explain why the order of some papers appears skewed.

Finally, it should be pointed out that structural parameter identification and adaptive control are not included in the present literature review.

9.1 ISSUES IN MODELING AND CONTROL DESIGN

9.1.1 Actuator and Sensor Dynamics

The issue of actuator and sensor dynamics is discussed very infrequently in the technical literature, although in some cases the effect can be significant. It is addressed to a varying degree by Hughes and Abdel-Rahman [H23],

Kosut et al. [K11, K12], Goh and Caughey [G8], Meirovitch and Quinn [M47] and Quinn and Meirovitch [Q1].

9.1.2 Actuator and Sensor Locations

Gevarter [G4] provides an introductory discussion of elastic vehicle and control system interactions, with emphasis on the stability implications of sensor and actuator locations. Martin [M7] uses relaxed control theory to determine the best location of a finite number of controllers at each time so as to minimize the cost functional for distributed-parameter systems. Juang and Rodriguez [J13] use the steady-state solutions of the optimal control and estimation problems to study the actuator and sensor locations for space structures. Wu et al. [W13] concentrate on the sensor and actuator locations for controlling space structures. A reduced-order model is selected in [W13] for linear control design and controllability and observability are considered to determine the minimum number of actuators and sensors. Within the framework of modal control, Chang and Soong [C8] advance an approach for the determination of optimal locations of a limited number of actuators such that the total control energy is minimized. Arbel [A14] uses modal equations to study the problem of actuator locations for undamped structures. Using certain controllability measures, he develops a design procedure for optimal actuator placement. Johnson [J6] develops criteria for the number, type and location of sensors and actuators for distributed structures. Horner [H17] presents a technique for determining the optimum damper locations and damping rates for a flexible structure. Then, he develops a technique for determining the best actuator locations, actuator gains and number of actuators in a planar structure [H18]. Schulz and Heimbold [S16] propose a method allowing an integrated determination of actuator/sensor positions and feedback gains for control of flexible structures. The method is based on maximization of energy dissipation due to control action. Barker and Jacquot [B19] determine optimal sensor and actuator locations in conjunction with direct output feedback control. Observation spillover is minimized in [B19] by optimal placement of several sensors and then total spillover is minimized by placement of one actuator. The subject is also discussed by Creedon and Lindgren [C15], Hughes and Skelton [H24, H25], Lin [L10], Meirovitch and Öz [M23], Soong and Chang [S47], Benhabib et al. [B24], Kosut et al. [K11, K12], Mahesh et al. [M3] and Lim and Junkins [L8].

9.1.3 Control and Observation Spillover

In an early paper on modal control, Creedon and Lindgren [C15] call attention to the problem of control spillover into the uncontrolled modes and propose ways of minimizing it. Balas [B4, B5] demonstrates that a combination of control and observation spillover due to the residual (uncontrolled) modes in a distributed-parameter systems controlled by a finite number of actuators and sensors can lead to instability. Using the orthogonality property of modes in a structure, Coradetti [C13] develops an algorithm for the design of optimal low-order feedback control of a set of controlled modes independently of the residual modes and with zero control spillover into the residual modes. Joshi and Groom [J9] use the Lyapunov method to derive two bounds on spectral norms of control and observation spillover terms in space structures. Satisfaction of any of the two bounds assures asymptotic stability. Balas [B10] proposes to eliminate the observation spillover from critical residual modes by adding an innovation feedthrough term to the control law and a residual mode aggregation term to the state estimator. Three steps are proposed by Lin [L10] to alleviate control and observation spillover problems in space structures. They consist of proper actuator and sensor placement, proper actuator and sensor influences synthesis and proper filtering of the control inputs and sensor outputs. Calico and Janiszewski [C2] suggest a way of handling observation and control spillover through a suppression transformation. Meirovitch and Baruh [M27] demonstrate that a minimal amount of inherent damping can at times eliminate observation spillover instability in distributed structures. In a later paper, Meirovitch and Baruh [M32] propose ways of preventing observation spillover in self-adjoint distributed-parameter systems. Additional reference to the subject is made by Balas [B3], Longman [L15], Juang et al. [J15], Meirovitch and Öz [M23] and Ulsoy [U3].

9.1.4 Controllability and Observability

Hughes and Skelton [H24, H25] investigate the controllability and observability of linear matrix second-order systems. Simple criteria provide insights into the modal behavior of the system and furnish information on the number and positioning of sensors and actuators. Juang and Balas [J14] consider controllability and observability conditions to determine minimum actuator and sensor requirements for the modal control of spinning space structures. Other related work is by Klein and Hughes [K6], Rappaport and Silverman [R3], Knowles [K7], Krabs [K13], Kuo [K15], Wu et al. [W13], Juang et al. [J15], Meirovitch and Öz [M23] and Hughes and Skelton [H26].

9.1.5 Discrete-Time Systems

Dorato and Levis [D10] present the optimal discrete-time version of the linear-quadratic-regulator problem. The optimization equations are derived in [D10] by means of dynamic programming [B23]. Rappaport and Silverman [R3] examine the optimization of discrete-time linear systems with respect to general quadratic performance indices. A complete stability theory is obtained in [R3] based on the concept of perfect observability. Kuo et al. [K14] develop an approach to the digital control system design for a spinning flexible spacecraft. A continuous-data control system is first designed and then redesigned to provide a digital control system whose states are similar to those of the continuous system at the sampling instants. A Luenberger observer is considered by Leondes and Novak [L3] as an alternative to a Kalman filter for obtaining state estimates in linear discrete-time stochastic stystems. An IMSC-based procedure for the nonlinear control of spinning flexible spacecraft is presented by VanLandingham and Meirovitch [V4]. Precision attitude control, as well as active damping of flexible modes, is accomplished in [V4] with digital processing and quantized control. Kuo [K15] presents a method allowing the design of digital control for output regulation in a system with constant disturbances. The controllability condition for the digital control systems is derived. Öz et al. [O4] contrast two approaches to digital control of continuous-time systems 1) time discretization of a previously designed continuous-time control system and 2) direct design in the discrete-time domain. Balas [B14] discusses the discrete-time stability of direct velocity feedback control, as well as other controls designed in continuous time. Then he presents [B15] a mathematical framework for discrete-time finite-dimensional control of distributed parameter systems. McClamroch [M10] presents results which can serve as guidelines for choice of feedback gains and sampling period to guarantee that a discrete-time-controlled structure is stable. Constant gain output velocity feedback control is used. Meirovitch and Öz [M37] extend the IMSC method to the discrete-time stochastic control of distributed-parameter systems. A discrete-time control methodology is developed and applied to structural control by Rodellar et al. [R8, R9]. The basic idea of the method in [R8, R9] is that the control input to the plant is the one that makes the predicted plant output equal to a dynamic desired plant output. Meirovitch and France [M53] present a discrete-time control of a spacecraft with maneuvering flexible antennas using decentralized control. The subject is also discussed by Gran et al. [G12], Benhabib [B25], Johnson [J3] and Pu and Hsu [P7].

9.1.6 Distributed Models, Controllers and Observers

Athans [A19] outlines the problems encountered in the control of distributed parameter systems and suggests some approaches to the problem. A monograph by Komkov [K9] presents a general discussion of the problem of optimal control of distributed-parameter systems, such as vibrating beams and plates, including the derivation of Pontryagin's maximum principle[†]. Balas [B3] uses modal equations to obtain feedback control of a finite number of modes of a distributed-parameter system and treats the problem of control spillover into the uncontrolled modes. The control of vibrating elastic systems, particularly the optimal feedback control of beams and plates transverse vibration, is investigated by Köhne [K8]. General develop-

---

[†] The difference in the title from that used in this book is due to a sign difference in the definition of the Hamiltonian.

ments both for distributed and discretized models are presented. Lions [L13] presents a general mathematical theory for the control of distributed-parameter systems. Gibson [G5] obtains a sequence of modal control laws from the solution of a sequence of matrix Riccati equations by modeling more and more modes of a flexible structure, and studies the convergence of this sequence of control laws. Krabs [K13] studies the problem of controlling a distributed-parameter vibrating system, such as a string or a beam, by means of a controller at the boundary. Akulenko and Bolotnik [A10, A11] investigate the problem of rotating a free-free elastic shaft through a given angle without vibration, where the shaft has a rigid flywheel at one end and a control torque at the other end. Breakwell [B35] presents the design of an optimal steady-state regulator for a distributed system. He shows a means for obtaining a distributed gain function for the implementation of the optimal control. De Silva [D7] develops a gradient algorithm for the optimal design of discrete passive dampers in the vibration control of a class of distributed flexible structures. Gibson [G7] considers the convergence and stability of the sequence of modal control laws corresponding to a sequence of models of increasing dimension. Meirovitch and Baruh [M30] develop an optimal IMSC for self-adjoint distributed-parameter systems and introduce the concept of modal filters. They also present a sensitivity analysis in the presence of variations in the natural frequencies. Juang and Dwyer [J16] describe the application of operator approximation theory to the optimal control of distributed-parameter structures. Meirovitch and Silverberg [M35] demonstrate that IMSC provides globally optimal control for self-adjoint distributed-parameter systems. Skaar [S34] presents an approach to the control of distributed-parameter elastic systems whereby the control solutions have the form of a convergent series of time-varying terms. Distributed-parameter systems are also considered by Gould and Murray-Lasso [G11], Berkman and Karnopp [B26], Creedon and Lindgren [C15], Takahashi et al. [T2], Klein and Hughes [K6], Balas [B4–B11, B13–B18], Balas and Ginter [B12], Martin [M7], Yang and Giannopoulos [Y4], Caglayan et al. [C1], Gibson [G6], Gran [G14], Lutze and Goff [L17], Meirovitch and Baruh [M27, M31, M32, M41], Hale and Rahn [H2], Baruh and Silverberg [B20], Meirovitch [M36, M46], Öz and Meirovitch [O5], Meirovitch and Öz [M37], Meirovitch and Silverberg [M39, M40], Sadek et al. [S1–S4], Bailey and Hubbard [B1], Crawley and De Luis [C14], Lisowski and Hale [L14], Barker and Jacquot [B19], Meirovitch and Bennighof [M43], Carotti et al. [C5], Hagedorn [H1], Mace [M1], Plump et al. [P2], Wie and Bryson [W9] and Meirovitch and Thangjitham [M57].

9.1.7 Nonlinear Plant and Control

Knowles [K7] considers the bang-bang control of certain distributed-parameter systems, and discusses the relationship between these results and the approximate controllability of the system. Turner and Junkins [T6] present results providing a basis for systematic solution for optimal large-angle single-axis flexible spacecraft rotational maneuvers. They propose and demonstrate a method for solving the nonlinear two-point boundary-value problems of attitude dynamics. Dehghanyar et al. [D6] essentially use decentralized direct output feedback control of nonlinear structures in which control pulses opposing the velocity are applied whenever the velocity reaches a maximum. Vander Velde and He [V1] use on-off thrusters to design control systems for space structures. The design in [V1] is based on an approximation to an optimal control formulation which minimizes a weighted combination of response time and fuel consumption. Dodds and Williamson [D9] develop a bang-bang control for simultaneous control of the attitude and a number of flexible modes of a flexible spacecraft. Meirovitch et al. [M38] demonstrate experimentally the effectiveness of the IMSC method in controlling the motion of a free-free beam using on-off modal control laws, where on-off modal controls translate into quantized actual controls. Turner and Chun [T7] use distributed control to solve the problem of [T6]. Balas [B16] considers a flexible structure subjected to small disturbances depending nonlinearly on the state. A theory is developed by McClamroch [M11] for the multivariable displacement control of an arbitrary nonlinear structure controlled by multiple electrohydraulic servo actuators. Shenhar and Meirovitch [S26] demonstrate how the IMSC method can be used to control a space structure with a minimum amount of fuel, which implies nonlinear control. Skaar et al. [S36] discuss the problem of reorienting a flexible spacecraft by means of on-off actuators. The object in [S36] is to select the control switch times so as to minimize postmaneuver vibrational energy. Abdel-Rohman and Nayfeh [A7, A8] consider active and passive means to control a nonlinear structure modeled by a single mode. Reinhorn et al. [R6] present methodology for the control of civil structures undergoing inelastic deformations through the use of an active pulse force system. Singh et al. [S32] consider the problem of time-optimal, rest-to-rest slewing of a flexible spacecraft through a large angle. The control histories in [S32] are bang-bang with multiple switches in each control variable. Other work related to the subject is by Meirovitch et al. [M20, M34], Van-Landingham and Meirovitch [V4], Meirovitch and Öz [M21, M23–M25, M37], Meirovitch [M36], Meckl and Seering [M15], Horta et al. [H19], Chun et al. [C11], Thompson et al. [T4] and Meirovitch and France [M53].

9.1.8 Reduced-Order Models, Controllers and Observers

Davison [D1] suggests a method for reducing appreciably the order of a dynamical system. The principle of the method in [D1] is to retain in the reduced model only the dominant eigenvalues of the original system. Rogers and Sworder [R11] present a procedure for evaluating a simple model of a complicated linear dynamical system. The model in [R11] is selected to insure that the actual system response is as close as possible to that attained by the optimally controlled model. Skelton and Likins [S37] propose to accommodate model errors in linear control problems by focusing on the construction of the model. To this end, they use three models of different fidelity. Gran and Rossi [G13] present a technique for reducing the order of the model for a structure on the basis of the closed-loop dynamics. Johnson and Lin [J5] develop an aggregation technique for model reduction based on the influence of the actuators and sensors, as an alternative to model reduction based on the open-loop modes. Longman [L15] presents methodology for generating reduced-order controllers for the shape and attitude control of space structures. Both annihilation and the more limited objective of suppression of control and observation spillover are discussed. Skelton [S40] develops a model reduction problem related to the control problem by means of a model quality index which measures the performance of the higher-order system when the control is based upon a lower-order model. Balas [B11] gives conditions under which stable control of a finite element model of a distributed structure will produce stable control of the actual structure. Gibson [G6] considers the stability implications of using a finite number of actuators to control a distributed-parameter system. A parameter optimization algorithm for designing optimal low-order controllers for high-order systems is applied by Martin and Bryson [M6] to the problem of attitude control of a flexible spacecraft. A hierarchy of dynamical models is developed by Hughes and Skelton [H26], where at each reduction level the model is expressed in terms of natural modes. Criteria for mode deletion include inertial completeness, frequency relationship, controllability and observability and modal cost. Hyland [H29] addresses the problem of inaccuracies in structural modeling arising from the high dimensionality of dynamical systems. A minimum set of a priori data needed for modeling fidelity is first identified and then a complete probability assignment is induced by a maximum entropy principle. Kabamba and Longman [K1] present a formalism for the optimal design of controllers of arbitrary prescribed order using a quadratic performance measure with additional penalties on the controller parameter magnitudes. Velman [V8] addresses the problem of using a low-order filter to estimate an approximation to a desired linear function of the state in conjunction with a high-order design model. The filter output in [V8] is characterized in terms of the transfer function of the state estimator, thus permitting the use of classical control design methods. A model reduction procedure for space structures based on aggregation with respect to sensor and actuator influences, rather than modes, is presented by Yam et al. [Y1, Y2]. A method of synthesizing reduced-order optimal feedback control laws for a high-order system is developed by Mukhopadhyay et al. [M65]. The method is applied in [M65] to the synthesis of an active flutter-suppression control law for an aeroelastic wind tunnel wing model. The concept of suboptimality is applied by Sezer and Siljak [S23] to test the validity of reduced-order models in the design of feedback control for large-scale systems. The approach is applied in [S23] to the control design of a space structure. Skelton et al. [S42] use the approach of [S41] to obtain a reduced-order model of a space structure. Meirovitch [M36] discusses modeling implications in the modal control of distributed structures by means of finite-dimensional controllers and observers designed on the basis of finite-dimensional models. A method for reducing the order of the model of multistory structures is presented by Wang et al. [W4]. The active control of the structural vibration using the pole placement method is outlined. Hyland and Bernstein [H30] are concerned with the design of quadratically optimal fixed-order (i.e., reduced-order) dynamic compensation for a plant subject to stochastic disturbances and nonsingular measurement noise. Hale and Rahn [H2] distinguish among designing control forces for a reduced-order model of a distributed structure, implementing the control forces with a finite number of actuators and estimating the distributed state from a finite number of sensors. By introducing three corresponding projection operations, they study the effects of using a reduced-order controller on the actual closed-loop eigenvalues. Bhaya and Desoer [B28] consider noncollocated actuators and sensors in a lumped model of a structure and state necessary and sufficient conditions for stabilizing a certain number of modes. Balas [B17, B18] develops two approaches based on the Galerkin method for the generation of finite-dimensional controllers for distributed-parameter systems. The first approximates the open-loop distributed-parameter system and then generates the controller from this approximation. The second approximates an infinite-dimensional controller. Gruzen and Vander Velde [G16] describe the capabilities of optimal projection/maximum entropy design methodology for the development of an active vibration damper. Hu et al. [H21] present a complete modal cost analysis for reducing the order of beam-like structures for control design purposes. Additional reference to the subject is made by Leondes and Novak [L3], Larson and Likins [L1], Martin [M7], Skelton and Likins [S38], Skelton [S39], Sesak et al. [S18], Wu et al. [W13], Meirovitch and Öz [M23], Skelton and Hughes [S41], Kosut et al. [K11, K12], Schaechter [S11], Yedavalli and Skelton [Y12], Hale and Lisowski [H3], Hwang and Pi [H28], Yousuff and Skelton [Y18], Lisowski and Hale [L14] and Greene [G15].

9.1.9 Sensitivity and Robustness

Davison and Goldenberg [D4] are concerned with the problem of finding a robust controller for a system so that asymptotic tracking, in the presence of disturbance, occurs independently of perturbation of the plant parameters and gains of the system. Doyle and Stein [D12] describe an adjustment procedure for observer-based linear control systems which asymptotically achieve the same loop transfer functions (and hence the same relative stability, robustness and disturbance rejection properties) as full-state feedback control implementations. Arbel and Gupta [A15, A16] are concerned with finding an output feedback matrix yielding a closed-loop stable system, where the actuators and sensors are collocated. Gran [G14] discusses the problem of controlling a self-adjoint distributed system by means of a finite-dimensional controller so as to exhibit insensitivity with respect to disturbances, stability and insensitivity with respect to variation in the system parameters. Kosut et al. [K11, K12] develop frequency-domain measures quantifying the stability and robustness properties of control systems for flexible spacecraft. Problems examined in [K11, K22] are: controlling an infinite-dimensional system with a reduced-order controller, the effect of actuators and sensors location and number and uncertain system parameters. Lehtomaki et al. [L2] consider the robustness of control systems with respect to model uncertainty using simple frequency domain criteria. Hallauer and Barthelemy [H7] examine the sensitivity of the independent modal-space control to imperfect modal response measurement. A sensitivity analysis based on modal equations is presented by Kida et al. [K3]. The eigenvalue, eigenvector and cost function sensitivity are derived in [K3] for a closed-loop system designed by the pole placement method. Ashkenazi and Bryson [A17] present an algorithm for designing controllers of specified order to minimize the response to zero-mean random disturbances with variations in some plant parameter specified. Meirovitch and Baruh [M31] demonstrate the robustness of the IMSC method with respect to variations in the mass and stiffness distributions. Schaechter [S11] develops a technique for computing the closed-loop performance sensitivity to parameter variations of a dynamic system with a reduced-order controller. Baruh and Silverberg [B22] discuss the robustness of the independent modal-space control method with respect to parameter uncertainties and errors due to spatial discretization. Blelloch and Mingori [B29] develop a procedure for dealing with performance and robustness issues in lightly damped structures. The procedure represents a modified version of the loop transfer recovery (LTR) design method. Joshi [J11] investigates the robustness of two types of controllers for space structures using collocated sensors and actuators in the presence of uncertainties, an attitude controller using attitude and rate feedback and a damping enhancement controller using velocity feedback. Greene [G15] is concerned with the robustness of control design for flexible structures in view of the necessity of model truncation and the variability of the structure itself. Hefner and Mingori [H13] describe a method for designing a controller with improved robustness with respect to truncated flexible modes. The approach involves minimization of a quadratic performance index subject to frequency-domain constraints. The subjects are also discussed by Van de Vegte and Hladun [V3], Sesak et al. [S18], Iwens et al. [I2], Benhabib et al. [B24], Meirovitch and Baruh [M30], Breakwell and Chambers [B36], Bernstein and Hyland [B27], Gruzen and Vander Velde [G16] and Schmidt and Chen [S14].

9.1.10 Time-Varying Plants

The subject of time-varying plants is covered in a general way in many textbooks on controls. Quite often, such systems are treated through discretization in time. Among papers addressing specifically this problem, we single out the work by Friedland et al. [F5], who propose to design suboptimal control laws for linear time-varying systems by solving the algebraic Riccati equation for the optimal control law at each instant of time, a design referred to as "adiabatic approximations." Time-varying plants occur at times as a result of a perturbation technique, such as in Meirovitch and Sharony [M49, M54] and Sharony and Meirovitch [S24, S25], or in systems with varying configuration, such as in Meirovitch and Kwak [M50, M55] and Meirovitch and France [M53].

9.1.11 Uncertainties and Modeling Errors

Larson and Likins [L1] examine the consequences of modeling errors in the observer for an elastic spacecraft by setting up various evaluation models against which to measure these consequences. They discuss the advantages of using frequency-domain analysis. Doyle and Stein [D13] investigate the issue of feedback control design in the face of uncertainties and generalize single-input, single-output statements and constraints of the design problem to multi-input, multi-output cases. They essentially develop the procedure known as linear-quadratic-Gaussian/loop transfer recovery (LQR/LTR). Uncertainties and/or modeling errors are also considered by Skelton and Likins [S37, S38], Sesak et al. [S18–S21], Kosut et al. [K11, K12], Yedavalli and Skelton [Y12], Meirovitch [M36], Yedavalli [Y13] and Joshi et al. [J11, J12].

9.1.12 Classical versus Modern Control

Classical control has been the backbone of control design for many years. Applications have been mainly for single-input, single-output systems. One of the advantages of classical control is that it affords good physical insight. In the case of multi-input, multi-output systems, difficulties are encountered if the design is for one loop at a time. Modern control has the advantage of being better able to cope with multi-input, multi-output systems, but some physical insight is lost. There has been a significant effort toward extending many of the concepts and techniques of classical control to multi-input, multi-output systems. Application of feedback control to structures has provided added impetus to this effort. Kosut [K10] presents a method for determining the transfer function matrix for a system described by a matrix second-order equation, such as in the case of an aircraft. Perkins and Cruz [P1] review several results relating frequency-domain characterization of optimal linear regulators, stability, return difference, comparison sensitivity and sensitivity performance indices. The monograph by Rosenbrock [R14] discusses ways of using classical techniques for multi-input, multi-output systems, albeit of limited order. An approach is developed by MacFarlane and Kouvaritakis [M2] for the design of linear multivariable feedback control systems based on a manipulation of eigenvalues and eigenvectors of an open-loop transfer-function matrix. Doyle [D11] presents a new approach to the frequency-domain analysis of multiloop linear feedback systems. Herrick [H15, H16] examines the application of the root locus techniques to multi-input, multi-output systems. Lutze and Goff [L17] use classical techniques to design controls for distributed structures. In particular, they use decentralized direct output feedback, whereby sensor signals with appropriate gains are fed back directly to the actuators at the same locations. Williams [W11] emphasizes the fact that the transmission zeros of a system are as important as its poles. For a space structure with collocated sensors and actuators, these zeros are shown to lie in a region of the complex plane defined by its natural frequencies and damping factors. Classical techniques in conjunction with multi-input, multi-output systems are also considered by Gould and Murray-Lasso [G11], Ohkami and Likins [O1], Larson and Likins [L1], Yocum and Slafer [Y14], Kosut et al. [K11, K12], Gupta [G19], Lehtomaki et al. [L2], Velman [V8], Kida et al. [K4], Schmidt and Chen [S14], Wie [W8] and Wie and Bryson [W9].

9.2 METHODS, PROCEDURES AND APPROACHES

9.2.1 Control of Domains and Local Control

The necessary condition for the solution of the linear quadratic optimal control problem with the constraint of local state feedback is derived by Schaechter [S9], where local control is defined as a control law that includes feedback of only those state variables that are physically near a particular actuator. Luzzato [L18] uses optimal control to develop active protection of certain domains of vibrating distributed structures. Protection criteria such as deflection, spatial gradient and velocity at given points, as well as perturbing energy are used. Luzzato and Jean [L19, L20] extend the concepts of [L18] to viscoelastic distributed structures.

9.2.2 Decentralized Control

West-Vukovich et al. [W6] consider the decentralized control problem for space structures using point force actuators and point displacement/displacement-rate sensors. They show that a solution to the decentralized control problem exists if it exists for the centralized control problem. Young [Y17] describes a framework for the design of decentralized control of space structures integrating finite element modeling and control design. Decentralized control is used by Siljak [S29], Sesak et al. [S18, S21] and Meirovitch and France [M53].

9.2.3 Decomposition-Aggregation

In the context of the decomposition-aggregation method, Siljak [S28] uses a feedback control scheme for multilevel stabilization of large-scale linear systems to maximize the stability region of a structural parameter which couples the wobble and spin motions of a flexible spinning spacecraft. The approach is used by Siljak [S29] and Yam et al. [Y1, Y2].

9.2.4 Disturbance Accommodation

Johnson [J1, J2] presents the theory of disturbance-accommodating controllers, where disturbances are defined as inputs which are not known accurately a priori. Balas [B8] considers disturbance accommodation for distributed-parameter systems subjected to external disturbances of known waveform but unknown amplitude. Johnson [J3] presents a discrete-time version of the theory developed in [J1, J2]. Burdess and Metcalfe [B39] demonstrate how steady periodic vibration in structures can be reduced by feedback control. Burdess [B40] shows how a control law can be derived which will minimize a quadratic performance index based upon the steady state system response in the presence of persistent harmonic disturbances. Burdess and Metcalfe [B41] consider the vibration control of a mass-damper-spring system subjected to an arbitrary, unmeasurable disturbance. Chun et al. [C10] consider the problem of maneuvering a flexible spacecraft through a large angle, where the disturbance-accommodating feedback control tracks a desired output state. The desired output state is obtained from an open-loop solution for the linear model. The approach is also used by Sharony and Meirovitch [S24, S25] and Meirovitch and Sharony [M49, M54].

9.2.5 Eigenstructure Assignment, Pole Allocation

Wonham [W12] shows that controllability of an open-loop system is equivalent to the possibility of assigning an arbitrary set of poles to the transfer matrix of the closed-loop system, formed by means of suitable linear feedback of the state. Davison [D2] provides a simple proof for the proposition that it is always possible to assign (place) the closed-loop poles of a linear time-invariant system. Simon and Mitter [S31] present a complete theory of modal control. They investigate questions of existence and uniqueness of modal controllers, as well as recursive algorithms for the realization of modal controllers. Davison [D3] gives the conditions for approximate placing of the closed-loop poles in output feedback. A design approach using pole placement techniques for a class of space structures is developed by Wu et al. [W14]. They examine numerical problems of the pole placement algorithm when used on large dimensional systems with extremely low-frequency eigenvalues. Andry et al. [A13] consider the use of output feedback to obtain desired natural frequencies and/or to reshape mode shapes of a structure rather than altering the mass and stiffness distributions. Meirovitch [M46] demonstrates that, when decentralized direct output feedback is used to control distributed structures, poles cannot be placed arbitrarily. A comprehensive treatment of pole allocation can be found in the book by Porter and Crossley [P5]. The subject is also considered by Balas [B4, B5], Tseng and Mahn [T5], Wu et al. [W13], Juang et al. [J15], Kida et al. [K3], Abdel-Rohman [A4], Wang et al. [W4], Bodden and Junkins [B30], Garrard and Liebst [G2] and Liebst et al. [L6, L7].

9.2.6 Filters (Modal, Orthogonal)

The discussion is confined here to a procedure for approximating model errors (orthogonal filters) and a procedure for extracting modal states from distributed measurements (modal filters). Skelton and Likins [S38] propose to improve the convergence properties of state estimators for nonrigid spacecraft. To this end, they approximate the modal error vector by orthogonal functions over given time intervals. Meirovitch and Baruh [M41] discuss aspects of implementation of modal filters, where the filters are capable of producing estimates of modal states from distributed measurements of the states. Modal filters were proposed originally by Meirovitch and Baruh [M30] and used later by Öz and Meirovitch [O5] and Meirovitch and Öz [M37].

9.2.7 Frequency Shaping

Gupta [G19] extends the linear-quadratic-Gaussian method for feedback control design to include frequency-shaped weighting matrices in the quadratic cost functional, thus combining classical design requirements with automated computational procedures of modern control. The approach is used by Chun et al. [C11].

9.2.8 The Independent Modal-Space Control (IMSC) Method

Gould and Murray-Lasso [G11] essentially present a Laplace transform version of independent modal-space control for distributed systems. Implementation is carried out on the assumption that the output and the forcing functions have negligible eigenfunction content beyond a finite number. Creedon and Lindgren [C15] present a modal approach similar to that in [G11] for controlling the vibration of a plate by means of a finite number of actuators acting over given controlled surfaces through pads. The size and location of the actuators are chosen in [C15] so as to minimize the control spillover into the uncontrolled modes. Takahashi et al. [T2] introduce some preliminary ideas involved in the independent modal-space control method. Meirovitch et al. [M20] develop the independent modal-space control method for spinning flexible spacecraft. Caglayan et al. [C1] essentially use the independent modal-space control method to show that the problem of control of a distributed structure can be replaced by a set of uncoupled problems. Hallauer and Barthelemy [H6] use independent modal-space control in conjunction with modal on-off control for active vibration control in structures. Meirovitch and Öz [M23] develop further the IMSC method for flexible gyroscopic systems by considering a reduced-order modal observer. The questions of sensor and actuator locations and controllability and observability are also discussed. Meirovitch and Öz [M24] discuss the use of the IMSC method to control civil structures. An IMSC dual-level control scheme for controlling rigid-body motions and elastic motions of a space structure is presented by Meirovitch and Öz [M25]. Öz and Meirovitch [O3] develop an optimal control theory for large-order flexible gyroscopic systems. The approach is based on the IMSC method and it requires the solution of $n/2$ decoupled 2×2 matrix Riccati equations instead of a single $n \times n$ matrix Riccati equation. Meirovitch and Baruh [M28] extend the optimal IMSC developed in [O3] to damped flexible gyroscopic systems. Rajaram [R2] describes an IMSC method for the optimal control of a flexible structure, including integral feedback of modal coordinates. VanLandingham et al. [V5] use the decoupling procedure suggested by IMSC in conjunction with matrix pseudo-inverses to implement suboptimal modal control of flexible systems with a reduced number of sensors and actuators. Meirovitch and Baruh [M30] develop an optimal IMSC for self-adjoint distributed-parameter systems and introduce the concept of modal filters. They also present a system sensitivity analysis in the presence of variations in the natural frequencies. Meirovitch and Baruh [M31] demonstrate the robustness of the IMSC method developed in [M30] with respect to variations in the mass and stiffness distributions. Meirovitch and Silverberg [M35] show that IMSC provides globally optimal control for self-adjoint distributed-parameter systems. Öz and Meirovitch [O5] develop a stochastic independent modal-space state estimation and control for distributed-parameter system. Inman [I1] presents necessary conditions permitting application of the independent modal-space control method to a large class of structures. Lindberg and Longman [L12] propose to reduce the number of actuators required by IMSC by using the same approach as that suggested in [V5]. Meirovitch and Silverberg [M40] extend the IMSC method to non-self-adjoint distributed-parameter systems. The method in [M40] is capable of handling viscous damping forces, circulatory forces and aerodynamic forces. Baruh [B21] develops duality relations between two methods for designing controls for distributed parameter systems, direct output feedback with collocated sensors and actuators and independent modal-space control. The IMSC method is also considered by VanLandingham and Meirovitch [V4], Meirovitch and Öz [M21, M37], Hallauer and Barthelemy [H7], Meirovitch and Silverberg [M33, M39, M40], Meirovitch et al. [M34, M38], Hale and Rahn [H2], Baruh and Silverberg [B20],

Skidmore et al. [S43, S44], Hallauer et al. [H9], Meirovitch and Bennighof [M43], Shenhar and Meirovitch [S26], McLaren and Slater [M12], Meirovitch and Ghosh [M45], Meirovitch and Quinn [M48], Mesquita and Kamat [M58], Pu and Hsu [P7] and Quinn and Meirovitch [Q1].

9.2.9 Input-Output Decoupling

Falb and Wolovich [F1] determine necessary and sufficient conditions for the decoupling of an m-input, m-output time-invariant linear system using state variable feedback.

9.2.10 Linear-Quadratic-Gaussian/Loop Transfer Recovery (LQG/LTR)

The LQG/LTR procedure is essentially developed by Doyle and Stein in [D13]. Stein and Athans [S49] provide a tutorial overview of the LQG/LTR procedure for linear multivariable feedback systems. The method is applied to the synthesis of an attitude control system for a space structure by Joshi et al. [J12] and Sundararajan et al. [S51]. A modified version is presented by Blelloch and Mingori [B29].

9.2.11 Modal Control (Coupled)

Modal control represents a family of techniques whose basic idea is to induce asymptotic stability in a system by altering the open-loop eigenvalues. Basic work on modal control in coupled form was carried out by Wonham [W12], Simon and Mitter [S31], Davison [D2, D3] and Porter and Crossley [P5], as discussed above in connection with the pole allocation method. The subject of (coupled) modal control in conjunction with structures is considered by Martin and Soong [M5], Balas [B3–B5], Abdel-Rohman et al. [A2, A4], Chang and Soong [C8], Juang and Balas [J14], Breakwell [B34], Vilnay [V10], Wu et al. [W14], Meirovitch et al. [M34], Meirovitch [M36], Radcliffe and Mote [R1], Hale and Rahn [H2], Liebst et al. [L6, L7] and Meirovitch and Thangjitham [M57].

9.2.12 Modal Cost Analysis, Component Cost Analysis

Skelton [S39] discusses the construction of reduced-order models by means of cost sensitivity calculations of truncation indices to indicate which modes are to be truncated from the higher-order model to form the low-order estimator design model. Skelton and Hughes [S41] show that the decomposition of quadratic cost indices into the sum of contributions from each modal coordinate can be used to rank the importance of modes in the model and in the control problem. The contribution of each mode to the cost is referred to as modal cost and may be used as a modal truncation criterion.

Component cost analysis is used by Yousuff and Skelton [Y18] to develop a method for controller reduction based on the participation of the controller states in the value of a quadratic performance metric. The approach is used by Hughes and Skelton [H26], Skelton et al. [S42] and Hu et al. [H21].

9.2.13 Model Error Sensitivity Suppression (MESS)

Sesak et al. [S18] advance a model error sensitivity suppression (MESS) method to solve the problems of sensitivity to modeling errors and the use of low-order state estimators. The method is based on the decentralized control concept. Sesak and Likins [S19] consider the capability of MESS [S18] to handle the truncation of known vibration modes. Sesak [S20] advances a suppressed mode damping scheme in conjunction with MESS [S18] based on direct feedback of sensor signals. Sesak and Halstenberg [S21] apply the MESS method [S18] to the decentralization of attitude and vibration control in a space structure.

9.2.14 Multi-Objective Optimization

Hale et al. [H3–H5] consider the problem of optimal integrated structure-control design for maneuvering space structures by means of a reduced-order model. Sadek and Adali [S1] treat the problem of controlling the vibration of a membrane with a minimum amount of expenditure of force. The problem is formulated as a multi-objective control problem involving the deflection, velocity and force spent, and the necessary condition for optimality is given in the form of a maximum principle. Bodden and Junkins [B30] develop an eigenspace optimization approach for the design of feedback controllers for the maneuver of and vibration suppression in flexible structures. The approach developed in [B30] is applicable to structural and control parameter optimization. Lisowski and Hale [L14] use reduced-order models to study the unified problem of synthesizing a control and structural configuration for open-loop rotational maneuvers of flexible structures. Venkayya and Tischler [V9] study the effect of structural optimization on optimal control design. Structural optimization is treated in [V9] as a problem of mass minimization with constraint on the open-loop frequency. Khot et al. [K2] propose an approach for the vibration control of space structures by simultaneously integrating the structure and control design to reduce the response to disturbances. The formulation in [K2] consists of structural modification of some finite element model, which is controlled in an optimal fashion by a linear regulator. Sadek et al. [S2] discuss the control of a damped Timoshenko beam by the approach of [S1]. Mesquita and Kamat [M58] consider the effect of structural optimization on the control of stiffened laminated structures. The structural optimization in [M58] consists of maximizing the natural frequencies subject to frequency separation constraints and an upper bound on weight; the control is by the

IMSC method. Miller and Shim [M59] consider the problem of combined structural and control optimization for flexible structures. They minimize the sum of structural mass and control energy terms with respect to structural and control parameters using gradient searches. Onoda and Haftka [O2] present an approach to the simultaneous optimal design of a structure and control system for flexible spacecraft whereby the combined cost of structure and control system is minimized subject to a constraint on the magnitude of the response of the structure. Sadek [S3] seeks approximate solutions for the distributed control of vibrating beams based on the approach of [S1]. Sadek et al. [S4] extend the approach of [S1] to a damped two-span beam. Lim and Junkins [L8] discuss the robustness optimization of structural and controller parameters. Three cost functions, total mass, stability robustness and eigenvalue sensitivity, are optimized with respect to a set of design parameters that include structural and control parameters and actuator locations.

9.2.15 Optimal Projection/Maximum Entropy (OP/ME)

Bernstein and Hyland [B27] present a review of optimal projection/maximum entropy (OP/ME), an approach to designing low-order controllers for high-order systems with parameter uncertainties. The basic ideas are introduced by Hyland [H29] and Hyland and Bernstein [H30]. The method is used by Gruzen and Vander Velde in [G16].

9.2.16 Output Feedback

A technique for determining optimal linear constant output feedback controls is developed by Levine and Athans [L4]. This feedback control law does not require state estimation. Canavin [C3] and Balas [B6] suggest controlling the vibration of a flexible spacecraft by means of multivariable output feedback. The approach of [L4], [C3] and [B6] is known as direct output feedback control, which implies that sensor outputs are multiplied by a gain matrix to produce actuator commands. Balas [B7] discusses the approach of [B6], in which output signals from velocity sensors alone are fed back individually to collocated actuators. Elliott et al. [E2] demonstrate that the output feedback controller suggested in [C3] is capable of high performance. Lin et al. [L9] examine four output feedback control methods for possible application to vibration control in space structures. An output feedback control design capable of increasing the damping ratio and the frequency of selected modes in space structures is proposed by Lin and Lin [L11]. Joshi and Groom [J10] formulate the problem of selecting velocity feedback gains for the control of space structures as an optimal output feedback regulator problem, and derive the necessary conditions for minimizing a quadratic performance index. Hegg [H14] demonstrates that the linear equation for the output feedback control gain matrix for reduced-order structural models is algebraically consistent regardless of the rank of the reduced-order observation matrix. Bossi and Price [B33] present a study of "mode residualization" for reducing the order of compensators obtained by LQG design and an extension of an output feedback technique, developed to alter the compensator in order to overcome instability induced by the residualization. Goh and Caughey [G8] discuss direct velocity feedback and show that actuator dynamics can cause instability; they suggest instead the use of positive position feedback. The procedure is also discussed by Sesak [S20], Arbel and Gupta [A15, A16], Lutze and Goff [L17], Balas [B14], Hallauer et al. [H8], Delghanyar et al. [D6], McClamroch [M10], Schäfer and Holzach [S13], Baruh [B21], Barker and Jacquot [B19], Joshi [J11], Joshi et al. [J12], von Flotow and Schäfer [V12], McLaren and Slater [M12], Meirovitch [M46] and Pu and Hsu [P7].

9.2.17 Perturbation Approaches

Aubrun [A21] uses a perturbation approach to determine the closed-loop eigenvalues and eigenvectors resulting from a low-authority controller, (i.e., a controller imparting small forces to the structure). Williams [W10] presents a perturbation method for the prediction of modal dynamics resulting from moderate amounts of velocity and/or position feedback applied to space structures. Sharony and Meirovitch [S24, S25] are concerned with the control of perturbations experienced by a flexible spacecraft during a minimum-time maneuver. The controller is divided into an optimal finite-time linear quadratic regulator and a disturbance-accommodating control. Meirovitch and Sharony [M54] present a general review of the problem of flexible space structure maneuvering and control with emphasis on a perturbation technique. Perturbation approaches are also used by Bodden and Junkins [B30], Meirovitch and Quinn [M48] and Quinn and Meirovitch [Q1].

9.2.18 Positivity

Iwens et al. [I2] discuss a technique for designing robust control systems for space structures based on the positivity of operators. Benhabib et al. [B24] propose a technique for the design of stable and robust control systems for space structures based on the positivity of transfer matrices. Benhabib [B25] extends the concept of positivity discussed in [B24] to discrete-time systems. McLaren and Slater [M12] examine robust multivariable control of space structures designed on the basis of a reduced-order model using positivity concepts. Three controller designs are compared: ordinary multivariable control, independent modal-space control and decentralized direct output feedback control. The concept is also used by Champetier [C6] and Takahashi and Slater [T1].

9.2.19 Proportional-Plus-Derivative-Plus-Integral Control

Hughes and Abdel-Rahman [H23] consider the linear attitude control of flexible spacecraft in conjunction with a proportional-plus-derivative-plus-integral feedback control law. Simple models of sensor and actuator dynamics are included. The control law is considered also by Rajaram [R2], Meirovitch and Sharony [M49] and Meirovitch and Kwak [M50].

9.2.20 Pulse Control

An active control method for flexible structures is presented by Masri et al. [M8, M9] whereby an open-loop control pulse is applied as soon as some specified response threshold has been exceeded. The optimum pulse characteristics are determined analytically so as to minimize a nonnegative performance index related to the structure energy. Udwadia and Tabaie [U1] consider the feasibility of using on-line, pulsed, open-loop control for reducing the oscillation of structural and mechanical systems modeled as single-degree-of-freedom oscillators. Then they extend the approach to multi-degree-of-freedom structures [U2]. The approach is used also by Champetier [C6] and Reinhorn et al. [R6].

9.2.21 Sequential Linear Optimization

Horta et al. [H19] develop an optimization approach using a continuation procedure in conjunction with linear programming for controller design in flexible structures. Using first-order sensitivity of the system eigenvalues, the method converts a nonlinear optimization problem into a maximization problem with linear inequality constraints.

9.2.22 Traveling Wave Control

Vaughan [V7] considers wave propagation concepts to control the bending vibration in beams. Van de Vegte [V2] examines the use of wave reflection concepts in the control of the vibration of slender uniform beams. Meirovitch and Bennighof [M43] are concerned with the use of the IMSC method to control traveling waves in a structure. Von Flotow [V11] describes the elastic motion in a space structure in terms of traveling waves. To achieve control, the energy is shunted into unimportant portions of the structure or is absorbed by an active wave absorber. Von Flotow and Schäfer [V12] describe the experimental implementations of the wave absorbers introduced in [V11]. They discuss the similarity between wave-absorbing compensator and direct velocity output feedback. Hagedorn [H1] applies the traveling wave concept to the control problem is distributed structures. He simplifies the control problem by dropping the boundary condition in the first stage of the design, instead of discretizing the boundary-avalue problem, thus avoiding control spillover. Mace [M1] considers active control of disturbances propagating along a waveguide by means of a finite number of point sensors and actuators. Miller et al. [M60] use feedback compensators, based on spatially local models, to actively modify wave transmission and reflection characteristics of a structure. Redman-White et al. [R5] present results from an experiment designed to reduce the magnitude of flexural waves propagating in elastic structures by active control.

9.2.23 Uniform Damping Control

Silverberg [S30] introduces the concept of uniform damping control of flexible spacecraft. He also shows that the uniform damping control solution represents a first-order approximation to a special globally optimal control problem. The method is used by Meirovitch and Quinn [M47], Quinn and Meirovitch [Q1] and Meirovitch and France [M53].

9.3 AIRCRAFT AND ROTORCRAFT CONTROL

Smith the Lum [S45] use linear optimal control theory for design of active structural bending control to improve ride qualities and structural fatigue life of high-performance aircraft. Analytical and experimental results demonstrating the viability of using feedback control to suppress flutter instability in a thin cantilevered plate are presented by Moon and Dowell [M62]. The basic physical relationships involved in control of a flexible aircraft disturbed by random wind gusts are used by Johnson et al. [J4] in formulating the aerodynamic surface location problem as one in optimal control of a distributed system by a limited number of point actuators. Nissim [N4] and Nissim and Lottati [N5] treat the problem of flutter suppression from an energy point of view, whereby the energy dissipated per cycle is reduced to a quadratic form involving a diagonal matrix of eigenvalues of an aerodynamic energy matrix. Abel et al. [A9] consider the use of classical control theory in conjunction with the aerodynamic energy method [N4, N5] to increase the flutter velocity by 20%. Balakrishnan [B2] suggests the use of a linear-quadratic-regulator for active control of flutter and for gust alleviation in flexible flight vehicles. Edwards et al. [E1] describe the application of generalized unsteady aerodynamic theory to the problem of active flutter control. They investigate the controllability of flutter modes. Newsom [N1] describes a study investigating the use of optimal control for the synthesis of an active flutter-suppression control law. As a design application, he considers a high-aspect-ratio cantilever wind-tunnel wing model. Noll and Huttsell [N6] present an analytical study of an active flutter suppression wind-tunnel model which has both a leading-edge and a trailing-edge control structure. Schulz [S15] designs a control law for the compensation of vibration induced by the rotor of a helicopter and transmitted to the cabin. Ham et al. [H10, H11] describe a technique for active control of helicopters, whereby each blade is individually controlled in the rotating frame over a wide range of frequencies; test results are discussed. McLean [M13] proposes a method for reducing weight and fuel consumption of a transport aircraft, while still being able to respond to the same maneuver commands, by using a structural load alleviation control. Sensburg et al. [S17] provide a summary of active control concepts for flutter prevention and vibration alleviation, where the flutter and vibration are caused by gusts. Mahesh et al. [M3] describe the application of LQG methodology to the design of active control for suppression of aerodynamic flutter. Results of control surface and sensor position optimization are presented. Garrard et al. [G1] seek to provide additional insights into the use of the method developed in [D13] in aeroelastic control problems by applying the method to a flutter control problem. Gupta and Du Val [G20] design a helicopter vibration controller by optimizing a cost functional placing a large penalty on fuselage accelerations at vibration frequencies. Chang [C9] develops a state-space model for motion of an airfoil in a two-dimensional unsteady flow of an inviscid, incompressible fluid and then uses a linear quadratic regulator for flutter suppression. A modal method for the active vibration suppression of a cantilever wing is presented by Meirovitch and Silverberg [M39], in which a modal feedback control law relates the motion of the control surfaces to the modes controlled. Noll et al. [N7] establish the feasibility of using active feedback control to prevent aeroelastic instabilities associated with forward swept wings. Garrard and Liebst [G2] examine the application of eigenspace techniques (closed-loop eigenvalues and eigenvectors shaping) and optimal control to the preliminary design of an active flutter suppression system for a flight test vehicle. Newsom et al. [N3] describe the design of a flutter-suppression system for a remotely piloted research aircraft, with major emphasis on use of optimal control. An eigenvalue placement is used by Liebst et al. [L6] to design a full state controller for flutter suppression. The controller is shown to satisfy specifications on rms control surface activity and gain and phase margins over a range of flight conditions. In addition, the use of eigenvector shaping for gust load alleviation is examined. Schmidt and Chen [S14] synthesize a simple robust control law for flutter suppression with classical control techniques, aided with computer graphics, thus demonstrating that such methods are not too slow for higher-order systems. Takahashi and Slater [T1] examine the use of positive real feedback for the design of a flutter mode control design. Freymann [F4] presents analytical and experimental results in the field of active control of aircraft structures. He investigates the interaction between active control systems and structures and calls attention to possible dangers from the coupling of rigid-body and elastic modes. The techniques used in [L6] are used by Liebst [L7] for the design of a flutter suppression/gust load alleviation system for a hypothetical research drone.

Additional pertinent work is presented by Roger et al. [R10] and Mukhopadhyay et al. [M65].

9.4 CONTROL OF CIVIL STRUCTURES

In an early paper, Yao [Y11] attempts to stimulate interest in the use of active control to suppress the vibration of civil structures subjected to wind and earthquake excitations. Roorda [R12] examines some of the concepts involved in active damping in structures by feedback controls by way of an analysis of a tendon-control scheme for tall flexible structures. Yang [Y3] studies the feasibility of applying optimal control theory to control the vibration of civil engineering structures under stochastic dynamic loadings, such as earthquakes and wind loads. Modal control as formulated by Porter and Crossley [P5] is used by Martin and Soong [M5] to control civil engineering structures. McNamara [M14] proposes the use of a tuned mass damper, also known as a vibration absorber, as an energy-absorbing system to reduce wind-induced structural response of buildings in the elastic range. Abdel-Rohman and Leipholz [A1] propose to control the lowest three modes of a simply-supported beam, simulating a bridge, by means of a collocated actuator/sensor pair. Sae-Ung and Yao [S5] propose to determine a feasible feedback comfort control function for building structures subjected to wind forces. Yang and Giannopoulos [Y4] present a method for active control of distributed-parameter civil structures based on the transfer matrix technique. A dynamic analysis of a two-cable-stayed bridge subject to active feedback control is presented by the same authors [Y5, Y6]. They show that the bridge vibration can be reduced appreciably and the flutter speed can be raised significantly. Abdel-Rohman et al. [A2] investigate the use of optimal control to structures of the type considered in [A1]. The same authors [A3] consider the design of a reduced-order observer for the structure investigated in [A1]. Balas [B9] discusses the applicability of control techniques developed for space structures to civil engineering structures. Chang and Soong [C7] propose to augment a tuned mass damper for reducing the vibration of a civil structure with an added optimal feedback control. The pole assignment method is used by Juang et al. [J15] to control the critical modes of civil structures. Controllability and observability and control and observation spillover are discussed. Meirovitch and Öz [M24] discuss the use of the IMSC method to control civil structures. Soong and Chang [S47] use modal control to establish a simple criterion of optimal design of the control matrix under energy constraint. Guidelines are provided for placing the actuators so as to minimize the energy. Soong and Skinner [S48] report results of an experimental investigation of active structural control in a wind tunnel. The control device used in the experiment is an aerodynamic appendage located at the top of the structure. Vilnay [V10] studies the characteristics of the modal participation matrix in modal control of structures. Abdel-Rohman [A4] attempts to identify an optimal control gain matrix among gain matrices determined by the pole allocation method. A method of open-loop optimal critical-mode control for tall buildings subjected to earthquake excitation is presented by Yang [Y7] and Yang and Lin [Y8, Y10], where the control is implemented by means of 1) an active mass damper and 2) active tendons. Abdel-Rohman and Leipholz [A5] demonstrate the advantages of controlling the vibration of tall structures by means of tendons, as opposed to using tuned mass dampers. Hrovat et al. [H20] propose to control wind-induced vibrations in tall buildings by means of a semi-active tuned mass damper using a small amount of external power to modulate the damping. Meirovitch and Silverberg [M33] discuss the use of optimal IMSC to control civil structures subjected to seismic excitation. Methodology for the active control of tall buildings subjected to strong wind gusts is presented by Yang and Samali [Y9]. They show that both active mass dampers and active tendons are capable of reducing the vibration significantly. Abdel-Rohman and Leipholz [A6] suggest the use of direct output velocity feedback to control a distributed structure. Samali et al. [S6] investigate the possible application of an active mass damper control system to tall buildings excited by strong wind turbulence. Samali et al. [S7] explore the possible application of both active tendon and active mass damper control to buildings excited by earthquakes. Carotti et al. [C5] suggest an active control system to protect a pipeline suspension bridge from excessive wind-induced vibration. Meirovitch and Ghosh [M45] discuss the use of the IMSC method to suppress the unstable flutter mode in a suspension bridge, while leaving the other modes unchanged. Chung et al. [C12] describe an experimental study concerning the possible application of active structural control by means of tendon using linear quadratic regulator theory. Pu and Hsu [P7] present an IMSC approach to the optimal linear discrete-time output feedback control of multistory civil structures. Civil structures are also considered by Roorda [R13], Masri et al. [M8, M9], Udwadia and Tabaie [U1, U2], Wang et al. [W4], Sirlin et al. [S33], Abdel-Rohman and Nayfeh [A7, A8], Reinhorn et al. [R6], Rodellar et al. [R8, R9] and Shinozuka et al. [S27].

9.5 SOUND RADIATION SUPPRESSION

Walker and Yaneske [W1, W2] discuss the use of feedback control to suppress the vibration of plates in the interest of damping the vibration of an acoustic barrier. Guicking and Karcher [G17] develop a system for active impedance control, generalizing the concept of active sound absorption. Vyalyshev et al. [V13] investigate the increase in the acoustic reduction achieved through the application of an auxiliary (control) force on a plate. The parameters of the auxiliary force are determined so as to achieve a reduction of the sound transmission coefficient in a given direction and of the acoustic power transmitted through the plate. Fuller and Jones [F6] investigate experimentally the performance of an active control system for different sensor and actuator locations. The object is to minimize the internal sound level in a closed cylindrical shell excited by an external acoustic source. Analytical and experimental studies of the use of active vibration control for reducing sound fields enclosed in elastic cylinders are presented by Jones and Fuller [J7, J8]. Meirovitch and Thangjitham [M57] are concerned with the problem of suppressing the acoustic radiation pressure generated by a vibrating structure. The approach is to control the modes of the structure most responsible for the radiation pressure. Fuller [F7] presents an analytical study of active control of sound radiation from vibrating plates by oscillating forces applied directly to the structure. Optimal control is achieved by minimizing a cost function proportional to the radiated acoustic power.

9.6 MANEUVERING OF SPACE STRUCTURES. ROBOTICS

Book et al. [B31, B32] consider modeling and control of flexible manipulator arms. The approach uses both a frequency-domain representation and a state-variable representation of the arm model. Alfriend and Longman [A12] use the concept of degree of controllability to choose actuator placement and suppress control spillover in the slewing of a flexible spacecraft. Breakwell [B34] presents a method for maneuvering a flexible spacecraft while leaving an arbitrary number of modes inactive at the end of the maneuver. Baruh and Silverberg [B20] describe a method for the simultaneous large angle maneuver and vibration suppression of flexible spacecraft. The control is divided into two parts, one carrying out the maneuver and the other suppressing the elastic motion by the independent modal-space control method. Judd and Falkenburg [J19] present the kinematic and dynamic analysis of an n-link robot with flexible members. Meckl and Seering [M15] explore ways of using open-loop control to eliminate residual vibration of a three-axis robotic manipulator at the end of a move. Vibration control is achieved using either bang-bang control for minimum-time response or a forcing function designed to avoid resonance throughout the move. Goldenberg and Rakhsha [G10] use the computed torque technique to control the rotary joint of a single-link flexible robot. The technique is implemented by calculating, for a given trajectory, the nominal torque from the rigid-body model of the robot. Skaar and Tucker [S35] treat the problem of maneuvering a flexible manipulator arm by an approach similar to that in Skaar [S34]. Chun et al. [C11] consider the problem of maneuvering a flexible spacecraft through large angles in finite time. The control problem consists of an open-loop nominal solution for the nonlinear rigid-body motion and a feedback control for the linearized flexible body response about several points along the rigid-body nominal solution. Gebler [G3] is concerned with the modeling and control of flexible industrial robots so that, for a given trajectory, the tip of the robot arm will not oscillate. Juang and Horta [J18] introduce air drag forces into the equations for the slewing control maneuver of a flexible structure to assess the effects of atmosphere on the controller design. Meirovitch and Quinn [M47] are concerned with the problem of slewing a flexible structure in space and suppressing the elastic vibration. A perturbation method permits a maneuver strategy independent of the vibration control. The same authors [M48] derive the equations of motion for maneuvering flexible spacecraft. A perturbation approach is presented in which the quantities defining the rigid-body maneuver are regarded as the unperturbed motion and the elastic motions and deviations from the rigid-body motions are regarded as the perturbed motion. Meirovitch and Sharony [M49] are concerned with simultaneous maneuver and vibration control of a flexible spacecraft. A perturbation approach is used, whereby the "rigid-body" slewing of the spacecraft is carried out in minimum time and the vibration control is carried out by a time-dependent linear quadratic regulator including integral feedback and prescribed convergence rate. Meirovitch and Kwak [M50] are concerned with the problem of reorienting the line of sight of a given number of flexible antennas in a spacecraft. The maneuver of the antennas is carried out according to a bang-bang control law, and the control of the elastic vibration and of the rigid-body motions caused by the maneuver is implemented by means of a proportional-plus-integral control law. Quinn and Meirovitch [Q1] are concerned with the maneuver and vibration control of the Spacecraft Control Laboratory Experiment (SCOLE). The approach is based on a perturbation method permitting a maneuver strategy independent of the vibration control. Actuator dynamics is considered both in [M47] and [Q1]. Meirovitch and Kwak [M52] use the developments of [M51] to derive general hybrid (ordinary and partial) differential equations of motion describing multi-body maneuvering in flexible spacecraft. Meirovitch and Sharony [M54] present a general review of the problem of flexible space structure maneuvering and control with emphasis on a perturbation technique. Meirovitch and Kwak [M55] are concerned with the dynamics and control of spacecraft consisting of a rigid platform and a given number of maneuvering antennas. The control strategy consists of stabilizing the platform relative to an inertial space and maneuvering the antennas relative to the platform, while suppressing the vibration of the antennas. Thompson et al. [T4] present a near-minimum time, open-loop optimal control formulation for slewing flexible structures. The control is saturation-bounded and is parametrized by two independent and arbitrary constants such that the control can be shaped to meet specified hardware constraints or mission requirements. The subject is also treated by Turner and Junkins [T6], Hale and Lisowski [H3–H5], Chun et al. [C10], Juang et al. [J17], Skaar et al. [S36], Turner and Chun [T7], Soga et al. [S46], Meirovitch and France [M53] and Singh et al. [S32].

9.7 CONTROL OF SPACE STRUCTURES AND LAUNCH VEHICLES

Fisher [F2] develops a linear-quadratic regulator to control one rigid-body and three bending modes of a flexible rocket booster. Swaim [S52] uses a state variable formulation to synthesize a control system for a launch vehicle with severe interaction between the rigid-body and the elastic modes. A control for a flexible launch vehicle to minimize a performance index is designed by Maki and Van de Vegte [M4] using a small number of sensors and constant feedback gains. The feedback gains and the sensor positions required for minimization of the performance index are determined by parameter optimization. Poelaert [P3] describes some of the problems involved in the control of a nonrigid spacecraft and proposes ways of approaching them. Ohkami and Likins [O1] present a method for determining the poles and zeros of the transfer function describing the attitude dynamics of a flexible spacecraft. The problem is reduced to that of finding the eigenvalues of matrices constructed by simple manipulations of the inertia and modal parameter matrices. Based on pole allocation, Tseng and Mahn [T5] propose a design approach permitting direct control over the closed-loop eigenvalues of a flexible spacecraft. Yocum and Slafer [Y14] discuss the problem of control system design in the presence of severe structural/control loop interactions by presenting the case history of the Orbiting Solar Observatory-8 despin control. Gran et al. [G12] are concerned with the optimal control of a space structure using rigid-body control only. Both continuous- and discrete-time control systems are presented. Meirovitch and Öz [M21] apply the theory developed in Meirovitch et al. [M20] to dual-spin flexible spacecraft. Balas and Ginter [B12] use optimal control to design a controller capable of stabilizing the attitude and suppressing the vibration of a flexible spacecraft. Reddy et al. [R4] consider the dynamics, stability and control of an orbiting flexible platform. Four different control techniques are discussed. Sesak et al. [S22] develop a filter-accommodated optimal control methodology to cope with unknown high-frequency vibration modes in space structures. A method for designing antenna-feed attitude control systems for deployable spaceborne antenna systems with flexible booms is proposed by Wang [W3]. The method is based on mechanical decoupling of the antenna-feed from the boom. Yedavalli and Skelton [Y12] and Yedavalli [Y13] present a method for the delineation of critical parameters of a space structure for a given performance objective in linear-quadratic-Gaussian regulators with uncertain parameters. A control design study of a space-deployable antenna is presented by Wang and Cameron [W5]. Three controller-estimator systems are designed and optimized based on requirements of focused communications missions, disturbances and hardware configurations. Goh and Caughey [G9] explore the theory of stiffness modification to the vibration suppression of a space structure. Stability is guaranteed by virtue of positive definite rate of energy decay. Kida et al. [K4] present a method for determining numerically poles and zeros of nonspinning flexible space structure consisting of one rigid part and elastic appendages. Joshi et al. [J12] investigate the application of the LQG/LTR procedure to the problem of synthesizing an attidude control system for a space structure. Sundararajan et al. [S51] use the LQG/LTR method to synthesize an attitude control system for a flexible space antenna. Preliminary results of pole-zero modeling and active control synthesis for a space structure are presented by Wie [W8]. He demonstrates the simplicity and practicality of a classical transfer-function approach enhanced by the nonminimum-phase filtering concept. Generic models of space structures are investigated by Wie and Bryson [W9] from the infinite discrete-spectrum viewpoint of distributed-parameter systems. Transfer functions of the various generic models are derived analytically and their pole-zeros patterns are investigated. Control of such structures is also considered by Gevarter [G4], Kuo et al. [K14], Siljak [S28], Larson and Likins [L1], Meirovitch et al. [M20], Canavin [C3], Skelton and Likins [S38], VanLandingham and Meirovitch [V4], Balas [B6, B7, B11, B14], Hughes and Abdel-Rahman [H23], Joshi and Groom [J9, J10], Juang and Rodriguez [J13], Lin et al. [L9–L11], Longman [L15], Sesak et al. [S18], Wu et al. [W13, W14], Juang and Balas [J14], Martin and Bryson [M6], Meirovitch and Öz [M23, M25], Sesak and Halstenberg [S21], Turner and Junkins [T6], Arbel and Gupta [A15, A16], Benhabib et al. [B24], Benhabib [B25], Breakwell [B34], Calico and Janiszewski [C2], Kosut et al. [K11, K12], Meirovitch and Baruh [M28], Velman [V8], Yam et al. [Y1, Y2], Sezer and Siljak [S23], Skelton et al. [S42], Baruh and Silverberg [B20], Cannon and Rosenthal [C4], Hale et al. [H3–H5], Turner and Chun [T7], West-Vukovich et al. [W6], Bhaya and Desoer [B28], Chun et al. [C10, C11], Goh and Caughey [G8], Lisowski and Hale [L14], Joshi [J11], Shenhar and Meirovitch [S26], Williams [W10, W11], Meirovitch and Quinn [M47, M48], Meirovitch and Sharony [M49, M54], Meirovitch and Kwak [M50, M52, M55], Meirovitch [M51], Meirovitch and France [M53], Sharony and Meirovitch [S24, S25], Singh et al. [S32] and Thompson et al. [T4].

9.8 MISCELLANEOUS APPLICATIONS

Klein and Hughes [K6] are concerned with the longitudinal torsion control in seagoing ships. Topics discussed include modeling, observability and state estimation. Van de Vegte and Hladun [V3] present optimal control techniques for design of passive systems, such as linear dampers, rotary dampers and dynamic absorbers, to minimize the vibration of slender uniform beams. Mote et al. [M63, M64] show analytically and experimentally that control of the axisymmetric temperature distribution in a circular saw can be an effective means of reducing blade vibration. Ellis and Mote [E3] present the analysis, design and testing of a feedback controller which increases the transverse stiffness and damping of a circular saw. Radcliffe and Mote [R1] discuss a method for the active control of the transverse displacement of a rotating disk. The approach is based on the identification of the dominant mode of the disk vibration and application of electromagnetic forces to suppress the amplitude of that mode. Ulsoy [U3] discusses the design of an active controller for a mechanical system considering the velocity of rotation and translation, as well as observation and control spillover. Sirlin et al. [S33] and Shinozuka et al. [S27] demonstrate experimentally that active control of floating structures can substantially reduce wave-induced motion at frequencies lower than the structure's natural frequency. Optimal control theory is used by Yoshimura et al. [Y15, Y16] to design an active suspension system to control the vertical and lateral vibration of a track vehicle system. Lewis and Allaire [L5] address the problem of using feedback control to limit vibration. A rotor model is considered in which a control force is imposed between the rotor and the base supporting the rotor bearings.

9.9 EXPERIMENTAL WORK AND FLIGHT TEST

Roger et al. [R10] describe the flight test of a flutter control system permitting an aircraft to fly faster than the flutter speed. Roorda [R13] presents several experiments in the feedback control of structures by such devices as electrohydraulic and electromagnetic actuators. Hallauer et al. [H8] describe experimental implementation of direct output velocity feedback control for the active vibration control of a structure. Schaechter [S10] and Schaechter and Eldred [S12] describe the analytical work performed in support of an experimental facility, the final design specifications, control law synthesis and some preliminary results. Breakwell and Chambers [B36] describe the hardware setup and experimental results of the TOYSAT structural control experiment. The experiment is used to test LQG optimal estimation and control algorithms. Aubrun et al. [A22] describe an experiment designed to examine problems in hardware, as opposed to evaluating control synthesis theories. Cannon and Rosenthal [C4] describe an experimental apparatus for investigating control laws for flexible spacecraft. The experiments are intended to demonstrate the difficulties associated with the control of space structures, particularly when the sensors and actuators are noncollocated. Skidmore et al. [S43, S44] present a theory of multiple-actuator excitation and/or control of individual vibration modes. Then, they describe an experiment in which the independent modal-space control method is applied to a beam-cable structure. Bailey and Hubbard [B1] discuss an active vibration damper for a cantilever beam using a distributed piezoelectric actuator. Crawley and De Luis [C14] discuss the use of distributed segmented piezoelectric actuators bonded to the structure so as to control the deformations of the structure. Hallauer et al. [H9] describe the experimental implementation of the independent modal-space control for the active damping of a plane grid. Schäfer and Holzach [S13] describe a hardware experiment designed to study the active vibration control (low-authority control) of a flexible beam, where the design methodology is based on direct output velocity feedback control. Soga et al. [S46] propose a control method for the slewing manuever of flexible spacecraft using open-loop control for the rigid-body mode and closed-loop feedback control for the elastic modes. The approach is demonstrated by means of an experiment using a single-axis air-bearing table. Juang et al. [J17] describe an experimental apparatus for the slewing control of flexible structures. Plump et al. [P2] describe a nonlinear active vibration damper using a spatially distributed piezoelectric actuator to control a cantilever beam. Dimitriadis and Fuller [D8] investigate the possibility of controlling sound transmission/radiation from a vibrating plate using as actuators piezoelectric elements bounded to the plate surface. Other experimental work is reported by Moon and Dowell [M62], Mote et al. [M63, M64], Ellis and Mote [E3], Ham et al. [H10, H11], Breakwell [B34], Soong and Skinner [S48], Radcliffe and Mote [R1], Meirovitch et al. [M38], Juang and Horta [J18], Redman-White et al. [R5], Chung et al. [C12] and Jones and Fuller [J8].

REFERENCES

A1. Abdel-Rohman, M. and Leipholz, H. H., "Structural Control by Pole Assignment Method." *ASCE Journal of the Engineering Mechanics Division*, Vol. 104, No. EM5, 1978, pp. 1159–1175.

A2. Abdel-Rohman, M., Quintana, V. H. and Leipholz, H. H., "Optimal Control of Civil Engineering Structures." *ASCE Journal of the Engineering Mechanics Division*, Vol. 106, No. EM1, 1980, pp. 57–73.

A3. Abdel-Rohman, M., Leipholz, H. H. E. and Quintana, V. H., "Design of Reduced-Order Observers for Structural Control Systems." In *Structural Control*, H. H. E. Leipholz (Editor), North-Holland Publishing Co., Amsterdam, 1980, pp. 57–77.

A4. Abdel-Rohman, M., "Active Control of Large Structures." *ASCE Journal of the Engineering Mechanics Division*, Vol. 108, No. EM5, 1982, pp. 719–730.

A5. Abdel-Rohman, M. and Leipholz, H. H., "Active Control of Tall Buildings." *ASCE Journal of Structural Engineering*, Vol. 109, No. 3, 1983, pp. 628–645.

A6. Abdel-Rohman, M. and Leipholz, H. H. E., "Optimal Feedback Control of Elastic, Distributed Parameter Structures." *Computers and Structures*, Vol. 19, No. 5/6, 1984, pp. 801–805.

A7. Abdel-Rohman, M. and Nayfeh, A. H., "Active Control of Nonlinear Oscillations in Bridges." *ASCE Journal of Engineering Mechanics*, vol. 113, No. 3, 1987, pp. 335–348.

A8. Abdel-Rohman, M. and Nayfeh, A. H., "Passive Control of Nonlinear Oscillations in Bridges." *ASCE Journal of Engineering Mechanics*, Vol. 113, No. 11, 1987, pp. 1694–1708.

A9. Abel, I., Perry, B., III and Murrow, H. N., "Two Synthesis Techniques Applied to Flutter Suppression on a Flight Research Wing." *Journal of Guidance and Control*, Vol. 1, No. 5, 1978, pp. 340–346.

A10. Akulenko, L. D. and Bolotnik, N. N., "On the Control of Systems with Elastic Elements." *Journal of Applied Mathematics and Mechanics* (PMM), Vol. 44, 1980, pp. 13–18.

A11. Akulenko, L. D., "Reduction of an Elastic System to a Prescribed State by Use of a Boundary Force." *Journal of Applied Mathematics and Mechanics* (PMM), Vol. 45, 1981, pp. 827–833.

A12. Alfriend, K. T. and Longman, R. W., "Rotational Maneuvers of Large Flexible Spacecraft." *Proceedings of the AAS Annual Rocky Mountain Guidance and Control Conference*, 1980, pp. 453–475.

A13. Andry, A. N., Jr., Shapiro, E. Y. and Sobel, K. M., "Modal Control and Vibrations." In *Frequency Domain and State Space Methods for Linear Systems*, C. I. Byrnes and A. Lindquist (Editors), North-Holland Publishing Co., Amsterdam, 1986, pp. 185–199.

A14. Arbel, A., "Controllability Measures and Actuator Placement in Oscillatory Systems." *International Journal of Control*. Vol. 33, No. 3, 1981, pp. 565–574.

A15. Arbel, A. and Gupta, N. K., "Robust Colocated Control for Large Flexible Space Structures." *Journal of Guidance and Control*, Vol. 4, No. 5, 1981, pp. 480–486.

A16. Arbel, A. and Gupta, N., "Robust Inverse Optimal Control for Space Structures." *Journal of Optimization Theory and Applications*, Vol. 35, No. 3, 1981, pp. 403–416.

A17. Ashkenazi, A. and Bryson, A. E., Jr., "Control Logic for Parameter Insensitivity and Disturbance Attenuation." *Journal of Guidance and Control*, Vol. 5, No. 4, 1982, pp. 383–388.

A18. Athans, M. and Falb, P. L., *Optimal Control*, McGraw-Hill, New York, 1966.

A19. Athans, M. "Toward a Practical Theory for Distributed Parameter Systems." *IEEE Transactions on Automatic Control*, Vol. AC-15, April 1970, pp. 245–247.

A20. Atluri, S. N. and Amos, A. K. (Editors), *Large Space Structures: Dynamics and Control*, Springer-Verlag, Berlin, 1988.

A21. Aubrun, J. N., "Theory of the Control of Structures by Low-Authority Controllers." *Journal of Guidance and Control*, Vol. 3, No. 5, 1980, pp. 444–451.

A22. Aubrun, J. N., Ratner, M. J. and Lyons, M.G., "Structural Control for a Circular Plate." *Journal of Guidance, Control, and Dynamics*, Vol. 7, No. 5, 1984, pp. 535–545.

B1. Bailey, T. and Hubbard, J. E., Jr., "Distributed Piezoelectric Polymer Active Vibration Control of a Cantilever Beam." *Journal of Guidance, Control, and Dynamics*, Vol. 8, No. 5, 1985, pp. 605–611.

B2. Balakrishnan, A. V., "Active Control of Airfoils in Unsteady Aerodynamics." *Applied Mathematics and Optimization*, Vol. 4, 1978, pp. 171–195.

B3. Balas, M. J., "Modal Control of Certain Flexible Dynamic Systems." *SIAM Journal on Control and Optimization*, Vol. 16, No. 3, 1978, pp. 450–462.

B4. Balas, M. J., "Active Control of Flexible Systems." *Journal of Optimization Theory and Applications*, Vol. 25, No. 3, 1978, pp. 415–436.

B5. Balas, M. J., "Feedback Control of Flexible Systems." *IEEE Transactions on Automatic Control*, Vol. AC-23, No. 4, 1978, pp. 673–679.

B6. Balas, M. J., "Direct Output Feedback Control of Large Space Structures." *Journal of Astronautical Sciences*, Vol. 27, No. 2, 1979, pp. 157–180.

B7. Balas, M. J., "Direct Velocity Feedback Control of Large Space Structures." *Journal of Guidance and Control*, Vol. 2, No. 3, 1979, pp. 252–253.

B8. Balas, M. J., "Disturbance-Accommodating Control of Distributed Parameter Systems: An Introduction." *Journal of Interdisciplinary Modeling and Simulation*, Vol. 3, No. 1, 1980, pp. 63–81.

B9. Balas, M. J., "Active Control of Large Civil Engineering Structures: A Naive Approach." In *Structural Control*, H. H. E. Leipholz (Editor), North-Holland Publishing Co., Amsterdam, 1980, pp. 107–126.

B10. Balas, M. J., "Enhanced Modal Control of Flexible Structures Via Innovations Feedthrough." *International Journal of Control*, Vol. 32, No. 6, 1980, pp. 983–1003.

B11. Balas, M. J., "Finite Element Models and Feedback Control of Flexible Aerospace Structures." *Proceedings of the Joint Automatic Control Conference*, San Francisco, California, 1980, Paper FP1-D.

B12. Balas, M. and Ginter, S., "Attitude Stabilization of Large Flexible Spacecraft." *Journal of Guidance and Control*, Vol. 4, No. 5, 1981, pp. 561–564.

B13. Balas, M. J., "Trends in Large Space Structure Control Theory: Fondest Hopes, Wildest Dreams." *IEEE Transactions on Automatic Control*, Vol. AC-27, No. 3, 1982, pp. 522–535.

B14. Balas, M. J., "Discrete-Time Stability of Continuous-Time Controller Designs for Large Space Structures." *Journal of Guidance and Control*, Vol. 5, No. 5, 1982, pp. 541–543.

B15. Balas, M. J., "The Structure of Discrete-Time Finite-Dimensional Control of Distributed Parameter Systems." *Journal of Mathematical Analysis and Applications*, Vol. 102, 1984, pp. 519–538.

B16. Balas, M. J., "Distributed Parameter Control of Nonlinear Flexible Structures with Linear Finite-Dimensional Controllers." *Journal of Mathematical Analysis and Applications*, Vol. 108, 1985, pp. 528–545.

B17. Balas, M. J., "Finite-Dimensional Control of Distributed Parameter Systems by Galerkin Approximation of Infinite Dimensional Controllers." *Journal of Mathematical Analysis and Applications*, Vol. 114, 1986, pp. 17–36.

B18. Balas, M. J., "Exponentially Stabilizing Finite-Dimensional Controllers for Linear Distributed Parameter Systems: Galerkin Approximation of Infinite Dimensional Controllers." *Journal of Mathematical Analysis and Applications*, Vol. 117, 1986, pp. 358–384.

B19. Barker, D. S. and Jacquot, R. G., "Spillover Minimization in the Control of Self-Adjoint Distributed Parameter Systems." *Journal of Astronautical Sciences*, Vol. 34, No. 2, 1986, pp. 133–146.

B20. Baruh, H. and Silverberg, L. M., "Maneuver of Distributed Spacecraft." *Proceedings of the AIAA Guidance and Control Conference*, 1984, pp. 637–647.

B21. Baruh, H., "On Duality Relations in the Control of Distributed Systems." *Proceedings of the Annual 19th Conference on Information Science and Systems*, 1985, pp. 623–629.

B22. Baruh, H. and Silverberg, L. M., "Robust Natural Control of Distributed Systems." *Journal of Guidance, Control, and Dynamics*, Vol. 8, No. 6, 1985, pp. 717–724.

B23. Bellman, R., *Dynamic Programming*, Princeton University Press, Princeton, New Jersey, 1957.

B24. Benhabib, R. J., Iwens, R. P. and Jackson, R. L., "Stability of Large Space Structure Control Systems Using Positivity Concepts." *Journal of Guidance and Control*, Vol. 4, No. 5, 1981, pp. 487–494.

B25. Benhabib, R. J., "Discrete Large Space Structure Control System Design Using Positivity." *Proceedings of the 20th IEEE Conference on Decision and Control*, 1981, pp. 715–724.

B26. Berkman, F. and Karnopp, D., "Complete Response of Distributed Systems Controlled by a Finite Number of Linear Feedback Loops." *ASME Journal of Engineering for Industry*, Vol. 91, November 1969, pp. 1063–1068.

B27. Bernstein, D. S. and Hyland, D. C., "The Optimal Projection/Maximum Entropy Approach to Designing Low-Order, Robust Controllers for Flexible Structures." *Proceedings of the 24th IEEE Conference on Decision and Control*, 1985, pp. 745–752.

B28. Bhaya, A. and Desoer, C. A., "On the Design of Large Flexible Space Structures (LFSS)." *IEEE Transactions on Automatic Control*, Vol. AC-30, No. 11, 1985, pp. 1118–1120.

B29. Blelloch, P. A. and Mingori, D. L., "Modified LTR Robust Control for Flexible Structures." *Proceedings of the AIAA Guidance, Navigation and Control Conference*, 1986, pp. 314–318.

B30. Bodden, D. S. and Junkins, J. L., "Eigenvalue Optimization Algorithms for Structure/Controller Design Iterations." *Journal of Guidance, Control, and Dynamics*, Vol. 8, No. 6, 1985, pp. 697–706.

B31. Book, W. J., Maizza-Neto, O. and Whitney, D. E., "Feedback Control of Two Beam, Two Joint Systems with Distributed Flexibility." *ASME Journal of Dynamic Systems, Measurement, and Control*, Vol. 97, December 1975, pp. 424–431.

B32. Book, W. J. and Majette, M., "Controller Design for Flexible, Distributed Parameter Mechanical Arms Via Combined State Space and Frequency Domain Techniques." *ASME Journal of Dynamic Systems, Measurement, and Control*, Vol. 105, December 1983, pp. 245–254.

B33. Bossi, J. A. and Price, G. A., "A Flexible Structure Controller Design Method Using Mode Residualization and Output Feedback." *Journal of Guidance, Control, and Dynamics*, Vol. 7, No. 1, 1984, pp. 125–127.

B34. Breakwell, J. A., "Optimal Feedback Slewing of Flexible Spacecraft." *Journal of Guidance and Control*, Vol. 4, No. 5, 1981, pp. 472–479.

B35. Breakwell, J. A., "Optimal Control of Distributed Systems." *Journal of Astronautical Sciences*, Vol. 29, No. 4, 1981, pp. 343–372.

B36. Breakwell, J. A. and Chambers, G. J., "The Toysat Structural Control Experiment." *Journal of Astronautical Sciences*, Vol. 31, No. 3, 1983, pp. 441–454.

B37. Brogan, W. L., *Modern Control Theory*, Quantum Publishers, New York, 1974.

B38. Bryson, A. E., Jr., "New Concepts in Control Theory, 1959–1984." *Journal of Guidance, Control, and Dynamics*, Vol. 8, No. 4, 1985, pp. 417–425.

B39. Burdess, J. S. and Metcalfe, A. V., "Active Control of Forced Harmonic Vibration in Finite Degree of Freedom Structures with Negligible Natural Damping." *Journal of Sound and Vibration*, Vol. 91, No. 3, 1983, pp. 447–459.

B40. Burdess, J. S., "The Control of Linear Multivariable Systems in the Presence of Harmonic Disturbances." *ASME Journal of Dynamic Systems, Measurement, and Control*, Vol. 105, March 1983, pp. 48–49.

B41. Burdess, J. S. and Metcalfe, A. V., "The Active Control of Forced Vibration Produced by Arbitrary Disturbances." *ASME Journal of Vibration, Acoustics, Stress, and Reliability in Design*, Vol. 107, January 1985, pp. 33–37.

C1. Caglayan, A. K., VanLandingham, H. F. and Sathe, S. G., "A Procedure for Optimal Control of Flexible Surface Vibrations. *Proceedings of the Second VPI&SU/AIAA Symposium on Dynamics and Control of Large Flexible Spacecraft*, L. Meirovitch (Editor), 1979, pp. 129–144.

C2. Calico, R. A. and Janiszewski, A. M., "Control of a Flexible Satellite via Elimination of Observation Spillover." *Proceedings of the Third VPI&SU/AIAA Symposium on Dynamics and Control of Large Flexible Spacecraft*, L. Meirovitch (Editor), 1981, pp. 15–33.

C3. Canavin, J. R., "The Control of Spacecraft Vibrations Using Multivariable Output Feedback." *AIAA/AAS Astrodynamics Conference*, Palo Alto, California, Aug. 7–9, 1978, Paper 78-1419.

C4. Cannon, R. H., Jr. and Rosenthal, D. E., "Experiments in Control of Flexible Structures with Noncolocated Sensors and Actuators." *Journal of Guidance, Control, and Dynamics*, Vol. 7, No. 5, 1984, pp. 546–553.

C5. Carotti, A., De Miranda, M. and Turci, E., "An Active Protection System for Wind Induced Vibrations on Pipe-Line Suspension Bridges." In *Structural Control*, H. H. E. Leipholz (Editor), Martinus Nijhoff Publishers, Dordrecht, The Netherlands, 1987, pp. 76–104.

C6. Champetier, C., "Impulse Control of Flexible Structures" (in French). *Proceedings of the Third IFAC Symposium on Control of Distributed Parameter Systems*, Toulouse, France, 1982, pp. V.14–V.19.

C7. Chang, J. C. H. and Soong, T. T., "Structural Control Using Active Tuned Mass Dampers." *ASCE Journal of the Engineering Mechanics Division*, Vol. 106, No. EM6, 1980, pp. 1091–1098.

C8. Chang, M. I. J. and Soong, T. T., "Optimal Controller Placement in Modal Control of Complex Systems." *Journal of Mathematical Analysis and Applications*, Vol. 75, 1980, pp. 340–358.

C9. Chang, S., "Modeling and Control of Aircraft Flutter Problem." *Proceedings of the 23rd IEEE Conference on Decision and Control*, 1984, pp. 1163–1168.

C10. Chun, H. M., Turner, J. D. and Juang, J. N., "Disturbance-Accommodating Tracking Maneuvers of Flexible Spacecraft." *Journal of Astronautical Sciences*, Vol. 33, No. 2, 1985, pp. 197–216.

C11. Chun, H. M., Turner, J. D. and Juang, J. N., "Frequency-Shaped Large-Angle Maneuvers." *AIAA 25th Aerospace Sciences Meeting*, Reno, Nevada, Jan. 12–15, 1987, Paper 87-0174.

C12. Chung, L. L., Reinhorn, A. M. and Soong, T. T., "Experiments on Active Control of Seismic Structures." *ASCE Journal of Engineering Mechanics*, Vol. 114, No. 2, 1988, pp. 241–256.

C13. Coradetti, T., "Orthogonal Subspace Reduction of Optimal Regulator Order." *Proceedings of the AIAA Guidance and Control Conference*, 1979, pp. 352–358.

C14. Crawley, E. F. and De Luis, J., "Use of Piezo-Ceramics as Distributed Actuators in Large Space Structures." *Proceedings of the AIAA/ASME/ASCE/AHS 26th Structures, Structural Dynamics, and Materials Conference*, 1985, pp. 126–133.

C15. Creedon, J. F. and Lindgren, A. G., "Control of the Optical Surface of a Thin, Deformable Primary Mirror with Application to an Orbiting Astronomical Observatory." *Automatica*, Vol. 6, 1970, pp. 643–660.

C16. Croopnick, S. R., Lin, Y. H. and Strunce, R. R., "A Survey of Automatic Control Techniques for Large Space Structures." *Proceedings of 8th IFAC Symposium on Automatic Control in Space*, 1979, pp. 275–284.

D1. Davison, E. J., "A Method for Simplifying Linear Dynamic Systems." *IEEE Transactions on Automatic Control*, Vol. AC-11, No. 1, 1966, pp. 93–101.

D2. Davison, E. J., "On Pole Assignment in Multivariable Linear Systems." *IEEE Transactions on Automatic Control*, Vol. AC-13, No. 6, 1968, pp. 747–748.

D3. Davison, E.J., "On Pole Assignment in Linear Systems with Incomplete State Feedback." *IEEE Transactions on Automatic Control*, Vol. AC-15, June 1970, pp. 348–351.

D4. Davison, E. J. and Goldenberg, A., "Robust Control of a General Servomechanism Problem: The Servo Compensator." *Automatica*, Vol. 11, 1975, pp. 461–471.

D5. D'Azzo, J. J. and Houpis, C. H., *Linear Control System Analysis and Design*, McGraw-Hill, New York, 1981.

D6. Dehghanyar, T. J., Masri, S. F., Miller, R. K., Bekey, G. A. and Caughey, T. K., "An Analytical and Experimental Study into the Stability and Control of Nonlinear Flexible Structures." *Proceedings of the Fourth VPI&SU/AIAA Symposium on Dynamics and Control of Large Structures*, L. Meirovitch (Editor), 1983, pp. 291–310.

D7. de Silva, C. W., "An Algorithm for the Optimal Design of Passive Vibration Controllers for Flexible Systems." *Journal of Sound and Vibration*, Vol. 74, No. 4, 1981, pp. 495–502.

D8. Dimitriadis, E. K. and Fuller, C. R., "Investigation on Active Control of Sound Transmission Through Elastic Plates Using Piezoelectric Actuators."

AIAA 27th Aerospace Sciences Conference, Reno, Nevada, Jan. 8–12, 1989, Paper 89-106.

D9. Dodds, S. J. and Williamson, S. E., "A Signed Switching Time Bang-Bang Attitude Control Law for Fine Pointing of Flexible Spacecraft." *International Journal of Control*, Vol. 40, No. 4, 1984, pp. 795–81.

D10. Dorato, P. and Levis, A. H., "Optimal Linear Regulators: The Discrete Time Case." *IEEE Transactions on Automatic Control*, Vol. AC-16, No. 6, 1971, pp. 613–620.

D11. Doyle, J. C., "Robustness of Multiloop Linear Feedback Systems." *Proceedings of the 17th IEEE Decision and Control Conference*, 1978, pp. 12–18.

D12. Doyle, J. C. and Stein, G., "Robustness with Observers." *IEEE Transactions on Automatic Control*, Vol. AC-24, No. 4, 1979, pp. 607–611.

D13. Doyle, J. C. and Stein, G., "Multivariable Feedback Design: Concepts for a Classical/Modern Synthesis." *IEEE Transactions on Automatic Control*, Vol. AC-26, No. 1, 1981, pp. 4–16.

E1. Edwards, J. W., Breakwell, J. V. and Bryson, A. E., Jr., "Active Flutter Control Using Generalized Unsteady Aerodynamic Theory." *Journal of Guidance and Control*, Vol. 1, No. 1, 1978, pp. 32–40.

E2. Elliott, L. E., Mingori, D. L. and Iwens, R. P., "Performance of Robust Output Feedback Controller for Flexible Spacecraft." *Proceedings of the Second VPI&SU/AIAA Symposium on Dynamics and Control of Large Flexible Spacecraft*, L. Meirovitch (Editor), 1979, pp. 409–420.

E3. Ellis, R. W. and Mote, C. D., Jr., "A Feedback Vibration Controller for Circular Saws." *ASME Journal of Dynamic Systems, Measurement, and Control*, Vol. 101, March 1979, pp. 44–49.

F1. Falb, P. L. and Wolovich, W. A., "Decoupling in the Design and Synthesis of Multivariable Control Systems." *IEEE Transactions on Automatic Control*, Vol. AC-12, No. 6, 1967, pp. 651–659.

F2. Fisher, E. E., "An Application of the Quadratic Penalty Function Criterion to the Determination of a Linear Control for a Flexible Vehicle." *AIAA Journal*, Vol. 3, No. 7, 1965, pp. 1262–1267.

F3. Franklin, G. F. and Powell, J. D., *Digital Control of Dynamic Systems*, Addison-Wesley, Reading, Massachusetts. 1980.

F4. Freymann, R., "Interactions Between an Aircraft Structure and Active Control Systems." *Journal of Guidance, Control, and Dynamics*, Vol. 10, No. 5, 1987, pp. 447–452.

F5. Friedland, B., Richman, J. and Williams, D. E., "On the 'Adiabatic Approximation' for Design of Control Laws for Linear, Time-Varying Systems." *IEEE Transactions on Automatic Control*, Vol. AC-32, No. 1, 1987, pp. 62–63.

F6. Fuller, C. R. and Jones, J. D., "Influence of Sensor and Actuator Location on the Performance of Active Control Systems." *ASME Winter Annual Meeting*, Boston, Massachusetts, Dec. 13–18, 1987, Paper 87-WA/NCA-9.

F7. Fuller, C. R., "Active Control of Sound Transmission/Radiation from Elastic Plates by Vibration Inputs. I. Analysis." *Journal of Sound and Vibration* (submitted for publication).

G1. Garrard, W. L., Mahesh, J. K., Stone, C. R. and Dunn, H. J., "Robust

Kalman Filter Design for Active Flutter Suppression Systems." *Journal of Guidance and Control*, Vol. 5, No. 4, 1982, pp. 412–414.

G2. Garrard, W. L. and Liebst, B. S., "Active Flutter Suppression Using Eigenvalue and Linear Quadratic Design Techniques." *Journal of Guidance, Control, and Dynamics*, Vol. 8, No. 3, 1985, pp. 304–311.

G3. Gebler, B., "Modeling and Control of an Industrial Robot with Elastic Components." In *Structural Control*, H. H. E. Leipholz (Editor), Martinus Nijhoff Publishers, Dordrecht, The Netherlands, 1987, pp. 221–234.

G4. Gevarter, W. B., "Basic Relations for Control of Flexible Vehicles." *AIAA Journal*, Vol. 8, No. 4, 1970, pp. 666–672.

G5. Gibson, J. S., "Convergence and Stability in Linear Modal Regulation of Flexible Structures." *Proceedings of the Second VPI&SU/AIAA Symposium on Dynamics and Control of Large Flexible Spacecraft*, L. Meirovitch (Editor), 1979, pp. 51–64.

G6. Gibson, J. S., "A Note on Stabilization of Infinite Dimensional Linear Oscillators by Compact Linear Feedback." *SIAM Journal on Control and Optimization*, Vol. 18, No. 3, 1980, pp. 311-316.

G7. Gibson, J. S., "An Analysis of Optimal Modal Regulation: Convergence and Stability." *SIAM Journal on Control and Optimization*, Vol. 19, No. 5, 1981, pp. 686–707.

G8. Goh, C. J. and Caughey, T. K., "On the Stability Problem Caused by Finite Actuator Dynamics in the Collocated Control of Large Space Structures." *International Journal of Control*, Vol. 41, No. 3, 1985, pp. 787–802.

G9. Goh, C. J. and Caughey, T. K., "A Quasi-Linear Vibration Suppression Technique for Large Space Structures Via Stiffness Modification." *International Journal of Control*, Vol. 41, No. 3, 1985, pp. 803–811.

G10. Goldenberg, A. A. and Rakhsha, F., "Feedforward Control of a Single-Link Flexible Robot." *Mechanism and Machine Theory*, Vol. 21, No. 4, 1986, pp. 325–335.

G11. Gould, L. A. and Murray-Lasso, M. A., "On the Modal Control of Distributed Systems with Distributed Feedback." *IEEE Transactions on Automatic Control*, Vol. AC-11, No. 4, 1966, pp. 729–737.

G12. Gran, R., Rossi, M. and Moyer, H. G., "Optimal Digital Control of Large Space Structures." *Journal of Astronautical Sciences*, Vol. 27, No. 2, 1979, pp. 115–130.

G13. Gran, R. and Rossi, M., "Closed Loop Order Reduction for Large Structures Control." *Proceedings of the Second VPI&SU/AIAA Symposium on Dynamics and Control of Large Flexible Spacecraft*, L. Meirovitch (Editor), 1979, pp. 443–457.

G14. Gran, R., "Finite-Dimensional Controllers for Hyperbolic Systems." *Proceedings of the Third VPI&SU/AIAA Symposium on Dynamics and Control of Large Flexible Spacecraft*," L. Meirovitch (Editor), 1981, pp. 301–317.

G15. Greene, M., "Robustness of Active Modal Damping of Large Flexible Structures." *International Journal of Control*, Vol. 46, No. 6, 1987, pp. 1009–1018.

G16. Gruzen, A. and Vander Velde, W. E., "Robust Reduced-Order Control of Flexible Structures Using the Optimal Projection/Maximum Entropy Design Methodology." *Proceedings of the AIAA Guidance, Navigation and Control Conference*, 1986, pp. 319–327.

G17. Guicking, D. and Karcher, K., "Active Impedance Control for One-Dimensional Sound." *ASME Journal of Vibration, Acoustics, Stress, and Reliability in Design*, Vol. 106, July 1984, pp. 393–396.

G18. Guicking, D., *Active Noise and Vibration Control Reference Bibliography*, 3rd ed. University of Göttingen, West Germany, 1988.

G19. Gupta, N. K., "Frequency-Shaped Cost Functionals: Extension of Linear-Quadratic-Gaussian Design Methods." *Journal of Guidance and Control*, Vol. 3, No. 6, 1980, pp. 529–535.

G20. Gupta, N. K. and Du Val, R. W., "A New Approach for Active Control of Rotorcraft Vibration." *Journal of Guidance and Control*, Vol. 5, No. 2, 1982, pp. 143–150.

H1. Hagedorn, P., "On a New Concept of Active Vibration Damping of Elastic Structures." In *Structural Control*, H. H. E. Leipholz (Editor), Martinus Nijhoff Publishers, Dordrecht, The Netherlands, 1987, pp. 261–277.

H2. Hale, A. L. and Rahn, G. A., "Robust Control of Self-Adjoint Distributed-Parameter Structures." *Journal of Guidance, Control, and Dynamics*, Vol. 7, No. 3, 1984, pp. 265–273.

H3. Hale, A. L. and Lisowski, R. J., "Reduced-Order Modeling Applied to Optimal Design of Maneuvering Flexible Structures." *Proceedings of the American Control Conference*, San Diego, California, 1984, pp. 1685–1690.

H4. Hale, A. L., Lisowski, R. J. and Dahl, W. E., "Optimal Simultaneous Structural and Control Design of Maneuvering Flexible Spacecraft." *Journal of Guidance, Control, and Dynamics*, Vol. 8, No. 1, 1985, pp. 86–93.

H5. Hale, A. L. and Lisowski, R. J., "Characteristic Elastic Systems of Time-Limited Optimal Maneuvers." *Journal of Guidance, Control, and Dynamics*, Vol. 8, No. 5, 1985, pp. 628–636.

H6. Hallauer, W. L., Jr. and Barthelemy, J. F. M., "Active Damping of Modal Vibrations by Force Apportioning." *Proceedings of the AIAA/ASME/ASCE/AHS 21st Structures, Structural Dynamics, and Materials Conference*, 1980, pp. 863–873.

H7. Hallauer, W. L., Jr. and Barthelemy, J. F. M., "Sensitivity of Modal-Space Control to Nonideal Conditions." *Journal of Guidance and Control*, Vol. 4, No. 5, 1981, pp. 564–566.

H8. Hallauer, W. L., Jr., Skidmore, G. R. and Mesquita, L. C., "Experimental-Theoretical Study of Active Vibration Control." *Proceedings of the First International Modal Analysis Conference*, Orlando, Florida, 1982, pp. 39–45.

H9. Hallauer, W. L., Jr., Skidmore, G. R. and Gehling, R. N., "Modal-Space Active Damping of a Plane Grid: Experiment and Theory." *Journal of Guidance, Control, and Dynamics*, Vol. 8, No. 3, 1985, pp. 366–373.

H10. Ham, N. D., "A Simple System for Helicopter Individual-Blade-Control Using Modal Decomposition." *Vertica*, Vol. 4, 1980, pp. 23–28.

H11. Ham, N. D., Behal, B. L. and McKillip, R. M., Jr., "Hellicopter Rotor Lag Damping Augmentation through Individual-Blade-Control." *Vertica*, Vol. 7, No. 4, 1983, pp. 361–371.

H12. Hashemipour, H. R. and Laub, A. J., "Kalman Filtering for Second-Order Models." *Journal of Guidance, Control, and Dynamics*, Vol. 11, No. 2, 1988, pp. 181–186.

H13. Hefner, R. D. and Mingori, D. L., "Robust Controller Design Using Frequency Domain Constraints." *Journal of Guidance, Control, and Dynamics*, Vol. 10, No. 2, 1987, pp. 158–165.

H14. Hegg, D. R., "Extensions of Suboptimal Output Feedback Control with Application to Large Space Structures." *Journal of Guidance and Control*, Vol. 4, No. 6, 1981, pp. 637–641.

H15. Herrick, D. C., "Application of a Root Locus Technique to the Design of Compensated Feedback Controllers for a Resonant Plant." *Proceedings of the Second VPI&SU/AIAA Symposium on Dynamics and Control of Large Flexible Spacecraft*, L. Meirovitch (Editor), 1979, pp. 161–175.

H16. Herrick, D. C., "Application of a Root Locus Technique to Structural Control." *Proceedings of the AAS/AIAA Astrodynamics Conference*, 1979, pp. 675–686.

H17. Horner, G. C., "Optimum Damping Locations for Structural Vibrations Control." *Proceedings of the AIAA/ASME/ASCE/AHS 23rd Structures, Structural Dynamics, and Materials Conference*, 1982, pp. 29–34.

H18. Horner, G. C., "Optimum Actuator Placement, Gain, and Number for a Two-Dimensional Grillage." *Proceedings of the AIAA/ASME/ASCE/AHS 24th Structures, Structural Dynamics, and Materials Conference*, 1983, Part 2, pp. 179–184.

H19. Horta, L. G., Juang, J. N. and Junkins, J. L., "A Sequential Linear Optimization Approach for Controller Design." *Journal of Guidance, Control, and Dynamics*, Vol. 9, No. 6, 1986, pp. 699–703.

H20. Hrovat, D., Barak, P. and Rabins, M., "Semi-Active versus Passive or Active Tuned Mass Dampers for Structural Control." *ASCE Journal of Engineering Mechanics*, Vol. 109, No. 3, 1983, pp. 691–705.

H21. Hu, A., Skelton, R. E. and Yang, T. Y., "Modeling and Control of Beam-Like Structures." *Journal of Sound and Vibration*, Vol. 117, No. 3, 1987, pp. 475–496.

H22. Huebner, K. H. and Thornton, E. A., *The Finite Element Method for Engineers*, 2nd ed., Wiley, New York, 1982.

H23. Hughes, P. C. and Abdel-Rahman, T. M., "Stability of Proportional-Plus-Derivative-Plus-Integral Control of Flexible Spacecraft." *Journal of Guidance and Control*, Vol. 2, No. 6, 1979, pp. 499–503.

H24. Hughes, P. C. and Skelton, R. E., "Controllability and Observability of Linear Matrix-Second-Order Systems." *ASME Journal of Applied Mechanics*, Vol. 47, June 1980, pp. 415–420.

H25. Hughes, P. C. and Skelton, R. E., "Controllability and Observability for Flexible Spacecraft." *Journal of Guidance and Control*, Vol. 3, No. 5, 1980, pp. 452–459.

H26. Hughes, P. C. and Skelton, R. E., "Modal Truncation for Flexible Spacecraft." *Journal of Guidance and Control*, Vol. 4, No. 3, 1981, pp. 291–297.

H27. Hurty, W. C., "Vibrations of Structural Systems by Component-Mode Synthesis." *ASCE Journal of the Engineering Mechanics Division*, Vol. 86, August 1960, pp. 51–69.

H28. Hwang, C. and Pi, W. S., "Optimal Control Applied to Aircraft Flutter Suppression." *Journal of Guidance, Control, and Dynamics*, Vol. 7, No. 3, 1984, pp. 347–354.

H29. Hyland, D. C., "Active Control of Large Flexible Spacecraft: A New Design Approach Based on Minimum Information Modelling of Parameter Uncertainties." *Proceedings of the Third VPI&SU/AIAA Symposium on Dynamics and Control of Large Flexible Spacecraft*, L. Meirovitch (Editor), 1981, pp. 631–646.

H30. Hyland, D. C. and Bernstein, D. S., "The Optimal Projection Equations for Fixed-Order Dynamic Compensation." *IEEE Transactions on Automatic Control*, Vol. AC-29, No. 11, 1984, pp. 1034–1037.

I1. Inman, D. J., "Modal Decoupling Conditions for Distributed Control for Flexible Structures." *Journal of Guidance, Control, and Dynamics*, Vol. 7, No. 6, 1984, pp. 750–752.

I2. Iwens, R. P., Benhabib, R. J. and Jackson, R. L., "A Unified Approach to the Design of Large Space Structure Control Systems." *Proceedings of the Joint Automatic Control Conference*, San Francisco, California, 1980, Paper FP1-A.

J1. Johnson, C. D., "Theory of Disturbance-Accommodating Controllers." In *Control and Dynamic Systems*, C. T. Leondes (Editor), Academic Press, New York, 1976, Vol. 12, pp. 387–489.

J2. Johnson, C. D., "Disturbance-Accommodating Control; An Overview of the Subject." *Journal of Interdisciplinary Modeling and Simulation*, Vol. 3, No. 1, 1980, pp. 1–29.

J3. Johnson, C. D., "A Discrete-Time Disturbance-Accommodating Control Theory for Digital Control of Dynamical Systems." In *Control and Dynamic Systems*, C. T. Leondes (Editor), Academic Press, New York, 1982, Vol. 18, pp. 223–315.

J4. Johnson, T. L., Athans, M. and Skelton, G. B., "Optimal Control-Surface Locations for Flexible Aircraft." *IEEE Transactions on Automatic Control*, Vol. AC-16, No. 4, 1971, pp. 320–331.

J5. Johnson, T. L. and Lin, J. G., "An Aggregation Method for Active Control of Large Space Structures." *Proceedings of the 18th IEEE Conference on Decision and Control*, 1979, pp. 1–3.

J6. Johnson, T. L., "Principles of Sensor and Actuator Location in Distributed Systems." *Proceedings of an International Symposium on Engineering Science and Mechanics*, Tainan, Taiwan, Dec. 29–31, 1981, pp. 1–14.

J7. Jones, J. D. and Fuller, C. R., "Reduction of Interior Sound Fields in Flexible Cylinders by Active Vibration Control." *Proceedings of the Sixth International Modal Analysis Conference*, London, United Kingdom, 1988, pp. 315–327.

J8 Jones, J. D. and Fuller, C. R., "Active Control of Structurally-Coupled Sound Fields in Elastic Cylinders by Vibrational Force Inputs." *Proceedings of the Seventh International Modal Analysis Conference*, Las Vegas, Nevada, 1989.

J9 Joshi, S. M. and Groom, N. J., "Stability Bounds for the Control of Large Space Structures." *Journal of Guidance and Control*, Vol. 2, No. 4, 1979, pp. 349–351.

J10 Joshi, S. M. and Groom, N. J., "Optimal Member Damper Controller Design for Large Space Structures." *Journal of Guidance and Control*, Vol. 3, No. 4, 1980, pp. 378–380.

J11 Joshi, S. M., "Robustness Properties of Collocated Controllers for Flexible Spacecraft." *Journal of Guidance, Control and Dynamics*, Vol. 9, No. 1, 1986, pp. 85–91.

J12 Joshi, S. M., Armstrong, E. S. and Sundararajan, N., "Application of LQG/LTR Technique to Robust Controller Synthesis for a Large Flexible Space Antenna." *NASA Technical Publication 2560*, September 1986, 61p.

J13 Juang, J. N. and Rodriguez, G., "Formulations and Applications of Large Structure Actuator and Sensor Placements." *Proceedings of the Second VPI&SU/AIAA Symposium on Dynamics and Control of Large Flexible Spacecraft*, L. Meirovitch (Editor), 1979, pp. 247–262.

J14 Juang, J. N. and Balas, M. J., "Dynamics and Control of Large Spinning Spacecraft." *Journal of Astronautical Sciences*, Vol. 28, No. 1, 1980, pp. 31–48.

J15 Juang, J. N., Sae-Ung. S. and Yang, J. N., "Active Control of Large Building Structures." In *Structural Control*, H. H. E. Leipholz (Editor), North-Holland Publishing Co., Amsterdam, 1980, pp. 663–676.

J16 Juang, J. N. and Dwyer, T. A. W., III, "First Order Solution of the Optimal Regulator Problem for a Distributed Parameter Elastic System." *Journal of Astronautical Sciences*, Vol. 31, No. 3, 1983, pp. 429–439.

J17 Juang, J. N., Horta, L. G. and Robertshaw, H. H., "A Slewing Control Experiment for Flexible Structures." *Journal of Guidance, Control, and Dynamics*, Vol. 9, No. 5, 1986, pp. 599–607.

J18 Juang, J. N. and Horta, L. G., "Effects of Atmosphere on Slewing Control of a Flexible Structure." *Journal of Guidance, Control, and Dynamics*, Vol. 10, No. 4, 1987, pp. 387–392.

J19 Judd, R. P. and Falkenburg, D. R., "Dynamics of Nonrigid Articulated Robot Linkages." *IEEE Transactions on Automatic Control*, Vol. AC-30, No. 5, 1985, pp. 499–502.

K1 Kabamba, P. T. and Longman, R. W., "An Integrated Approach to Optimal Reduced Order Control Theory." *Proceedings of the Third VPI&SU/AIAA Symposium on Dynamics and Control of Large Flexible Spacecraft*, L. Meirovitch (Editor), 1981, pp. 571–585.

K2 Khot, N. S., Venkayya, V. B. and Eastep, F. E., "Optimal Structural Modifications to Enhance the Active Vibration Control of Flexible Structures." *AIAA Journal*, Vol. 24, No. 8, 1986, pp. 1368–1374.

K3 Kida, T., Okamoto, O. and Ohkami, Y., "On Modern Modal Controller for Flexible Space Structures: A Sensitivity Analysis." *Proceedings of an Interna-*

tional Symposium on Engineering Science and Mechanics, Tainan, Taiwan, Dec. 29–31, 1981, pp. 968–978.

K4. Kida, T., Ohkami, Y. and Sambongi, S., "Poles and Transmission Zeros of Flexible Spacecraft Control Systems." *Journal of Guidance, Control, and Dynamics*, Vol. 8, No. 2, 1985, pp. 208–213.

K5. Kirk, D. E., *Optimal Control Theory*, Prentice-Hall, Englewood Cliffs, New Jersey, 1970.

K6. Klein, R. E. and Hughes, R. O., "The Distributed Parameter Control of Torsional Bending in Seagoing Ships." *Proceedings of the Joint Automatic Control Conference*, St. Louis, Missouri, 1971, pp. 867–875.

K7. Knowles, G., "Some Problems in the Control of Distributed Systems and Their Numerical Solution." *SIAM Journal on Control and Optimization*, Vol. 17, No. 1, 1979, pp. 5–22.

K8. Köhne, M., "The Control of Vibrating Elastic Systems." In *Distributed Parameter Systems-Identification, Estimation, and Control*, W. H. Ray and D. G. Lainiotis (Editors), Marcel Dekker, New York, 1978, pp. 387–456.

K9. Komkov, V., *Optimal Control Theory for the Damping of Vibrations of Simple Elastic Systems*, Lecture Notes in Mathematics, No. 253. Springer-Verlag, New York, 1972.

K10. Kosut, R. L., "The Determination of the System Transfer Function Matrix for Flight Control Systems." *IEEE Transactions on Automatic Control*, Vol. AC-13, April 1968, p. 214.

K11. Kosut, R. L. and Salzwedel, H., "Stability and Robustness of Control Systems for Large Space Structures." *Proceedings of the Third VPI&SU/AIAA Symposium on Dynamics and Control of Large Flexible Spacecraft*, L. Meirovitch (Editor), 1981, pp. 343–364.

K12. Kosut, R. L., Salzwedel, H. and Emami-Naeini, A., "Robust Control of Flexible Spacecraft." *Journal of Guidance, Control, and Dynamics*, Vol. 6, No. 2, 1983, pp. 104–111.

K13. Krabs, W., "On Boundary Controllability of One-Dimensional Vibrating Systems." *Mathematical Methods in the Applied Sciences*, Vol. 1, 1979, pp. 322–345.

K14. Kuo, B. C., Seltzer, S. M., Singh, G. and Yackel, R. A., "Design of a Digital Controller for Spinning Flexible Spacecraft." *Journal of Spacecraft and Rockets*, Vol. 11, No. 8, 1974, pp. 584–589.

K15. Kuo, B. C., "Design of Digital Control Systems with State Feedback and Dynamic Output Feedback." *Journal of Astronautical Sciences*, Vol. 27, No. 2, 1979, pp. 207–214.

K16. Kwakernaak, H. and Sivan, R., *Linear Optimal Control Systems*, Wiley-Interscience, New York, 1972.

L1. Larson, V. and Likins, P. W., "Optimal Estimation and Control of Elastic Spacecraft." In *Control and Dynamic Systems*, C. T. Leondes (Editor), Academic Press, New York, 1977, pp. 285–322.

L2. Lehtomaki, N. A., Sandell, N. R., Jr. and Athans, M., "Robustness Results in Linear-Quadratic Gaussian Based Multivariable Control Designs." *IEEE Transactions on Automatic Control*, Vol. AC-26, No. 1, 1981, pp. 75–92.

L3. Leondes, C. T. and Novak, L. M., "Reduced-Order Observers for Linear Discrete Time Systems." *IEEE Transactions on Automatic Control*, Vol. AC-19, February 1974, pp. 42–46.

L4. Levine, W. S. and Athans, M., "On the Determination of the Optimal Constant Output Feedback Gains for Linear Multivariable Systems." *IEEE Transactions on Automatic Control*, Vol. AC-15, No. 1, 1970, pp. 44–48.

L5. Lewis, D. W. and Allaire, P. E., "Rotor to Base Control of Rotating Machinery to Minimize Transmitted Force." In *Structural Control*, H. H. E. Leipholz (Editor), Martinus Nijhoff Publishers, Dordrecht, The Netherlands, 1987, pp. 408–425.

L6. Liebst, B. S., Garrard, W. L. and Adams, W. M., "Design of an Active Flutter Suppression System." *Journal of Guidance, Control, and Dynamics*, Vol. 9, No. 1, 1986, pp. 64–71.

L7. Liebst, B. S., Garrard, W. L. and Farm, J. A., "Design of a Multivariable Flutter Suppression/Gust Load Alleviation System." *Journal of Guidance, Control, and Dynamics*, Vol. 11, No. 3, 1988, pp. 220–229.

L8. Lim, K. B. and Junkins, J. L., "Robustness Optimization of Structural and Controller Parameters." *Journal of Guidance, Control, and Dynamics*, Vol. 12, No. 1, 1989, pp. 89–96.

L9. Lin, J. G., Hegg, D. R., Lin, Y. H. and Keat, J. E., "Output Feedback Control of Large Space Structures: An Investigation of Four Design Methods." *Proceedings of the Second VPI&SU/AIAA Symposium on Dynamics and Control of Large Flexible Spacecraft*, L. Meirovitch (Editor), 1979, pp. 1–18.

L10. Lin, J. G., "Three Steps to Alleviate Control and Observation Spillover Problems of Large Space Structures." *Proceedings of the 19th IEEE Conference on Decision and Control*, 1980, pp. 438–444.

L11. Lin, Y. H. and Lin, J. G., "An Improved Output Feedback Control of Flexible Large Space Structures." *Proceedings of the 19th IEEE Conference on Decision and Control*, 1980, pp. 1248–1250.

L12. Lindberg, R. E., Jr. and Longman, R. W., "On the Number and Placement of Actuators for Independent Modal Space Control." *Journal of Guidance, Control, and Dynamics*, Vol. 7, No. 2, 1984, pp. 215–221.

L13. Lions, J.-L., "Remarks on the Theory of Optimal Control of Distributed Systems." In *Control Theory of Systems Governed by Partial Differential Equations*, A. K. Aziz, J. W. Wingate and M. J. Balas (Editors), Academic Press, New York, 1977, pp. 1–103.

L14. Lisowski, R. J. and Hale, A. L., "Optimal Design for Single Axis Rotational Maneuvers of a Flexible Structure." *Journal of Astronautical Sciences*, Vol. 33, No. 2, 1985, pp. 179–196.

L15. Longman, R. W., "Annihilation or Suppression of Control and Observation Spillover in the Optimal Shape Control of Flexible Spacecraft." *Journal of Astronautical Sciences*, Vol. 27, No. 4, 1979, pp. 381–399.

L16. Luenberger, D. G., "An Introduction to Observers." *IEEE Transactions on Automatic Control*, Vol. AC-16, No. 6, 1971, pp. 596–602.

L17. Lutze, F. H. and Goff, R. M. A., "Application of Classical Techniques to

Control of Continuous Systems." *Proceedings of the Third VPI&SU/AIAA Symposium on Dynamics and Control of Large Flexible Spacecraft*, L. Meirovitch (Editor), 1981, pp. 119–136.

L18. Luzzato, E., "Active Protection of Domains of a Vibrating Structure by Using Optimal Control Theory: A Model Determination." *Journal of Sound and Vibration*, Vol. 91, No. 2, 1983, pp. 161–180.

L19. Luzzato, E. and Jean, M., "Protection of Continuous Structures Against Vibrations by Active Damping." *ASME Journal of Vibration, Acoustics, Stress, and Reliability in Design*, Vol. 105, July 1983, pp. 374–381.

L20. Luzzato, E. and Jean, M., "Mechanical Analysis of Active Vibration Damping in Continuous Structures." *Journal of Sound and Vibration*, Vol. 86, No. 4, 1983, pp. 455–473.

M1. Mace, B. R., "Active Control of Flexural Vibrations." *Journal of Sound and Vibration*, Vol. 114, April 1987, pp. 253–270.

M2. MacFarlane, A. G. J. and Kouvaritakis, B., "A Design Technique for Linear Multivariable Feedback Systems." *International Journal of Control*, Vol. 25, No. 6, 1977, pp. 837–874.

M3. Mahesh, J. K., Stone, C. R., Garrard, W. L. and Dunn, H. J., "Control Law Synthesis for Flutter Suppression Using Linear Quadratic Gaussian Theory." *Journal of Guidance and Control*, Vol. 4, No. 4, 1981, pp. 415–422.

M4. Maki, M. C. and Van de Vegte, J., "Optimal and Constrained-Optimal Control of a Flexible Launch Vehicle." *AIAA Journal*, Vol. 10, No. 6, 1972, pp. 796–799.

M5. Martin, C. R. and Soong, T. T., "Modal Control of Multistory Structures." *ASCE Journal of the Engineering Mechanics Division*, Vol. 102, No. EM4, 1976, pp. 613–623.

M6. Martin, G. D. and Bryson, A. E., Jr., "Attitude Control of a Flexible Spacecraft." *Journal of Guidance and Control*, Vol. 3, No. 1, 1980, pp. 37–41.

M7. Martin, J.-C. E., "Optimal Allocation of Actuators for Distributed-Parameter Systems." *ASME Journal of Dynamic Systems, Measurement, and Control*, Vol. 100, February 1978, pp. 227–228.

M8. Masri, S. F., Bekey, G. A. and Caughey, T. K., "Optimum Pulse Control of Flexible Structures." *ASME Journal of Applied Mechanics*, Vol. 48, September 1981, pp. 619–626.

M9. Masri, S. F., Bekey, G. A. and Caughey, T. K., "On-Line Control of Nonlinear Flexible Structures." *ASME Journal of Applied Mechanics*, Vol. 49, December 1982, pp. 877–884.

M10. McClamroch, N. H., "Sampled Data Control of Flexible Structures Using Constant Gain Velocity Feedback." *Journal of Guidance, Control, and Dynamics*, Vol. 7, No. 6, 1984, pp. 747–749.

M11. McClamroch, N. H., "Displacement Control of Flexible Structures Using Electrohydraulic Servo-Actuators." *ASME Journal of Dynamic Systems, Measurement, and Control*, Vol. 107, March 1985, pp. 34–39.

M12. McLaren, M. D. and Slater, G. L., "Robust Multivariable Control of Large Space Structures Using Positivity." *Journal of Guidance, Control, and Dynamics*, Vol. 10, No. 4, 1987, pp. 393–400.

M13. McLean, D., "Effects of Reduced Order Mathematical Models on Dynamic Response of Flexible Aircraft with Closed-Loop Control." In *Structural Control*, H. H. E. Leipholz (Editor), North-Holland Publishing Co., Amsterdam, 1980, pp. 493–504.

M14. McNamara, R. J., "Tuned Mass Dampers for Buildings." *ASCE Journal of the Structural Division*, Vol. 103, No. ST9, 1977, pp. 1785–1798.

M15. Meckl, P. and Seering, W., "Active Damping in a Three-Axis Robotic Manipulator." *ASME Journal of Vibration, Acoustics, Stress, and Reliability in Design*, Vol. 107, January 1985, pp. 38–46.,

M16. Meirovitch, L., *Analytical Methods in Vibrations*, Macmillan, New York, 1967.

M17. Meirovitch, L., *Methods of Analytical Dynamics*, McGraw-Hill, New York, 1970.

M18. Meirovitch, L., "A New Method of Solution of the Eigenvalue Problem for Gyroscopic Systems." *AIAA Journal*, Vol. 12, No. 10, 1974, pp. 1337–1342.

M19. Meirovitch, L., "A Modal Analysis for the Response of Gyroscopic Systems." *Journal of Applied Mechanics*, Vol. 42, No. 2, 1975, pp. 446–450.

M20. Meirovitch, L., Van Landingham, H. F. and Öz, H., "Control of Spinning Flexible Spacecraft by Modal Synthesis." *Acta Astronautica*, Vol. 4, No. 9–10, 1977, pp. 985–1010.

M21. Meirovitch, L. and Öz, H., "Observer Modal Control of Dual-Spin Flexible Spacecraft." *Journal of Guidance and Control*, Vol. 2, No. 2, 1979, pp. 101–110.

M22. Meirovitch, L. and Öz, H., "An Assessment of Methods for the Control of Large Space Structures." *Proceedings of the Joint Automatic Control Conference*, Denver, Colorado, 1979, pp. 34–41.

M23. Meirovitch, L. and Öz, H., "Modal-Space Control of Distributed Gyroscopic Systems." *Journal of Guidance and Control*, Vol. 3, No. 2, 1980, pp. 140–150.

M24. Meirovitch, L. and Öz, H., "Active Control of Structures by Modal Synthesis." In *Structural Control*, H. H. E. Leipholz (Editor), North-Holland Publishing Co., Amsterdam, 1980, pp. 505–521.

M25. Meirovitch, L. and Öz, H., "Modal-Space Control of Large Flexible Spacecraft Possessing Ignorable Coordinates." *Journal of Guidance and Control*, Vol. 3, No. 6, 1980, pp. 569–577.

M26. Meirovitch, L., *Computational Methods in Structural Dynamics*, Sijthoff & Noordhoff, The Netherlands, 1980.

M27. Meirovitch, L. and Baruh, H., "Effect of Damping on Observation Spillover Instability." *Journal of Optimization Theory and Applications*, Vol. 35, No. 1, 1981, pp. 31–44.

M28. Meirovitch, L. and Baruh, H., "Optimal Control of Damped Flexible Gyroscopic Systems." *Journal of Guidance and Control*, Vol. 4, No. 2, 1981, pp. 157–163.

M29. Meirovitch, L. and Hale, A. L., "On the Substructure Synthesis Method." *AIAA Journal*, Vol. 19, No. 7, 1981, pp. 940–947.

M30. Meirovitch, L. and Baruh, H., "Control of Self-Adjoint Distributed-Parameter Systems." *Journal of Guidance and Control*, Vol. 5, No. 1, 1982, pp. 60–66.

M31. Meirovitch, L. and Baruh, H., "Robustness of the Independent Modal-Space Control Method." *Journal of Guidance, Control, and Dynamics*, Vol. 6, No. 1, 1983, pp. 20–25.

M32. Meirovitch, L. and Baruh, H., "On the Problem of Observation Spillover in Self-Adjoint Distributed-Parameter Systems." *Journal of Optimization Theory and Applications*, Vol. 39, No. 2, 1983, pp. 269–291.

M33. Meirovitch, L. and Silverberg, L. M., "Control of Structures Subjected to Seismic Excitation." *ASCE Journal of Engineering Mechanics*, Vol. 109, No. 2, 1983, pp. 604–618.

M34. Meirovitch, L., Baruh, H. and Öz, H., "A Comparison of Control Techniques for Large Flexible Systems." *Journal of Guidance, Control, and Dynamics*. Vol. 6, No. 4, 1983, pp. 302–310.

M35. Meirovitch, L. and Silverberg, L. M., "Globally Optimal Control of Self-Adjoint Distributed Systems." *Optimal Control Applications and Methods*, Vol. 4, 1983, pp. 365–386.

M36. Meirovitch, L., "Modeling and Control of Distributed Structures." In *Proceedings of the Workshop on Applications of Distributed Systems Theory to the Control of Large Space Structures*, G. Rodriguez (Editor), JPL Publication 83-46, 1983, pp. 1–30.

M37. Meirovitch, L. and Öz, H., "Digital Stochastic Control of Distributed-Parameter Systems." *Journal of Optimization Theory and Applications*, Vol. 43, No. 2, 1984, pp. 307–325.

M38. Meirovitch, L., Baruh, H., Montgomery, R. C. and Williams, J. P., "Nonlinear Natural Control of an Experimental Beam." *Journal of Guidance, Control, and Dynamics*, Vol. 7, No. 4, 1984, pp. 437–442.

M39. Meirovitch, L. and Silverberg, L. M., "Active Vibration Suppression of a Cantilever Wing." *Journal of Sound and Vibration*, Vol. 97, No. 3, 1984, pp. 489–498.

M40. Meirovitch, L. and Silverberg, L. M., "Control of Non-Self-Adjoint Distributed-Parameter Systems." *Journal of Optimization Theory and Applications*, Vol. 47, No. 1, 1985, pp. 77–90.

M41. Meirovitch, L. and Baruh, H., "The Implementation of Modal Filters for Control of Structures." *Journal of Guidance, Control, and Dynamics*, Vol. 8, No. 6, 1985, pp. 707–716.

M42. Meirovitch, L., *Introduction to Dynamics and Control*, Wiley, New York, 1985.

M43. Meirovitch, L. and Bennighof, J. K., "Modal Control of Traveling Waves in Flexible Structures." *Journal of Sound and Vibration*, Vol. 111, No. 1, 1986, pp. 131–144.

M44. Meirovitch, L., *Elements of Vibration Analysis*, 2nd ed. McGraw-Hill, New York, 1986.

M45. Meirovitch, L. and Ghosh, D., "Control of Flutter in Bridges." *ASCE Journal of Engineering Mechanics*, Vol. 113, No. 5, 1987, pp. 720–736.

M46. Meirovitch, L., "Some Problems Associated with the Control of Distributed Structures." *Journal of Optimization Theory and Applications*, Vol. 54, No. 1, 1987, pp. 1–21.

M47. Meirovitch, L. and Quinn, R. D., "Maneuvering and Vibration Control of

Flexible Spacecraft." *Journal of Astronautical Sciences*, Vol. 35, No. 3, 1987, pp. 301–328.

M48. Meirovitch, L. and Quinn, R. D., "Equations of Motion for Maneuvering Flexible Spacecraft." *Journal of Guidance, Control, and Dynamics*, Vol. 10, No. 5, 1987, pp. 453–465.

M49. Meirovitch, L. and Sharony, Y., "Optimal Vibration Control of a Flexible Spacecraft During a Minimum-Time Maneuver." *Proceedings of the Sixth VPI&SU/AIAA Symposium on Dynamics and Control of Large Structures*, L. Meirovitch (Editor), 1987, pp. 576–601.

M50. Meirovitch, L. and Kwak, M. K., "Control of Spacecraft with Multi-Targeted Flexible Antennas." *AIAA/AAS Astrodynamics Conference*, Minneapolis, Minnesota, Aug. 15–17, 1988, Paper 88-4268-CP.

M51. Meirovitch, L., "State Equations of Motion for Flexible Bodies in Terms of Quasi-Coordinates." In *Dynamics of Controlled Mechanical Systems*, G. Schweitzer and M. Mansour (Editors), Springer-Verlag, Berlin, 1989, pp. 37–48.

M52. Meirovitch, L. and Kwak, M. K., "State Equations for a Spacecraft with Flexible Appendages in Terms of Quasi-Coordinates." *Applied Mechanics Reviews*, Vol. 42, No. 11, Part 2, 1989, pp. 5161–5170.

M53. Meirovitch, L. and France, M. E. B., "Discrete-Time Control of a Spacecraft with Retargetable Flexible Antennas." *12th Annual AAS Guidance and Control Conference*, Keystone, Colorado, Feb. 4–8, 1989, Paper 89-007.

M54. Meirovitch, L. and Sharony, Y., "A Perturbation Approach to the Maneuvering and Control of Space Structures." In *Advances in Control and Dynamic Systems*, C. T. Leondes (Editor), Academic Press, San Diego California, (to be published).

M55. Meirovitch, L. and Kwak, M. K., "Dynamics and Control of Spacecraft with Retargeting Flexible Antennas." *Journal of Guidance, Control, and Dynamics* (to be published).

M56. Meirovitch, L. and Kwak, M. K., "On the Convergence of the Classical Rayleigh-Ritz Method and the Finite Element Method." *AIAA Journal* (to be published).

M57. Meirovitch, L. and Thangjitham, S., "Active Control of Sound Radiation Pressure." *ASME Journal of Vibration and Acoustics* (to be published).

M58. Mesquita, L. and Kamat, M. P., "Structural Optimization for Control of Stiffened Laminated Composite Structures." *Journal of Sound and Vibration*, Vol. 116, No. 1, 1987, pp. 33–48.

M59. Miller, D. F. and Shim, J., "Gradient-Based Combined Structural and Control Optimization." *Journal of Guidance, Control, and Dynamics*, Vol. 10, No. 3, 1987, pp. 291–298.

M60. Miller, D. W., von Flotow, A. H. and Hall, S. R., "Active Modification of Wave Reflection and Transmission in Flexible Structures." *Proceedings of the American Control Conference*, Minneapolis, Minnesota, 1987, pp. 1318–1324.

M61. Miller, R. K., Masri, S. F., Dehghanyar, T. J. and Caughey, T. K., "Active

Vibration Control of Large Civil Structures." *ASCE Journal of Engineering Mechanics*, Vol. 114, No. 9, 1988, pp. 1542–1570.

M62. Moon, F. C. and Dowell, E. H., "The Control of Flutter Instability in a Continuous Elastic System Using Feedback." *Proceedings of the AIAA/ASME 11th Structures, Structural Dynamics, and Materials Conference*, 1970, pp. 48–65.

M63. Mote, C. D., Jr. and Holoyen, S., "Feedback Control of Saw Blade Temperature with Induction Heating." *ASME Journal of Engineering for Industry*, Vol. 100, May 1978, pp. 119–126.

M64. Mote, C. D., Jr., Schajer, G. S. and Holoyen, S., "Circular Saw Vibration Control by Induction of Thermal Membrane Stresses." *ASME Journal of Engineering for Industry*, Vol. 103, February 1981, pp. 81–89.

M65. Mukhopadhyay, V., Newsom, J. R. and Abel, I., "Reduced-Order Optimal Feedback Control Law Synthesis for Flutter Suppression." *Journal of Guidance and Control*, Vol. 5, No. 4, 1982, pp. 389–395.

N1. Newsom, J. R., "Control Law Synthesis for Active Flutter Suppression Using Optimal Control Theory." *Journal of Guidance and Control*, Vol. 2, No. 5, 1979, pp. 388–394.

N2. Newsom, J. R., Adams, N. M., Jr., Mukhopadhyay, V., Tiffany, S. H. and Abel, I., "Active Controls: A Look at Analytical Methods and Associated Tools." *Proceedings of the 14th Congress of the International Council of the Aeronautical Sciences*, B. Laschka and R. Staufenbiel (Editors), 1984, pp. 230–242.

N3. Newsom, J. R., Pototzky, A. S. and Abel, I., "Design of a Flutter Suppression System for an Experimental Drone Aircraft." *Journal of Aircraft*, Vol. 22, No. 5, 1985, pp. 380–386.

N4. Nissim, E., "Flutter Suppression Using Active Controls Based on the Concept of Aerodynamic Energy." *NASA Technical Note D-6199*, March 1971, 112p.

N5. Nissim, E. and Lottati, I., "Active Control for Flutter Suppression and Gust Alleviation in Supersonic Aircraft." *Journal of Guidance and Control*, Vol. 3, No. 4, 1980, pp. 345–351.

N6. Noll, T. E. and Huttsell, L. J., "Wing Store Active Flutter Suppression – Correlation of Analysis and Wind-Tunnel Data." *Journal of Aircraft*, Vol. 16, No. 7, 1979, pp. 491–497.

N7. Noll, T. E., Eastep, F. E. and Calico, R. A., "Prevention of Forward Swept Wing Aeroelastic Instabilities with Active Controls." *Proceedings of the 14th Congress of the International Council of the Aeronautical Sciences*, B. Laschka and R. Staufenbiel (Editors), 1984, pp. 439–448.

N8. Nurre, D. S., Ryan, R. S., Scofield, H. N. and Sims, J. L., "Dynamics and Control of Large Space Structures." *Journal of Guidance, Control, and Dynamics*, Vol. 7, No. 5, 1984, pp. 514–526.

O1. Ohkami, Y. and Likins, P. W., "Determination of Poles and Zeros of Transfer Functions for Flexible Spacecraft Attitude Control." *International Journal of Control*, Vol. 24, No. 1, 1976, pp. 13–22.

O2. Onoda, J. and Haftka, R. T., "An Approach to Structure/Control Simultaneous Optimization for Large Flexible Spacecraft." *AIAA Journal*, Vol. 25, No. 8, 1987, pp. 1133–1138.

O3. Öz, H. and Meirovitch, L., "Optimal Modal-Space Control of Flexible Gyroscopic Systems." *Journal of Guidance and Control*, Vol. 3, No. 3, 1980, pp. 218–226.

O4. Öz, H., Meirovitch, L. and Johnson, C. R., Jr., "Some Problems Associated with Digital Control of Dynamical Systems." *Journal of Guidance and Control*, Vol. 3, No. 6, 1980, pp. 523–528.

O5. Öz, H. and Meirovitch, L., "Stochastic Independent Modal-Space Control of Distributed-Parameter Systems." *Journal of Optimization Theory and Applications*, Vol. 40, No. 1, 1983, pp. 121–154.

P1. Perkins, W. R. and Cruz, J. B., Jr., "Feedback Properties of Linear Regulators." *IEEE Transactions on Automatic Control*, Vol. AC-16, No. 6, 1971, pp. 659–664.

P2. Plump, J. M., Hubbard, J. E., Jr. and Bailey, T., "Nonlinear Control of a Distributed System: Simulation and Experimental Results." *ASME Journal of Dynamic Systems, Measurement, and Control*, Vol. 109, June 1987, pp. 133–139.

P3. Poelaert, D. H. I., "A Guideline for the Analysis and Synthesis of a Nonrigid-Spacecraft Control System." *ESA/ASE Scientific and Technical Review*, Vol. 1, 1975, pp. 203–218.

P4. Pontryagin, L. S., Boltyanskii, V., Gamkrelidze, R. and Mishchenko, E., *The Mathematical Theory of Optimal Processes*, Interscience, New York, 1962.

P5. Porter, B. and Crossley, R., *Modal Control, Theory and Applications*, Taylor & Francis, London, 1972.

P6. Potter, J. E., "Matrix Quadratic Solutions." *SIAM Journal of Applied Mathematics*, Vol. 14, No. 3, 1966, pp. 496–501.

P7. Pu, J. P. and Hsu, D. S., "Optimal Control of Tall Buildings." *ASCE Journal of Engineering Mechanics*, Vol. 114, No. 6, 1988, pp. 973–989.

Q1. Quinn, R. D. and Meirovitch, L., "Maneuver and Vibration Control of SCOLE." *Journal of Guidance, Control, and Dynamics*, Vol. 11, No. 6, 1988, pp. 542–553.

R1. Radcliffe, C. J. and Mote, C. D., Jr., "Identification and Control of Rotating Disk Vibration." *ASME Journal of Dynamic Systems, Measurement, and Control*, Vol. 105, March 1983, pp. 39–45.

R2. Rajaram, S., "Optimal Independent Modal Space Control of a Flexible System Including Integral Feedback." *Proceedings of an International Symposium on Engineering Science and Mechanics*, Tainan, Taiwan, Dec. 29–31, 1981, pp. 1031–1041.

R3. Rappaport, D. and Silverman, L. M., "Structure and Stability of Discrete-Time Optimal Systems." *IEEE Transactions on Automatic Control*, Vol. AC-16, No. 3, 1971, pp. 227–233.

R4. Reddy, A. S. S. R., Bainum, P. M., Krishna, R. and Hamer, H. A., "Control of a Large Flexible Platform in Orbit." *Journal of Guidance and Control*, Vol. 4, No. 6, 1981, pp. 642–649.

R5. Redman-White, W., Nelson, P. A. and Curtis, A. R. D., "Experiments on the

Active Control of Flexural Wave Power Flow." *Journal of Sound and Vibration*, Vol. 112, No. 1, 1987, pp. 187–191.
R6. Reinhorn, A. M., Manolis, G. D. and Wen, C. Y., "Active Control of Inelastic Structures." *ASCE Journal of Engineering Mechanics*, Vol. 113, No. 3, 1987, pp. 315–333.
R7. Robinson, A. C., "A Survey of Optimal Control of Distributed-Parameter Systems." *Automatica*, Vol. 7, 1971, pp. 371–388.
R8. Rodellar, J. and Martin-Sanchez, J., "Predictive Structural Control." In *Structural Control*, H. H. E. Leipholz (Editor), Martinus Nijhoff Publishers, Dordrecht, The Netherlands, 1987, pp. 580–593.
R9. Rodellar, J., Barbat, A. H. and Martin-Sanchez, J. M., "Predictive Control of Structures." *ASCE Journal of Engineering Mechanics*, Vol. 113, No. 6, 1987, pp. 797–812.
R10. Roger, K. L., Hodges, G. E. and Felt, L. R., "Active Flutter Suppression – A Flight Test Demonstration." *Journal of Aircraft*, Vol. 12, No. 6, 1975, pp. 551–556.
R11. Rogers, R. O. and Sworder, D. D., "Suboptimal Control of Linear Systems Derived from Models of Lower Dimension." *AIAA Journal*, Vol. 9, No. 8, 1971, pp. 1461–1467.
R12. Roorda, J., "Tendon Control in Tall Structures." *ASCE Journal of the Structural Division*, Vol. 101, No. ST3, 1975, pp. 505–521.
R13. Roorda, J., "Experiments in Feedback Control of Structures." *Solid Mechanics Archives*, Vol. 5, Issue 2, 1980, pp. 131–163.
R14. Rosenbrock, H. H., *Computer-Aided Control System Design*, Academic Press, New York, 1974.
S1. Sadek, I. S. and Adali, S., "Control of the Dynamic Response of a Damped Membrane by Distributed Forces." *Journal of Sound and Vibration*, Vol. 96, No. 3, 1984, pp. 391–406.
S2. Sadek, I. S., Sloss, J. M., Bruch, J. C., Jr. and Adali, S., "Optimal Control of a Timoshenko Beam by Distributed Forces." *Journal of Optimization Theory and Applications*, Vol. 50, No. 3, 1986, pp. 451–461.
S3. Sadek, I. S., "Approximate Methods for Multiobjective Distributed Control of a Vibrating Beam." In *Structural Control*, H. H. E. Leipholz (Editor), Martinus Nijhoff Publishers, Dordrecht, The Netherlands, 1987, pp. 594–611.
S4. Sadek, I. S., Adali, S., Sloss, J. M. and Bruch, J. C., Jr., "Optimal Distributed Control of a Continuous Beam with Damping." *Journal of Sound and Vibration*, Vol. 117, No. 2, 1987, pp. 207–218.
S5. Sae-Ung, S. and Yao, J. T. P., "Active Control of Building Structures." *ASCE Journal of the Engineering Mechanics Division*, Vol. 104, No. EM2, 1978, pp. 335–350.
S6. Samali, B., Yang, J. N. and Yeh, C. T., "Control of Lateral-Torsional Motion of Wind-Excited Buildings." *ASCE Journal of Engineering Mechanics*, Vol. 111, No. 6, 1985, pp. 777–796.
S7. Samali, B., Yang, J. N. and Liu, S. C., "Active Control of Seismic-Excited Buildings." *ASCE Journal of Structural Engineering*, Vol. 111, No. 10, 1985, pp. 2165–2180.

S8. Sandell, N. R., Jr., Varaiya, P., Athans, M. and Safonov, M. G., "Survey of Decentralized Control Methods for Large Scale Systems." *IEEE Transactions on Automatic Control*, Vol. AC-23, No. 2, 1978, pp. 108–128.

S9. Schaechter, D. B., "Optimal Local Control of Flexible Structures." *Journal of Guidance and Control*, Vol. 4, No. 1, 1981, pp. 22–26.

S10. Schaechter, D. B., "Hardware Demonstration of Flexible Beam Control." *Journal of Guidance and Control*, Vol. 5, No. 1, 1982, pp. 48–53.

S11. Schaechter, D. B., "Closed-Loop Control Performance Sensitivity to Parameter Variations." *Journal of Guidance, Control, and Dynamics*, Vol. 6, No. 5, 1983, pp. 399–402.

S12. Schaechter, D. B. and Eldred, D. B., "Experimental Demonstration of the Control of Flexible Structures." *Journal of Guidance, Control, and Dynamics*, Vol. 7, No. 5, 1984, pp. 527–534.

S13. Schäfer, B. E. and Holzach, H., "Experimental Research on Flexible Beam Model Control." *Journal of Guidance, Control, and Dynamics*, Vol. 8, No. 5, 1985, pp. 597–604.

S14. Schmidt, D. K. and Chen, T. K., "Frequency Domain Synthesis of a Robust Flutter Suppression Control Law." *Journal of Guidance, Control, and Dynamics*, Vol. 9, No. 3, 1986, pp. 346–351.

S15. Schulz, G., "Active Multivariable Vibration Isolation for a Helicopter." *Automatica*, Vol. 15, 1979, pp. 461–466.

S16. Schulz, G. and Heimbold, G., "Dislocated Actuator/Sensor Positioning and Feedback Design for Flexible Structures." *Journal of Guidance, Control, and Dynamics*, Vol. 6, No. 5, 1983, pp. 361–367.

S17. Sensburg, O., Becker, J. and Hönlinger, H., "Active Control of Flutter and Vibration of an Aircraft." In *Structural Control*, H. H. E. Leipholz (Editor), North-Holland Publishing Co., Amsterdam, 1980, pp. 693–722.

S18. Sesak, J. R., Likins, P. and Coradetti, T., "Flexible Spacecraft Control by Model Error Sensitivity Suppression." *Journal of Astronautical Sciences*, Vol. 27, No. 2, 1979, pp. 131–156.

S19. Sesak, J. R. and Likins, P., "Model Error Sensitivity Suppression: Quasi-Static Optimal Control for Flexible Structures." *Proceedings of the 18th IEEE Conference on Decision and Control*, 1979, pp. 207–213.

S20. Sesak, J. R., "Suppressed Mode Damping for Model Error Sensitivity Suppression of Flexible Spacecraft Controllers." *Proceedings of the AIAA Guidance and Control Conference*, 1980, pp. 27–32.

S21. Sesak, J. R. and Halstenberg, R. V., "Decentralized Elastic Body and Rigid Body Control by Modal Error Sensitivity Suppression." *Proceedings of the Joint Automatic Control Conference*, San Francisco, California, 1980, Paper FA1-D.

S22. Sesak, J. R., Halstenberg, R. V., Chang, Y. and Davis, M. M., "Filter-Accommodated Optimal Control of Large Flexible Space Systems." *Proceedings of the AIAA Guidance and Control Conference*, 1981, pp. 177–186.

S23. Sezer, M. E. and Siljak, D. D., "Validation of Reduced-Order Models for Control System Design." *Journal of Guidance and Control*, Vol. 5, No. 5, 1982, pp. 430–437.

S24. Sharony, Y. and Meirovitch, L., "Accommodation of Kinematic Disturbances During a Minimum-Time Maneuver of a Flexible Spacecraft." *AIAA/AAS Astrodynamics Conference*, Minneapolis, Minnesota, Aug. 15–17, 1988, Paper 88-4253-CP.

S25. Sharony, Y. and Meirovitch, L., "A Perturbation Approach to the Minimum-Time Slewing of a Flexible Spacecraft." *IEEE International Conference on Control and Applications*, Jerusalem, Israel, Apr. 3–6, 1989.

S26. Shenhar, L. and Meirovitch, L., "Minimum-Fuel Control of High-Order Systems." *Journal of Optimization Theory and Applications*, Vol. 48, No. 3, 1986, pp. 469–491.

S27. Shinozuka, M., Samaras, E. and Paliou, C. "Active Control of Floating Structures." In *Structural Control*, H. H. E. Leipholz (Editor), Martinus Nijhoff Publishers, Dordrecht, The Netherlands, 1987, pp. 651–668.

S28. Siljak, D. D., "Multilevel Stabilization of Large-Scale Systems: A Spinning Flexible Spacecraft." *Proceedings of the Sixth IFAC Congress*, Boston, Massachusetts, 1975, pp. 309–320.

S29. Siljak, D. D., *Large-Scale Dynamic Systems: Stability and Structure*, North-Holland Publishing Co., New York, 1978.

S30. Silverberg, L. M., "Uniform Damping Control of Spacecraft." *Journal of Guidance, Control, and Dynamics*, Vol. 9, No. 2, 1986, pp. 221–227.

S31. Simon, J. D. and Mitter, S. K., "A Theory of Modal Control." *Information and Control*, Vol. 13, 1968, pp. 316–353.

S32. Singh, G., Kabamba, P. T. and McClamroch, N. H., "Planar, Time-Optimal, Rest-To-Rest Slewing Maneuvers of Flexible Spacecraft." *Journal of Guidance, Control, and Dynamics*, Vol. 12, No. 1, 1989, pp. 71–81.

S33. Sirlin, S. W., Paliou, C., Longman, R. W., Shinozuka, M. and Samaras, E., "Active Control of Floating Structures." *ASCE Journal of Engineering Mechanics*, Vol. 112, No. 9, 1986, pp. 947–965.

S34. Skaar, S. B., "Closed Form Optimal Control Solutions for Continuous Linear Elastic Systems." *Journal of Astronautical Sciences*, Vol. 32, No. 4, 1984, pp. 447–461.

S35. Skaar, S. B. and Tucker, D., "Point Control of a One-Link Flexible Manipulator." *Journal of Applied Mechanics*, Vol. 53, March 1986, pp. 23–27.

S36. Skaar, S. B., Tang, L. and Yalda-Mooshabad, I., "On-Off Attitude Control of Flexible Satellites." *Journal of Guidance, Control, and Dynamics*, Vol. 9, No. 4, 1986, pp. 507–510.

S37. Skelton, R. E. and Likins, P. W., "On the Use of Model Error Systems in the Control of Large Scale Linearized Systems." *Proceedings of the IFAC Symposium on Large Scale Systems, Theory and Applications*, Udine, Italy, June 1976, pp. 641–650.

S38. Skelton, R. E. and Likins, P. W., "Orthogonal Filters for Model Error Compensation in the Control of Nonrigid Spacecraft." *Journal of Guidance and Control*, Vol. 1, No. 1, 1978, pp. 41–49.

S39. Skelton, R. E., "On Cost-Sensitivity Controller Design Method for Uncertain Dynamic Systems." *Journal of Astronautical Sciences*, Vol. 27, No. 2, 1979, pp. 181–205.

S40. Skelton, R. E., "Observability Measures and Performance Sensitivity in the Model Reduction Problem." *International Journal of Control*, Vol. 29, No. 4, 1979, pp. 541–556.

S41. Skelton, R. E. and Hughes, P. C., "Modal Cost Analysis for Linear Matrix-Second-Order Systems." *ASME Journal of Dynamic Systems, Measurement, and Control*, Vol. 102, September 1980, pp. 151–158.

S42. Skelton, R. E., Hughes, P. C. and Hablani, H. B., "Order Reduction for Models of Space Structures Using Modal Cost Analysis." *Journal of Guidance and Control*, Vol. 5, No. 4, 1982, pp. 351–357.

S43. Skidmore, G. R., Hallauer, W. L., Jr. and Gehling, R. N., "Experimental-Theoretical Study of Modal-Space Control." *Proceedings of the Second International Modal Analysis Conference*, Orlando, Florida, 1984, pp. 66–74.

S44. Skidmore, G. R. and Hallauer, W. L., Jr., "Modal-Space Active Damping of a Beam-Cable Structure: Theory and Experiment." *Journal of Sound and Vibration*, Vol. 101, No. 2, 1985, pp. 149–160.

S45. Smith, R. E. and Lum, E. L. S., "Linear Optimal Theory Applied to Active Structural Bending Control." *Journal of Aircraft*, Vol. 5, No. 5, 1968, pp. 479–485.

S46. Soga, H., Hirako, K., Ohkami, Y., Kida, T., and Yamaguchi, I., "Large-Angle Slewing of Flexible Spacecraft." *Joint AAS/Japanese Rocket Society Symposium on Space Exploitation and Utilization*, Honolulu, Hawaii, Dec. 15–19, 1985, Paper AAS 85-671.

S47. Soong, T. T. and Chang, M. I. J., "On Optimal Control Configuration in Theory of Modal Control." In *Structural Control*, H. H. E. Leipholz (Editor), North-Holland Publishing Co., Amsterdam, 1980, pp. 723–738.

S48. Soong, T. T. and Skinner, G. T., "Experimental Study of Active Structural Control." *ASCE Journal of the Engineering Mechanics Division*, Vol. 107, No. EM6, 1981, pp. 1057–1067.

S49. Stein, G. and Athans, M., "The LQG/LTR Procedure for Multivariable Feedback Control Design." *IEEE Transactions on Automatic Control*, Vol. AC-32, No. 2, 1987, pp. 105–114.

S50. Stewart, G. W., *Introduction of Matrix Computations*, Academic Press, New York, 1973.

S51. Sundararajan, N., Joshi, S. M. and Armstrong, E. S., "Robust Controller Synthesis for a Large Flexible Space Antenna." *Journal of Guidance, Control, and Dynamics*, Vol. 10, No. 2, 1987, pp. 201–208.

S52. Swaim, R. L., "Control System Synthesis for a Launch Vehicle with Severe Mode Interaction." *IEEE Transactions on Automatic Control*, Vol. AC-14, No. 5, 1969, pp. 517–523.

S53. Swaim, R. L., "Aircraft Elastic Mode Control." *Journal of Aircraft*, Vol. 8, No. 2, 1971, pp. 65–71.

T1. Takahashi, M. and Slater, G. L., "Design of a Flutter Mode Controller Using Positive Real Feedback." *Journal of Guidance, Control, and Dynamics*, Vol. 9, No. 3, 1986, pp. 339–345.

T2. Takahashi, Y., Rabins, M. J. and Auslander, D. M., *Control and Dynamic Systems*, Addison-Wesley, Reading, Massachusetts, 1970.

T3. Thompson, G. O. and Kass, G. J., "Active Flutter Suppression – An Emerging Technology." *Journal of Aircraft*, Vol. 9, No. 3, 1972, pp. 230–235.

T4. Thompson, R. C., Junkins, J. L. and Vadali, S. R., "Near-Minimum Time Open-Loop Slewing of Flexible Vehicles." *Journal of Guidance, Control, and Dynamics*, Vol. 12, No. 1, 1989, pp. 82–88.

T5. Tseng, G. T. and Mahn, R. H., Jr., "Flexible Spacecraft Control Design Using Pole Allocation Technique." *Journal of Guidance and Control*, Vol. 1, No. 4, 1978, pp. 279–281.

T6. Turner, J. D. and Junkins, J. L., "Optimal Large-Angle Single Axis Rotational Maneuvers of Flexible Spacecraft." *Journal of Guidance and Control*, Vol. 3, No. 6, 1980, pp. 578–585.

T7. Turner, J. D. and Chun, H. M., "Optimal Distributed Control of a Flexible Spacecraft During a Large-Angle Maneuver." *Journal of Guidance, Control, and Dynamics*, Vol. 7, No. 3, 1984, pp. 257–264.

U1. Udwadia, F. E. and Tabaie, S., "Pulse Control of Single Degree-of-Freedom System." *ASCE Journal of the Engineering Mechanics Division*, Vol. 107, No. EM6, 1981, pp. 997–1009.

U2. Udwadia, F. E. and Tabaie, S. "Pulse Control of Structural and Mechanical Systems." *ASCE Journal of the Engineering Mechanics Division*, Vol. 107, No. EM6, 1981, pp. 1011–1028.

U3. Ulsoy, A. G., "Vibration Control in Rotating or Translating Elastic Systems." *ASME Journal of Dynamic Systems, Measurement, and Control*, Vol. 106, March 1984, pp. 6–14.

V1. Vander Velde, W. E. and He, J., "Design of Space Structure Control Systems Using On-Off Thrusters." *Journal of Guidance, Control, and Dynamics*, Vol. 6, No. 1, 1983, pp. 53–60.

V2. Van de Vegte, J., "The Wave Reflection Matrix in Beam Vibration Control." *ASME Journal of Dynamic Systems, Measurement, and Control*, Vol. 93, June 1971, pp. 94–101.

V3. Van de Vegte, J. and Hladun, A. R., "Design of Optimal Passive Beam Vibration Controls by Optimal Control Techniques." *ASME Journal of Dynamic Systems, Measurement, and Control*, Vol. 95, December 1973, pp. 427–434.

V4. VanLandingham, H. F. and Meirovitch, L., "Digital Control of Spinning Flexible Spacecraft." *Journal of Guidance and Control*, Vol. 1, No. 5, 1978, pp. 347–351.

V5. VanLandingham, H. F., Caglayan, A. K. and Floyd, J. B., "Approximation Techniques for Optimal Modal Control of Flexible Systems." *Proceedings of the Third VPI&SU/AIAA Symposium on Dynamics and Control of Large Flexible Spacecraft*, L. Meirovitch (Editor), 1981, pp. 89–99.

V6. VanLandingham, H. F., *Introduction to Digital Control Systems*, Macmillan, New York, 1985.

V7. Vaughan, D. R., "Application of Distributed Parameter Concepts to Dynamic Analysis and Control of Bending Vibrations." *ASME Journal of Basic Engineering*, Vol. 90, June 1968, pp. 157–166.

V8. Velman, J. R., "Low Order Controllers for Flexible Spacecraft." *Proceedings of the AAS/AIAA Astrodynamics Conference*, 1981, pp. 893–910.

V9. Venkayya, V. B. and Tischler, V. A., "Frequency Control and Its Effect on the Dynamic Response of Flexible Structures." *AIAA Journal*, Vol. 23, No. 11, 1985, pp. 1768–1774.

V10. Vilnay, O., "Design of Modal Control of Structures." *ASME Journal of the Engineering Mechanics Division*, Vol. 107, No. EM5, 1981, pp. 907–915.

V11. von Flotow, A. H., "Traveling Wave Control for Large Spacecraft Structures." *Journal of Guidance, Control, and Dynamics*, Vol. 9, No. 4, 1986, pp. 462–468.

V12. von Flotow, A. H. and Schäfer, B., "Wave-Absorbing Controllers for a Flexible Beam." *Journal of Guidance, Control, and Dynamics*, Vol. 9, No. 6, 1986, pp. 673–680.

V13. Vyalyshev, A. I., Dubinin, A. I. and Tartakovskii, B. D., "Active Acoustic Reduction of a Plate." *Soviet Physics—Acoustics (English Translation)* Vol. 32, No. 2, 1986, pp. 96–98.

W1. Walker, L. A. and Yaneske, P. P., "Characteristics of an Active Feedback System for the Control of Plate Vibrations." *Journal of Sound and Vibration*, Vol. 46, No. 2, 1976, pp. 157–176.

W2. Walker, L. A. and Yaneske, P. P., "The Damping of Plate Vibrations by Means of Multiple Active Control Systems." *Journal of Sound and Vibration*, Vol. 46, No. 2, 1976, pp. 177–193.

W3. Wang, P. K. C., "Control of Large Spaceborne Antenna Systems with Flexible Booms by Mechanical Decoupling." *Journal of the Franklin Institute*, Vol. 315, No. 5/6, 1983, pp. 469–493.

W4. Wang, P. C., Kozin, F. and Amini, F., "Vibration Control of Tall Buildings." *Engineering Structures*, Vol. 5, October 1983, pp. 282–288.

W5. Wang, S. J. and Cameron, J. M., "Dynamics and Control of a Large Space Antenna." *Journal of Guidance, Control, and Dynamics*, Vol. 7, No. 1, 1984, pp. 69–76.

W6. West-Vukovich, G. S., Davison, E. J. and Hughes, P. C., "The Decentralized Control of Large Flexible Space Structures." *IEEE Transactions on Automatic Control*, Vol. AC-29, No. 10, 1984, pp. 866–879.

W7. Whittaker, E. T. and Watson, G. N., *A Course in Modern Analysis*, Cambridge University Press, Cambridge, U.K., 1962.

W8. Wie, B., "Active Vibration Control Synthesis for the Control of Flexible Structures Mast Flight System." *Journal of Guidance, Control, and Dynamics*, Vol. 11, No. 3, 1988, pp. 271–277.

W9. Wie, B. and Bryson, A. E., Jr., "Pole-Zero Modeling of Flexible Space Structures." *Journal of Guidance, Control, and Dynamics*, Vol. 11, No. 6, 1988, pp. 554–561.

W10. Williams, T., "Efficient Modal Analysis of Damped Large Space Structures." *Journal of Guidance, Control, and Dynamics*, Vol. 9, No. 6, 1986, pp. 722–724.

W11. Williams, T., "Transmission-Zero Bounds for Large Space Structures with Applications." *Journal of Guidance, Control, and Dynamics*, Vol. 12, No. 1, 1989, pp. 33–38.

W12. Wonham, W. M., "On Pole-Assignment in Multi-Input Controllable Linear Systems." *IEEE Transactions on Automatic Control*, Vol. AC-12, No. 6, 1967, pp. 660–665.

W13. Wu, Y. W., Rice, R. B. and Juang, J. N., "Sensor and Actuator Placement for Large Flexible Space Structures." *Proceedings of the Joint Automatic Control Conference*, Denver, Colorado, 1979, pp. 230–238.

W14. Wu, Y. W., Rice, R. B. and Juang, J. N., "Control of Large Flexible Space Structures Using Pole Placement Design Techniques." *Journal of Guidance and Control*, Vol. 4, No. 3, 1981, pp. 298–303.

Y1. Yam, Y., Johnson, T. L. and Lin, J. G., "Aggregation of Large Space Structure Dynamics with Respect to Actuator and Sensor Influences." *Proceedings of the 20th IEEE Conference on Decision and Control*, 1981, pp. 936–942.

Y2. Yam, Y., Johnson, T. L. and Lang, J. H., "Flexible System Model Reduction and Control System Design Based Upon Actuator and Sensor Influence Functions." *IEEE Transactions on Automatic Control*, Vol. AC-32, No. 7, 1987, pp. 573–582.

Y3. Yang, J. N., "Application of Optimal Control Theory to Civil Engineering Structures." *ASCE Journal of the Engineering Mechanics Division*, Vol. 101, No. EM6, 1975, pp. 819–838.

Y4. Yang, J. N. and Giannopoulos, F., "Active Tendon Control of Structures." *ASCE Journal of the Engineering Mechanics Division*, Vol. 104, No. EM3, 1978, pp. 551–568.

Y5. Yang, J. N. and Giannopoulos, F., "Active Control and Stability of Cable-Stayed Bridge." *ASCE Journal of the Engineering Mechanics Division*, Vol. 105, No. EM4, 1979, pp. 677–694.

Y6. Yang, J. N. and Giannopoulos, F., "Active Control of Two-Cable Stayed Bridge." *ASCE Journal of the Engineering Mechanics Division*, Vol. 105, No. EM5, 1979, pp. 795–810.

Y7. Yang, J. N., "Control of Tall Buildings Under Earthquake Excitation." *ASCE Journal of the Engineering Mechanics Division*, Vol. 108, No. EM5, 1982, pp. 833–849.

Y8. Yang, J. N. and Lin, M. J., "Optimal Critical-Mode Control of Building Under Siesmic Load." *ASCE Journal of the Engineering Mechanics Division*, Vol. 108, No. EM6, 1982, pp. 1167–1185.

Y9. Yang, J. N. and Samali, B., "Control of Tall Buildings in Along-Wind Motion." *ASCE Journal of Structural Engineering*, Vol. 109, No. 1, 1983, pp. 50–68.

Y10. Yang, J. N. and Lin, M. J., "Building Critical-Mode Control: Nonstationary Earthquake." *ASCE Journal of Engineering Mechanics*, Vol. 109, No. 6, 1983, pp. 1375–1389.

Y11. Yao, J. T. P., "Concept of Structural Control." *ASCE Journal of the Structural Division*, Vol. 98, No. ST7, 1972, pp. 1567–1573.

Y12. Yedavalli, R. K. and Skelton, R. E., "Determination of Critical Parameters in Large Flexible Space Structures with Uncertain Modal Data." *ASME Journal of Dynamic Systems, Measurement, and Control*, Vol. 105, December 1983, pp. 238–244.

Y13. Yedavalli, R. K., "Critical Parameter Selection in the Vibration Suppression of Large Flexible Space Structures." *Journal of Guidance, Control, and Dynamics*, Vol. 7, No. 3, 1984, pp. 274–278.

Y14. Yocum, J. F. and Slafer, L. I., "Control System Design in the Presence of Severe Structural Dynamics Interactions." *Journal of Guidance and Control*, Vol. 1, No. 2, 1978, pp. 109–116.

Y15. Yoshimura, T., Ananthanarayana, N. and Deepak, D., "An Active Vertical Suspension for Track/Vehicle Systems." *Journal of Sound and Vibration*, Vol. 106, No. 2, 1986, pp. 217–225.

Y16. Yoshimura, T., Ananthanarayana, N. and Deepak, D., "An Active Lateral Suspension to a Track/Vehicle System Using Stochastic Optimal Control." *Journal of Sound and Vibration*, Vol. 115, No. 3, 1987, pp. 473–482.

Y17. Young, K. D., "A Distributed Finite Element Modeling and Control Approach for Large Flexible Structures." *Proceedings of the AIAA Guidance, Navigation and Control Conference*, 1988, pp. 253–263.

Y18. Yousuff, A. and Skelton, R. E., "Controller Reduction by Component Cost Analysis." *IEEE Transactions on Automatic Control*, Vol. AC-29, No. 6, 1984, pp. 520–530.

AUTHOR INDEX

A

Abdel-Rahman, T. M., 354, 372, 380
Abdel-Rohman, M., 359, 366, 368, 375, 376
Abel, I., 354, 360, 373, 374, 375
Adali, S., 358, 369, 370
Adams, N. M., Jr., 354
Adams, W. M., 366, 368, 374
Akulenko, L. D., 358
Alfriend, K. T., 377
Allaire, P. E., 381
Amini, F., 361, 366, 376
Amos, A. K., 354
Ananthanarayana, N., 381
Andry, A. N., Jr., 365
Arbel, A., 355, 361, 371, 380
Armstrong, E. S., 363, 368, 371, 380
Ashkenazi, A., 362
Athans, M., 192, 209, 210, 213, 252, 353, 357, 362, 364, 368, 370, 373
Atluri, S. N., 354
Aubrun, J. N., 371, 381
Auslander, D. M., 358, 366

B

Bailey, T., 358, 381, 382
Bainum, P. M., 379
Balakrishnan, A. V., 373
Balas, M. J., 327, 328, 333, 353, 355, 356, 357, 358, 359, 360, 361, 365, 366, 368, 370, 371, 375, 379, 380
Barak, P., 376
Barbat, A. H., 357, 376
Barker, D. S., 355, 358, 371
Barthelemy, J. F. M., 362, 367
Baruh, H., 257, 328, 356, 358, 359, 362, 366, 367, 368, 371, 377, 380, 382
Becker, J., 374
Behal, B. L., 374, 382
Bekey, G. A., 359, 371, 372, 376
Bellman, R., 192, 356
Benhabib, R. J., 355, 357, 362, 371, 380
Bennighof, J. K., 358, 368, 372
Berkman, F., 358
Bernstein, D. S., 361, 362, 370
Bhaya, A., 361, 380
Blelloch, P. A., 362, 368
Bodden, D. S., 366, 369, 371
Bolotnik, N. N., 358
Boltyanskii, V., 192, 203
Book, W. J., 377
Bossi, J. A., 371
Breakwell, J. A., 358, 362, 368, 377, 380, 381, 382
Breakwell, J. V., 373
Brogan, W. L., 82, 183

B

Bruch, J. C., Jr., 358, 369, 370
Bryson, A. E., Jr., 354, 358, 360, 362, 364, 373, 380
Burdess, J. S., 365

C

Caglayan, A. K., 358, 367
Calico, R. A., 356, 374, 380
Cameron, J. M., 379
Canavin, J. R., 370, 380
Cannon, R. H., Jr., 380, 381
Carotti, A., 358, 376
Caughey, T. K., 354, 355, 371, 372, 376, 379, 380
Chambers, G. J., 362, 381
Champetier, C., 371, 372
Chang, J. C. H., 375
Chang, M. I. J., 355, 368, 375
Chang, S., 374
Chang, Y., 379
Chen, T. K., 362, 364, 374
Chun, H. M., 359, 365, 366, 377, 378, 380
Chung, L. L., 376, 382
Coradetti, T., 356, 361, 362, 363, 364, 369, 380
Crawley, E. F., 358, 381
Creedon, J. F., 355, 358, 366
Croopnick, S. R., 353
Crossley, R., 183, 185, 188, 366, 368, 375
Cruz, J. B., Jr., 363
Curtis, A. R. D., 373, 382

D

Dahl, W. E., 369, 380
Davis, M. M., 379
Davison, E. J., 183, 359, 361, 364, 365, 368, 380
D'Azzo, J. J., 175
Deepak, D., 381
Dehghanyar, T. J., 354, 359, 371
De Luis, J., 358, 381
De Miranda, M., 358, 376
de Silva, C. W., 358
Desoer, C. A., 361, 380
Dimitriadis, E. K., 382
Dodds, S. J., 359

Dorato, P., 356
Dowell, E. H., 373, 382
Doyle, J. C., 361, 363, 364, 368, 374
Dubinin, A. I., 376
Dunn, H. J., 355, 374
Du Val, R. W., 374
Dwyer, T. A. W., III, 358

E

Eastep, F. E., 369, 374
Edwards, J. W., 373
Eldred, D. B., 381
Elliott, L. E., 370
Ellis, R. W., 380, 382
Emami-Naeini, A., 355, 361, 362, 363, 364, 380

F

Falb, P. L., 192, 368
Falkenburg, D. R., 377
Farm, J. A., 366, 368, 374
Felt, L. R., 375, 381
Fisher, E. E., 379
Floyd, J. B., 367
France, M. E. B., 357, 359, 363, 364, 373, 378, 380
Freymann, R., 374
Friedland, B., 363
Fuller, C. R., 377, 382

G

Gamkrelidze, R., 192, 203
Garrard, W. L., 355, 366, 368, 374
Gebler, B., 377
Gehling, R. N., 368, 381
Gevarter, W. B., 355, 380
Ghosh, D., 368, 376
Giannopoulos, F., 358, 375
Gibson, J. S., 358, 360
Ginter, S., 358, 379
Goff, R. M. A., 358, 364, 371
Goh, C. J., 355, 371, 379, 380
Goldenberg, A. A., 361, 377
Gould, L. A., 358, 364, 366
Gran, R., 357, 358, 360, 361, 379
Greene, M., 361, 362
Groom, N. J., 356, 370, 380

Gruzen, A., 361, 362, 370
Guicking, D., 353, 354, 376
Gupta, N. K., 361, 364, 366, 371, 374, 380

H

Hablani, H. B., 360, 369, 380
Haftka, R. T., 370
Hagedorn, P., 358, 372
Hale, A. L., 300, 304, 358, 361, 367, 368, 369, 378, 380
Hall, S. R., 373
Hallauer, W. L., Jr., 362, 367, 368, 371, 381
Halstenberg, R. V., 363, 364, 369, 379, 380
Ham, N. D., 374, 382
Hamer, H. A., 379
Hashemipour, H. R., 248
He, J., 359
Hefner, R. D., 362
Hegg, D. R., 370, 380
Heimbold, G., 355
Herrick, D. C., 364
Hirako, K., 378, 382
Hladun, A. R., 362, 380
Hodges, G. E., 375, 381
Holoyen, S., 380, 382
Holzach, H., 371, 382
Hönlinger, H., 374
Horner, G. C., 355
Horta, L. G., 359, 372, 378, 382
Houpis, C. H., 175
Hrovat, D., 376
Hsu, D. S., 357, 368, 371, 376
Hu, A., 361, 369
Hubbard, J. E., Jr., 358, 381, 382
Huebner, K. H., 296
Hughes, P. C., 354, 355, 356, 360, 361, 364, 368, 369, 372, 380
Hughes, R. O., 356, 358, 380
Hurty, W. C., 300
Huttsell, L. J., 373
Hwang, C., 361
Hyland, D. C., 360, 361, 362, 370

I

Inman, D. J., 367

Iwens, R. P., 355, 362, 370, 371, 380

J

Jackson, R. L., 355, 362, 371, 380
Jacquot, R. G., 355, 358, 371
Janiszewski, A. M., 356, 380
Jean, M., 364
Johnson, C. D., 357, 365
Johnson C. R., Jr., 357
Johnson, T. L., 355, 360, 365, 373, 380
Jones, J. D., 377, 382
Joshi, S. M., 356, 362, 363, 368, 370, 371, 380
Juang, J. N., 355, 356, 358, 359, 361, 365, 366, 368, 372, 375, 377, 378, 380, 382
Judd, R. P., 377
Junkins, J. L., 355, 358, 359, 366, 369, 370, 371, 372, 378, 380

K

Kabamba, P. T., 359, 360, 378, 380
Kamat, M. P., 368, 369
Karcher, K., 376
Karnopp, D., 358
Kass, G. J., 353
Keat, J. E., 370, 380
Khot, N. S., 369
Kida, T., 362, 364, 366, 378, 380, 382
Kirk, D. E., 192, 205
Klein, R. E., 356, 358, 380
Knowles, G., 356, 358
Köhne, M., 357
Komkov, V., 357
Kosut, R. L., 355, 361, 362, 363, 364, 380
Kouvaritakis, B., 363
Kozin, F., 361, 366, 376
Krabs, W., 356, 358
Krishna, R., 379
Kuo, B. C., 356, 357, 380
Kwak, M. K., 287, 288, 298, 363, 372, 378, 380
Kwakernaak, H., 233, 237, 240, 247

L

Lang, J. H., 360, 365, 380

Larson, V., 361, 363, 364, 380
Laub, A. J., 248
Lehtomaki, N. A., 362, 364
Leipholz, H. H. E., 354, 368, 375, 376
Leondes, C. T., 357, 361
Levine, W. S., 252, 370
Levis, A. H., 356
Lewis, D. W., 381
Liebst, B. S., 366, 368, 374
Likins, P. W., 360, 361, 362, 363, 364, 366, 369, 379, 380
Lim, K. B., 355, 370
Lin, J. G., 355, 356, 360, 365, 370, 380
Lin, M. J., 376
Lin, Y. H., 353, 370, 380
Lindberg, R. E., Jr., 367
Lindgren, A. G., 355, 358, 366
Lions, J.-L., 358
Lisowski, R. J., 358, 361, 369, 378, 380
Liu, S. C., 376
Longman, R. W., 356, 360, 367, 376, 377, 380, 381
Lottati, I., 373
Luenberger, D. G., 234
Lum, E. L. S., 373
Lutze, F. H., 358, 364, 371
Luzzato, E., 364
Lyons, M. G., 381

M

Mace, B. R., 358, 373
MacFarlane, A. G. J., 363
Mahesh, J. K., 355, 374
Mahn, R. H., Jr., 366, 379
Maizza-Neto, O., 377
Majette, M., 377
Maki, M. C., 379
Manolis, G. D., 359, 372, 376
Martin, C. R., 368, 375
Martin, G. D., 360, 380
Martin, J.-C. E., 355, 358, 361
Martin-Sanchez, J. M., 357, 376
Masri, S. R., 354, 359, 371, 372, 376
McClamroch, N. H., 357, 359, 371, 378, 380
McKillip, R. M., Jr., 374, 382
McLaren, M. D., 368, 371
McLean, D., 374
McNamara, R. J., 375
Meckl, P., 359, 377

Meirovitch, L., 5, 16, 17, 41, 42, 55, 60, 62, 64, 78, 80, 86, 135, 136, 139, 141, 227, 228, 257, 260, 262, 273, 274, 278, 280, 281, 286, 287, 288, 296, 297, 298, 300, 301, 304, 310, 319, 320, 322, 323, 328, 329, 335, 338, 347, 350, 353, 354, 355, 356, 357, 358, 359, 361, 362, 363, 364, 365, 366, 367, 368, 371, 372, 373, 374, 375, 376, 377, 378, 379, 380, 382
Mesquita, L., 368, 369, 371, 381
Metcalfe, A. V., 365
Miller, D. F., 370
Miller, D. W., 373
Miller, R. K., 354, 359, 371
Mingori, D. L., 362, 368, 370
Mishchenko, E., 192, 203
Mitter, S. K., 183, 365, 368
Montgomery, R. C., 359, 367, 382
Moon, F. C., 373, 382
Mote, C. D., Jr., 368, 380, 381, 382
Moyer, H. G., 357, 379
Mukhopadhyay, V., 354, 360, 375
Murray-Lasso, M. A., 358, 364, 366
Murrow, N. H., 373

N

Nayfeh, A. H., 359, 376
Nelson, P. A., 373, 382
Newsom, J. R., 354, 360, 373, 374, 375
Nissim, E., 373
Noll, T. E., 373, 374
Novak, L. M., 357, 361
Nurre, D. S., 354

O

Ohkami, Y., 362, 364, 366, 378, 379, 380, 382
Okamoto, O., 362, 366
Onoda, J., 370
Öz, H., 257, 260, 262, 323, 329, 353, 355, 356, 357, 358, 359, 361, 366, 367, 368, 375, 379, 380

P

Paliou, C., 376, 381

Perkins, W. R., 363
Perry, B., III, 373
Pi, W. S., 361
Plump, J. M., 358, 382
Poelaert, D. H. I., 379
Pontryagin, L. S., 192, 203
Porter, B., 183, 185, 188, 366, 368, 375
Pototzky, A. S., 374
Potter, J. E., 197, 198
Price, G. A., 371
Pu, J. P., 357, 368, 371, 376

Q

Quinn, R. D., 355, 368, 371, 373, 378, 380
Quintana, V. H., 368, 375

R

Rabins, M., 376
Rabins, M. J., 358, 366
Radcliffe, C. J., 368, 381, 382
Rahn, G. A., 358, 361, 367, 368
Rajaram, S., 367, 372
Rakhsha, F., 377
Rappaport, D., 356
Ratner, M. J., 381
Reddy, A. S. S. R., 379
Redman-White, W., 373, 382
Reinhorn, A. M., 359, 372, 376, 382
Rice, R. B., 355, 356, 361, 365, 366, 368, 380
Richman, J., 363
Robertshaw, H. H., 378, 382
Robinson, A. C., 353
Rodellar, J., 357, 376
Rodriguez, G., 355, 380
Roger, K. L., 375, 381
Rogers, R. O., 359
Roorda, J., 375, 376, 381
Rosenbrock, H. H., 177, 363
Rosenthal, D. E., 380, 381
Rossi, M., 357, 360, 379
Ryan, R. S., 354

S

Sadek, I. S., 358, 369, 370
Sae-Ung, S., 356, 366, 375

Safonov, M. G., 353
Salzwedel, H., 355, 361, 362, 363, 364, 380
Samali, B., 376
Samaras, E., 376, 381
Sambongi, S., 364, 380
Sandell, N. R., Jr., 353, 362, 364
Sathe, S. G., 358, 367
Schaechter, D. B., 361, 362, 364, 381
Schäfer, B. E., 371, 372, 382
Schajer, G. S., 380, 382
Schmidt, D. K., 362, 364, 374
Schulz, G., 355, 373
Scofield, H. N., 354
Seering, W., 359, 377
Seltzer, S. M., 356, 380
Sensburg, O., 374
Sesak, J. R., 361, 362, 363, 364, 369, 371, 379, 380
Sezer, M. E., 360, 380
Shapiro, E. Y., 365
Sharony, Y., 363, 365, 371, 372, 378, 380
Shenhar, J., 227, 228, 359, 368, 380
Shim, J., 370
Shinozuka, M., 376, 381
Siljak, D. D., 360, 364, 365, 380
Silverberg, L. M., 322, 323, 358, 362, 367, 373, 374, 376, 377, 380
Silverman, L. M., 356
Simon, J. D., 183, 365, 368
Sims, J. L., 354
Singh, G., 359, 378, 380
Singh, G., 356, 380
Sirlin, S. W., 376, 381
Sivan, R., 74, 233, 237, 240, 247
Skaar, S. B., 358, 359, 377, 378
Skelton, G. B., 373
Skelton, R. E., 355, 356, 360, 361, 363, 366, 368, 369, 379, 380
Skidmore, G. R., 368, 371, 381
Skinner, G. T., 375, 382
Slafer, L. I., 364, 379
Slater, G. L., 368, 371, 374
Sloss, J. M., 358, 369, 370
Smith, R. E., 373
Sobel, K. M., 365
Soga, H., 378, 382
Soong, T. T., 355, 368, 375, 376, 382
Stein, G., 361, 363, 368, 374

Stone, C. R., 355, 374
Strunce, R. R., 353
Sundararajan, N., 363, 368, 371, 380
Swaim, R. L., 353, 379
Sworder, D. D., 359

T

Tabaie, S., 372, 376
Takahashi, M., 371, 374
Takahashi, Y., 358, 366
Tang, L., 359, 378
Tartakovskii, B. D., 376
Thangjitham, S., 358, 368, 377
Thompson, G. O., 353
Thompson, R. C., 359, 378, 380
Thornton, E. A., 296
Tiffani, S. H., 354
Tischler, V. A., 369
Tseng, G. T., 366, 379
Tucker, D., 377
Turci, E., 358, 376
Turner, J. D., 358, 359, 365, 366, 377, 378, 380

U

Udwadia, F. E., 372, 376
Ulsoy, A. G., 356, 381

V

Vadali, S. R., 359, 378, 380
Vander Velde, W. E., 359, 361, 362, 370
Van de Vegte, J., 362, 372, 379, 380
VanLandingham, H. F., 77, 257, 357, 358, 359, 367, 380
Varaiya, P., 353
Vaughan, D. R., 372
Velman, J. R., 360, 364, 380
Venkayya, V. B., 369

Vilnay, O., 368, 375
von Flotow, A. H., 371, 372, 373
Vyalyshev, A. I., 376

W

Walker, L. A., 376
Wang, P. C., 361, 366, 376
Wang, P. C. K., 379
Wang, S. J., 379
Watson, G. N., 56
Wen, C. Y., 359, 372, 376
West-Vukovich, G. S., 364, 380
Whitney, D. E., 377
Whittaker, E. T., 56
Wie, B., 358, 364, 380
Williams, D. E., 363
Williams, J. P., 359, 367, 382
Williams, T., 364, 371, 380
Williamson, S. E., 359
Wolovich, W. A., 368
Wonham, W. M., 183, 365, 368
Wu, Y. W., 355, 356, 361, 365, 366, 368, 380

Y

Yackel, R. A., 356, 380
Yalda-Mooshabad, I., 359, 378
Yam, Y., 360, 365, 380
Yamaguchi, I., 378, 382
Yaneske, P. P., 376
Yang, J. N., 356, 358, 366, 375, 376
Yang, T. Y., 361, 369
Yao, J. T. P., 375
Yedavalli, R. K., 361, 363, 379
Yeh, C. T., 376
Yokum, J. F., 364, 379
Yoshimura, T., 381
Young, K. D., 364
Yousuff, A., 361, 369

SUBJECT INDEX

A

Actuating signal, 130
Actuator, 47, 130
  dynamics, 354–355
  locations, 355
  noise, 240
Adiabatic approximations method, 363
Adjoint eigenvalue problem, 79
Admissible control, 192
Admissible functions, 276
  global, 289
  local, 289
Admissible trajectory, 192
Admittance function, 53
Aircraft control, 373–375
Algebraic eigenvalue problem, 70, 99–108
Amplitude, 51
Angular momentum, 7
  conservation of, 7
Assembling process (in the finite element method), 293
Assumed-modes method, 307–309
Asymptotically stable equilibrium, 69, 70
  for discrete-time systems, 89
Autocorrelation function, 242

B

Back substitution, 119

Bandwidth, 149
Bang-bang function, 208
Bang-bang principle, 208
Bang-off-bang function, 225
Biorthogonality, 79
Biorthonormality, 79
Block diagram, 46
Bode diagrams (or plots), 150–157
Borel's theorem, 62
Boundary conditions (in vibrations), 272, 275
  geometric, 276
  natural, 276
Boundary-value problems (in vibrations), 270–274
Brachistochrone problem, 205–206
Breakaway point(s), 139

C

Cayley–Hamilton theorem, 82
Characteristic equation, 70, 137, 275
Characteristic polynomial, 70
Chattering, 232
Cholesky decomposition, 101
Circulatory:
  coefficients, 96
  forces, 96, 99
  matrix, 96
Civil structures control, 375–376

SUBJECT INDEX

Closed-loop:
 control, 47, 130, 181, 314
 eigenvalue problem, 184
 eigenvalues, *see* Pole(s)
 eigenvectors, 184
 equation(s), 183, 314
 frequency response, 163–169
 poles, 182, 184
 transfer function, 131, 134, 138
Collocation method, 300
Command input, 181
Comparison functions, 276
Compensator:
 cascade, or series, 171
 feedback, or parallel, 171
 lag, 175
 lead, 172
 lead-lag, 175
Complete set of functions, 276
 in energy, 278
Component cost analysis, 368
Component-mode synthesis, 300
Configuration space, 22, 38, 64, 94
Configuration vector, 63, 64
Conservation of:
 angular momentum, 7
 energy, 10, 99
 Hamiltonian, 99
 linear momentum, 6
Conservative forces, 10, 34
Conservative systems:
 gyroscopic, 105–107
 nongyroscopic, 100–105
Constraints, 22
Continuous-time systems, 86
Control gain(s)
 frequency-preserving, 321
 matrix, 181
 operators, 314
Controllability, 81–83, 356
 complete, 81, 83
 matrix, 83
 modal, 317, 337
Control law:
 for optimal control:
  linear regulator, 194
  linear tracking, 203
  minimum time, 207–208
 for pole allocation:
  dyadic control, 186

 multi-input control, 187
 single-input control, 183
Control spillover, 327, 345, 355–356
Control vector, 63
Convolution integral, 61–62
 in state form, 76, 81
Convolution sum, 89
Coordinates:
 generalized, 22, 34, 94
 global, 292
 local, 292
 natural:
  in the finite element method, 292
  in modal analysis, 305
 normal, 305
Corner frequency, 153
Correlation:
 coefficient, 241
 function, 241
 matrix, 242
Costate equations, 193
Costate vector, 192
Coupled control, 323–333
Covariance:
 function, 241
 matrix, 242
  compound, 243
Cross-correlation:
 function, 242
 matrix, 243
Cross-covariance:
 function, 242
 matrix, 242
Cutoff frequency, 149
Cutoff rate, 149

D

D'Alembert's principle, 25
Damped structures, 309–312
 control of, 348–351
Damping:
 coefficients, 96, 349
 matrix, 96, 312, 349
 operator, 310, 314, 348
  closed-loop, 315
 proportional, 310, 336–338
Dead-zone function, 224
Decade, 151
Decentralized control, 364

Decibel, 151
Decomposition-aggregation method, 365
Deflated matrix, 114
Degrees of freedom, 22
Degree of stability, 158, 161
Delay time, 135
Delta function, 56
Difference equations, 88
Differential eigenvalue problem, 274, 280
Differential operator, 275
   self-adjoint, 277
   positive definite, 278
   positive semidefinite, 278
Dirac delta function, 56
Direct output feedback control:
   discretized structures, 345–348
   distributed structures, 333–336
   energy considerations, 253–254, 335
   lumped structures, 252–255
Discrete systems, 48
Discrete-time systems, 86–91
   response of, 127
Discretization in space, 284, 339–342
Dissipation function, 96
Distributed parameters, 47
Distribution (singularity function), 57
Disturbance accommodation, 365
Double pendulum 14, 31
Dynamical equations, 83
Dynamical path, 38
Dynamical system, 64
Dynamic potential, 34
Dynamic programming, 192

E

Eigenfunctions, 276
   orthogonal, 278
   orthonormal, 279
Eigenstructure assignment method, 365–366
Eigenvalue problem (algebraic), 70, 79–81, 99–108
   adjoint, 79
   closed-loop, 184
   for conservative gyroscopic systems, 105–107
   for conservative nongyroscopic systems, 100–105
   for general systems, 107–108
   for real symmetric matrices, 104–109
Eigenvalue problem (differential), 274–280
   general form, 275
Eigenvalues, 70, 79, 89
   closed-loop, 182, 184
   for discrete-time systems, 89
   for distributed systems, 276
   zero, 278
Eigenvectors, 70, 79, 89
   adjoint, 79
   closed-loop, 184
   left, 79
      for discrete-time systems, 89
   orthogonality, 102
   right, 79
      for discrete-time systems, 89
Element mass matrix, 293
Element stiffness matrix, 293
Energy:
   conservation of, 10, 99
   kinetic, 9, 93
   potential, 9
   total, 10
Energy norm, 278
Equilibrium points, 64, 94
   trivial, 68, 94
      asymptotically stable, 69
      stable, 68
      unstable, 69
   nontrivial, 94
Error, 130, 133
Error detector, 130
Euclidean norm, 68
Euclidean space, 94
Euler's moment equations, 20, 43
Expansion theorem:
   distributed systems, 279
   lumped systems, 109
Expected value, 240
Experiments on structural control, 381–382

F

Feedback control, 47, 129–132, 180–182
Feedback gain matrix, 181
Feedback signal, 130
Feedforward matrix, 181
Filters:
   modal, 318, 328, 366
   orthogonal, 366

Final-value theorem, 136
Finite element method, 289–298, 307
  hierarchical, 296
Finite elements, 290
Frequency(ies), 51
  of damped oscillation, 135
  natural, 103, 278
  of undamped oscillation, 135
Frequency response, 50–53
  closed-loop, 163–169
  magnitude of, 147
  phase angle of, 147
  plots, 146–150
    logarithmic, 150–157
Frequency shaping, 366
Full-order, or identity observer(s), 233, 234, 237–238

G

Gain(s):
  constant, 151
  logarithmic, 151
  margin, 159
  matrix, 181, 186, 195
    observer, 234, 246, 247
Gain-crossover frequency, 160
Gain-crossover point, 160
Galerkin method, 300
Gaussian elimination, 119
Gaussian process, 241
Generalized coordinates, 22, 34, 94, 306
Generalized derivative, 58
Generalized forces, 23, 307
Generalized function, 57
Generalized momenta, 39, 63, 94
Generalized velocities, 38
Givens' method, 116
Global:
  admissible functions, 289
  mass matrix, 293
  stiffness matrix, 293
Globally optimal control, 322
Gram–Schmidt orthogonalization process, 120
Group property, 78
Gyroscopic:
  coefficients, 96
  matrix, 96
Gyroscopic systems, 105–107

  the eigenvalue problem for, 106–107

H

Hamilton's equations (in dynamics), 40, 63, 193
Hamilton's principle, 30, 270
  extended, 29
Hamiltonian (in controls):
  general, 192, 204
  for minimum fuel, 222
  for minimum time, 206
  minimum value for, 204
    global, 205
  for quadratic performance measure, 194, 201
Hamiltonian (in dynamics), 39, 98
  conservation of, 99
Harmonic function, 51
Hermite cubics, 295
Hessenberg form, 116
Householder's method, 116

I

Impedance function, 52
Impulse:
  angular, 7
  linear, 6
  unit, 56
Impulse response, 59
Inclusion principle, 111, 287
Independent modal-space control (IMSC), 366–368
  discretized structures, 342–345
  distributed structures, 320–323
    optimal control, 321–323
    pole allocation, 321
    systems with proportional damping, 338
  lumped structures, 257–267
    optimal control, 261–267
Inner product:
  energy, 277
  of functions, 276
Input, 46, 133
Input-output decoupling, 368
Intensity of stochastic process, 244
Interpolation functions (in the finite element method):
  cubic (Hermite), 295

linear, 292
Inverse iteration, 118–119
Inverse Laplace transform, 55

J

Jacobi method, 116
Joint probability density function, 241, 243
Jordan form, 80

K

Kalman–Bucy filter, 240–252
   for second-order systems, 248
Kinetic energy, 9, 28, 34, 93, 270
Kinetic energy density, 270

L

Lagrange's equation(s), 35
   for distributed systems, 272
   linearized, 95
   in terms of quasi-coordinates, 42, 273
Lagrange's multipliers, 192, 270
Lagrangian, 29, 94
Lagrangian density, 271
Laplace domain, 53
Laplace transform, 53
Launch vehicled control, 379–380
Leverrier algorithm, 77, 177
Linear elements (in the finite element method), 290
Linearization, 63–67, 95
Linearized solution, 70
   asymptotically stable, 70
   stable, 70
   unstable, 70
Linearized system, 71
   with critical behavior, 71
   with significant behavior, 71
Linear momentum, 3, 6
   conservation of, 6
Linear-quadratic-Gaussian/loop transfer recovery (LQG/LTR) method, 363, 368
Linear regulator:
   minimum-fuel, 222–229
   minimum-time, 208–221
   quadratic performance measure, 193–195

Linear systems, 48
Linear tracking problem, 201–203
Logarithmic gain, 151
Log magnitude-phase diagram, 161–163, 167
Luenberger observer, 232, 239
   full-order, or identity, 233, 234, 237–238
Lumped parameters, 47, 93

M

Magnitude (of function), 138
Maneuvering of space structures, 377–378
Mass:
   center, 11
   coefficients, 95, 284, 306
   density, 272
   matrix, 96, 284
Mathematical expectation, *see* Expected value
Matrix iteration by the power method, 113–116
Maximum-minimum theorem (of Courant and Fischer), 110, 281
Maximum overshoot, 135
Mean square error, 276
Mean square value, 240
Mean value:
   of a function, 240
   of a vector, 242
Mean-value theorem, 56
Minimum-fuel control, 222–229
   of harmonic oscillator, 225–229
   normal, 224
Minimum-phase system, 144
Minimum principle of Pontryagin, 192, 203–205
Minimum-time control, 205–221
   of double integral plant, 210–215
   of harmonic oscillator, 215–221
   normal, 207
   singular, 207
Modal analysis, 79–81, 123–127
   general systems, 126–127
   undamped gyroscopic systems, 125–126
   undamped nongyroscopic systems, 124–125
Modal control(s), 183–191, 255–267, 316
   coupled control, 257, 323–333, 368

Modal control(s) (*Continued*)
  independent modal-space control,
    257–262, 320–323, 338, 342–344,
    366–368
  pole allocation, 183–191, 255
    minimum gain, 188
  pseudo-independent modal-space control,
    344, 367
Modal cost analysis, 368–369
Modal equations, 124, 256, 305, 315–317
  closed-loop, 318–320
  open-loop, 319
Modal filters, 318, 328
Modal forces, 306, 316
Modal matrix, 102
Modal measurements, 317
Modal observer, 260, 326
Modal participation matrix, 324
Modal Riccati equations, 262
Mode controllability, 317
Mode observability, 317–318
Model error sensitivity suppression
  (MESS), 369
Momentum:
  angular, 7
    conservation of, 7
    generalized, 94
  linear, 3, 6
    conservation of, 6
Motor constant, 133
Multi-objective optimization, 369–370

N

Natural control, *see* Independent modal-
  space control (IMSC)
Natural frequencies, 103, 278
Natural modes, 278
Natural system, 99
Necessary condition(s) for optimal control:
  general, 192–193, 204
  minimum-fuel, 223
  minimum-time, 207
  quadratic performance measure, 194, 201
Newtonian mechanics, 3
Newton's laws, 3
  first, 6
  second, 3
Nichols charts, 167
Nichols plots, 169

Nodal displacements (in the finite element
  method), 290
Nodal vector (in the finite element
  method), 293
Nodes (in the finite element method), 290
Noise:
  actuator, 240
  observation, 240
  sensor, 240
  state excitation, 240
Nonconservative forces, 10, 34
  generalized, 35
Nonlinear control, 215, 358–359
Nonlinear plant, 358–359
Nonlinear systems, 48
Non-self-adjoint systems, 299
Normalization:
  of functions, 276
  of vectors, 102
Nyquist criterion, 144
Nyquist method, 141–146
Nyquist path, 143
Nyquist plot, 143

O

Observability, 83–84, 356
  complete, 84
  matrix, 84
  modal, 317, 337
Observation noise, 240
Observation spillover, 328, 355–356
Observer, *see also* State observer(s)
  error vector, 234, 246, 327
  gain matrix, 234, 246, 247, 326
  poles, 234
  Riccati equation, 247, 248
Octave, 151
On-off control, 225, 229–232
  simplified, 229–231
    harmonic oscillator, 231–232
On-off function, 225
On-off principle, 225
Open-loop:
  control, 46, 130, 180, 314
  eigenvalues, 182, 184
  eigenvectors, 184
  poles, 138, 182
  transfer function, 131, 138
  zeros, 138

Optimal control, 191–195
  linear tracking, 201–205
  minimum-fuel, 222–229
  minimum-time, 205–221
  quadratic performance measure,
    193–195
Optimal projection/maximum entropy
  (OP/ME), 370
Optimal trajectory, 192
Orthogonal filters, 366
Orthogonality:
  of eigenvectors, 102
  of modal matrix, 103
Orthonormality:
  of eigenfunctions, 279
  of eigenvectors, 102
  of modal matrix, 103
Orthonormal transformation, 118
Output, 46, 133
  equations, 83–84
  feedback control, 181, 252, 370–371
  vector, 83, 84

P

Peak amplitude, 148
Peak resonance, 148
Peak time, 135
Percent overshoot, 135
Performance (of control system), 132–137
Performance criteria, 136
Performance measure, performance index,
    (in optimal control), 192
  minimum-fuel, 222
  minimum-time, 206
  quadratic, 193, 201
Performance measure for optimal observer,
  246
Perturbation approaches to control, 371
Perturbation theory, 85
Phase angle, 138
Phase-crossover frequency, 160
Phase-crossover point, 159
Phase margin, 160
Phase space, 38, 63
Phase vector, 38
Plant, 46, 130
Pole(s):
  closed-loop, 182, 184
  open-loop, 138

Pole allocation, pole assignment, pole
    placement, 183–191, 365–366
  dyadic control, 186–187, 191
  multi-input control, 187–189
  single-input control, 183–185, 189–191
Pontryagin's minimum principle, 192,
  203–205
Position control system, 132
Positivity concept in controls, 371
Potential energy, 9, 29, 34, 270
Potential energy density, 270
Potter's algorithm, 197
Power spectral density:
  function, 245
  matrix, 244
Principle of:
  the argument, 141
  conservation of:
    angular momentum, 7, 13
    energy, 10, 99
    linear momentum, 6, 12
  d'Alembert, 25
  Hamilton, 29, 30
  Pontryagin, 192
  virtual work, 23
  optimality, 192
Probability density function, 240
  joint, 241, 243
Proportional damping, 310, 336–338
Proportional-plus-derivative-plus-integral
    control, 372
Pulse control, 372

Q

QL method, 118
QR method, 117–118
Quasi-comparison functions, 287, 288
Quasi-coordinates, 42

R

Random function, 240
Random process(es), 240
  stationary, 243
    weakly, 243
    wide-sense, 243
  uncorrelated, 245
  white noise, 245
Rayleigh's dissipation function, 97

Rayleigh's principle, 110
Rayleigh's quotient:
  for distributed systems, 280–282
    stationarity of, 281, 284
  for lumped systems, 109
    stationarity of, 109
Rayleigh–Ritz method (classical), 282–289, 300, 307
  convergence of, 287
Reduced-order controllers, 359–361
Reduced-order models, 359–361
Reduced-order observer(s), 237, 238–239, 359–361
Reference input, 130, 133, 181
Relative stability, 158–161
Residual modes, 326
Resolvent, 77, 176
Resonant frequency, 148
Riccati equation, 195
  algebraic, 197
  algorithms for solution of, 195–201
  observer, 247–248
  steady-state, 197
  transient, 197
Riccati matrix, 195
Rigid bodies, 16
Rigid-body modes, 278
Rigid-body motions, 103
Rise time, 136
Rms value, 240
Robotics, 377–378
Robustness, 361–362
Roof functions, 290
Root-locus method, 137–141
Root-locus plot(s), 137
  asymptotes of, 139
    centroid of, 139
  breakaway points of, 139
  complementary, 138
Root mean square value, 240
Rotorcraft control, 373–375
Routh array, 72
Routh–Hurwitz criterion, 72, 73

S

Sampling period, 87
Sampling property, 56
Second-order joint moment:
  function, 241

matrix, 242
Sensitivity, 361–362
  of control system, 169–171
  of eigensolution, 84–86
Sensor, 47, 130
  dynamics, 354–355
  locations, 355
  noise, 240
Separation principle, 235, 327
Sequential linear optimization, 372
Settling time, 136
Shape functions, 292
Similarity transformation, 80
Similar matrices, 80
Simple pendulum, 5, 7, 49, 66
Single-input, single-output, 132
Singularity functions, 55–58
  response to, 59–61
Small motions about equilibrium, 94
Sound radiation suppression, 376–377
Space structures:
  control, 379–380
  maneuvering, 377–378
Stability, 68
  for discrete-time systems, 89
  in the sense of Liapunov, 68
Stable equilibrium, 68, 70
  for discrete-time systems, 89
Standard deviation, 241
State controllability, 81–83
  complete, 81, 83
State equations, 63–67
  Laplace transform solution, 175–177
State excitation noise, 240
State feedback control, 181
State observer(s), 232–252
  deterministic, or Luenberger, 232–239
    full-order, or identity, 233, 234, 237–238
    reduced-order, 237, 238–239
  optimal, 240–252
  stochastic, or Kalman–Bucy filter, 240–252
State space, 40, 63, 94
State vector, 40, 63, 94
Stationary random process, 243
  weakly, 243
  wide-sense, 243
Steady-state error, 136
Steady-state response, 136

Step response, 59, 135
Stiffness:
　coefficients, 95, 284, 306
　　elastic, 96
　　geometric, 96
　matrix, 96, 284
　operator, 275, 314
　　closed-loop, 315
Stochastic process, 240. *See also* Random processes
Structural damping, 311
Sturm sequence, 117
Sturm's theorem, 116–117
Subspace iteration, 119–121
Substructure synthesis, 300–304
Superposition integral, 62
Superposition principle, 48
Switching curve, 215
Switching function(s), 214, 229
Switching time, 211
Sylvester's criterion, 111
Systems of particles, 11

T

Time-invariant systems, 48
Time-varying systems, 48, 362–363
Total energy, 10
Trajectory, 39, 68
Transfer matrix, 177
Transfer function(s), 53–55
　closed-loop, 131, 134
　feedback-path, 131
　forward-path, 131, 134
　matrix of, 177
　open-loop, 131
Transition matrix, 74, 80, 127
　computation of, 76–78
　by Laplace transformation, 77
　response by, 74–76
Traveling wave control, 372

Tridiagonal form, 116
Trivial equilibrium, 68
Truncation (of model), 284

U

Uniform damping control, 373
Unitary transformation, 118
Unit impulse, 56
Unit ramp function, 58
Unit step function, 56, 134
Unstable equilibrium, 69, 70
　for discrete-time systems, 89

V

Variance, 240
Variational approach to optimal control, 192
Virtual displacement(s), 23, 270
　generalized, 23
Virtual work, 23, 34, 97, 270
　principle of, 23
Virtual work density, 270
Viscous damping:
　coefficients, 96
　factor, 135
　forces, 96, 99
　matrix, 96

W

Weighted residuals method, 298–300
　collocation method, 300
　Galerkin method, 300
White noise, 245
Work, 8

Z

Zero(s) of function, 138
Zero-order hold, 87